Lecture Notes in Computer Science 13147

More information about this subseries at https://link.springer.com/bookseries/7409

Satish Narayana Srirama · Jerry Chun-Wei Lin ·
Raj Bhatnagar · Sonali Agarwal ·
P. Krishna Reddy (Eds.)

Big Data Analytics

9th International Conference, BDA 2021 Virtual Event, December 15–18, 2021 Proceedings

Editors
Satish Narayana Srirama
University of Hyderabad
Hyderabad, India

Raj Bhatnagar
University of Cincinnati
Cincinnati, OH, USA

P. Krishna Reddy
International Institute of Information Technology
Hyderabad, India

Jerry Chun-Wei Lin
Western Norway University of Applied Sciences
Bergen, Norway

Sonali Agarwal
Indian Institute of Information Technology Allahabad
Prayagraj, India

ISSN 0302-9743 ISSN 1611-3349 (electronic)
Lecture Notes in Computer Science
ISBN 978-3-030-93619-8 ISBN 978-3-030-93620-4 (eBook)
https://doi.org/10.1007/978-3-030-93620-4

LNCS Sublibrary: SL3 – Information Systems and Applications, incl. Internet/Web, and HCI

This Springer imprint is published by the registered company Springer Nature Switzerland AG
The registered company address is: Gewerbestrasse 11, 6330 Cham, Switzerland

Preface

With the proliferation of structured and unstructured data from a wide range of sources, and the anticipated opportunities it is going to offer through proper analytics, the big data horizon has caught significant attention of both the research community in computer science and domain/application experts. The adoption of emerging technologies such as the Internet of Things (IoT), cloud computing, 5G, and blockchains, by businesses, governments and the general public, further heightened the data proliferation and the importance of big data analytics. Numerous applications of big data analytics are found in many important and diverse fields, such as e-commerce, finance, smart healthcare, smart cities, education, e-governance, media and entertainment, security and surveillance, telecommunications, agriculture, astronomy, and transportation.

With key characteristics of big data, i.e. the 6Vs (volume, variety, velocity, veracity, value, and variability), big data analytics raises several challenges such as how to handle massive volumes of data, process data in real time, and deal with complex, uncertain, heterogeneous, and streaming data, which are often stored in multiple remote repositories. To address these challenges, innovative big data analysis solutions must be designed by drawing expertise from recent advances in several fields such as data mining, database systems, statistics, distributed data processing, machine/deep learning, and artificial intelligence. There is also an important need to build data analysis systems for emerging applications such as healthcare, autonomous vehicles, and IoT.

The 9th International Conference on Big Data Analytics (BDA 2021) was held online during December 15–18, 2021, and organized by the Indian Institute of Information Technology Allahabad, India. This proceedings contains 16 peer-reviewed research papers and contributions from keynote speakers, invited speakers, and tutorial and workshop speakers. This year's program covered a wide range of topics related to big data analytics: gesture detection, networking, social media, search, information extraction, image processing and analysis, spatial, text, mobile and graph data analytics, machine learning, and healthcare. It is expected that the research papers, keynote speeches, invited talks, and tutorials presented at the conference will promote research in big data analytics and stimulate the development of innovative solutions to industry problems.

The conference received 41 submissions. The Program Committee (PC) consisted of researchers from both academia and industry based in a number of different countries or territories including India, Ireland, Norway, Canada, Estonia and the USA. Each submission was reviewed by at least two to three reviewers from the Program Committee (PC) and was discussed by the PC chairs before a decision was made. Based on the above review process, the Program Committee accepted 16 full papers. The overall acceptance rate was approximately 39%.

We would like to sincerely thank the members of the Program Committee and the external reviewers for their time, energy, and expertise in supporting BDA 2021. In addition, we would like to thank all the authors who considered BDA 2021 as a forum for publishing their research contributions. A number of individuals contributed to the success of the conference. We thank Joao Gama, Albert Biefet, and Asoke K Talukder for their insightful keynote presentations. Our sincere thanks also go to P Nagabhushan, Anil Kumar Vuppala, and Ashish Ghosh for their valuable contribution to BDA 2021. We also thank all the workshop and tutorial organizers for their efforts and support. We would like to thank the supporting organizations including the International Institute of Information Technology Hyderabad (IIITH), the University of Hyderabad, the Indraprastha Institute of Information Technology Delhi (IIITD), the Western Norway University of Applied Sciences and Ahmedabad University. The conference also received invaluable support from the authorities of the Indian Institute of Information Technology Allahabad (IIITA) for hosting the conference. In addition, thanks are also due to the faculty, staff, and student volunteers of IIIT Allahabad for their continuous cooperation and support.

November 2021

Satish Narayana Srirama
Jerry Chun-Wei Lin
Raj Bhatnagar
Sonali Agarwal

Organization

BDA 2021 was organized by Indian Institute of Information Technology Allahabad (IIITA), India.

Honorary Chair

P. Nagabhushan	IIIT Allahabad, India

General Chair

Sonali Agarwal	IIIT Allahabad, India

Steering Committee Chair

P. Krishna Reddy	IIIT Hyderabad, India

Steering Committee

S. K. Gupta	IIT Delhi, India
Srinath Srinivasa	IIIT Bangalore, India
Krithi Ramamritham	IIT Bombay, India
Sanjay Kumar Madria	Missouri University of Science and Technology, USA
Masaru Kitsuregawa	University of Tokyo, Japan
Raj K. Bhatnagar	University of Cincinnati, USA
Vasudha Bhatnagar	University of Delhi, India
Mukesh Mohania	IIIT Delhi, India
H. V. Jagadish	University of Michigan, USA
Ramesh Kumar Agrawal	Jawaharlal Nehru University, India
Divyakant Agrawal	University of California, Santa Barbara, USA
Arun Agarwal	University of Hyderabad, India
Jaideep Srivastava	University of Minnesota, USA
Anirban Mondal	Ashoka University, India
Sharma Chakravarthy	University of Texas at Arlington, USA
Sanjay Chaudhary	Ahmedabad University, India

Program Committee Chairs

Satish Narayana Srirama	University of Hyderabad, India
Jerry Chun-Wei Lin	Western Norway University of Applied Sciences, Norway
Raj Bhatnagar	University of Cincinnati, USA

Publication Chair

Sanjay Chaudhary	Ahmedabad University, India

Workshop Chairs

Partha Pratim Roy	IIT Roorkee, India
Joao Gama	University of Porto, Portugal
Pradeep Kumar	IIM Lucknow, India

Tutorial Chairs

Mukesh Prasad	University of Technology Sydney, Australia
Anirban Mondal	Asoka University, India
Pragya Dwivedi	MNNIT Allahabad, India

Panel Chairs

Srinath Srinivasa	IIIT Bangalore, India
Rajeev Gupta	Microsoft, India
Sanjay Kumar Madria	Missouri University of Science and Technology, USA

Publicity Chairs

M. Tanveer	IIT Indore, India
Deepak Gupta	NIT Arunachal Pradesh, India
Navjot Singh	IIIT Allahabad, India

Organizing Committee Members

Shekhar Verma	IIIT Allahabad, India
Pavan Chakraborty	IIIT Allahabad, India
Vrijendra Singh	IIIT Allahabad, India
Manish Kumar	IIIT Allahabad, India
Krishna Pratap Singh	IIIT Allahabad, India
Triloki Pant	IIIT Allahabad, India
S. Venkatesan	IIIT Allahabad, India
Kokila Jagadeesh	IIIT Allahabad, India
Narinder Singh Punn	IIIT Allahabad, India

Program Committee Members

Anirban Mondal	Ashoka University, India
Chinmaya Dehury	University of Tartu, Estonia
Dhirendra Kumar	Delhi Technological University, India
Divya Sardana	Nike, USA
Engelbert Mephu Nguifo	University of Clermont Auvergne, France
Gautam Srivastava	Brandon University, Canada
Gowtham Atluri	University of Cincinnati, USA
Himanshu Gupta	IBM Research, India
Jose Maria Luna	University of Córdoba, Spain
Ladjel Bellatreche	ISAE - ENSMA, France
P. Krishna Reddy	IIIT Hyderabad, India
Philippe Fournier-Viger	Harbin Institute of Technology, China
Praveen Rao	University of Missouri, USA
Ravi Jampani	Conduira Education, India
Sadok Ben Yahia	Tallinn University of Technology, Estonia
Sangeeta Mittal	Jaypee Institute of Information Technology, India
Sanjay Chaudhary	Ahmedabad University, India
Sebastián Ventura	University of Córdoba, Spain
Sharma Chakravarthy	University of Texas at Arlington, USA
Sonali Agarwal	IIIT Allahabad, India
Srinath Srinavasa	IIIT Bangalore, India
Uday Kiran	University of Aizu, Japan
Vasudha Bhatnagar	University of Delhi, India
Vikram Goyal	IIIT Delhi, India
Vikram Singh	NIT Kurukshetra, India

Supporting Institutions

Western Norway University of Applied Sciences, Norway
International Institute of Information Technology Hyderabad (IIITH), India
University of Hyderabad, India
Indraprastha Institute of Information Technology Delhi (IIITD), India

Contents

IoTs, Sensors, and Networks

Fundamentation

Pattern Mining and data Analytics

Medical and Health Applications

MAG-Net: Multi-task Attention Guided Network for Brain Tumor Segmentation and Classification

Sachin Gupta, Narinder Singh Punn(✉), Sanjay Kumar Sonbhadra, and Sonali Agarwal

Indian Institute of Information Technology Allahabad, Prayagraj, India
{mit2019075,pse2017002,rsi2017502,sonali}@iiita.ac.in

Abstract. Brain tumor is the most common and deadliest disease that can be found in all age groups. Generally, MRI modality is adopted for identifying and diagnosing tumors by the radiologists. The correct identification of tumor regions and its type can aid to diagnose tumors with the followup treatment plans. However, for any radiologist analysing such scans is a complex and time-consuming task. Motivated by the deep learning based computer-aided-diagnosis systems, this paper proposes multi-task attention guided encoder-decoder network (MAG-Net) to classify and segment the brain tumor regions using MRI images. The MAG-Net is trained and evaluated on the Figshare dataset that includes coronal, axial, and sagittal views with 3 types of tumors meningioma, glioma, and pituitary tumor. With exhaustive experimental trials the model achieved promising results as compared to existing state-of-the-art models, while having least number of training parameters among other state-of-the-art models.

Keywords: Attention · Brain tumor · Deep learning · Segmentation

1 Introduction

Brain tumor is considered as the deadliest and most common form of cancer in both children and adults. Determining the correct type of brain tumor in its early stage is the key aspect for further diagnosis and treatment process. However, for any radiologist, identification and segmentation of brain tumor via multi-sequence MRI scans for diagnosis, monitoring, and treatment, are complex and time-consuming tasks.

Brain tumor segmentation is a challenging task because of its varied behavior both in terms of structure and function. Furthermore, the tumor intensity of a person differs significantly from each other. MRI is preferred over other imaging modalities [4] for the diagnosis of brain tumor because of its non-invasive property that follows from without the exposure to ionizing radiations and superior image contrast in soft tissues.

S. Gupta, N. S. Punn, S. K. Sonbhadra, and S. Agarwal—Equal contribution.

S. N. Srirama et al. (Eds.): BDA 2021, LNCS 13147, pp. 3–15, 2021.
https://doi.org/10.1007/978-3-030-93620-4_1

Deep learning has shown advancement in various fields with promising performance especially in the area of biomedical image analysis [24]. The convolutional neural networks (CNN) [2] are the most widely used models in image processing. The CNNs involve combination of convolution, pooling and activation layers accompanied with the normalization and regularization operations to extract and learn the target specific features for desired task (classification, localization, segmentation, etc.). In recent years various techniques have been proposed for identification (classification and segmentation) of the brain tumor using MRI images that achieved promising results [13,30]. However, most of the approaches use millions of trainable parameters that result in slower training and analysis time, while also having high variance in results in case of limited data samples.

In order to overcome the aforementioned drawbacks, Ronneberger et al. [26] proposed U shaped network (U-Net) for biomedical image segmentation. The model follows encoder-decoder design with feature extraction (contraction path) and reconstruction phases (expansion path) respectively. In addition, skip connections are introduced to propagate the extracted feature maps to the corresponding reconstruction phase to aid upsample the feature maps. Finally, model produces segmentation mask in same dimensions as the input highlighting the target structure (tumor in our case). Following the state-of-the-art potential of the U-Net model, many U-Net variants are proposed to further improve the segmentation performance. Attention based U-Net model [19] is one such variant that tend to draw the focus of the model towards target features to achieve better segmentation results. The attention filters are introduced in the skip connections where each feature is assigned weight coefficient to highlight its importance towards the target features. Despite achieving the promising results, these models have millions of trainable parameter which can be reduced by optimizing the convolution operation. This can be achieved by incorporating the depthwise convolution operations [8] that is performed in two stages: depthwise and pointwise convolutions. The reduction in the number of the parameters and multiplications as compared to standard convolution operation can represented as $1/r + 1/f^2$, where r is the depth of the output feature map and f is the kernel height or width [21]. The achieved reduction in number of parameters and multiplications is ~80%. Following this context, attention guided network is proposed that uses depthwise separable convolution for real time segmentation and classification of the brain tumor using MRI imaging. The major contribution of the present research work is as follows:

- A novel model, Multi-task (segmentation and classification) attention guided network (MAG-Net) is proposed for brain tumor diagnosis.
- Optimization of training parameters using depthwise separable convolution. The training parameters of the MAG-Net reduced from 26.0M to 5.4M.
- MAG-Net achieved significant improvement in classification and segmentation as compared to the state-of-the-art models while having limited data samples.

The rest paper is organized as follows: Sect. 2 describes the crux of related work on brain tumor segmentation and classification. Section 3, talks about the proposed architecture, whereas Sect. 4 discuses the training and testing environment with experimental and comparative analysis. Finally, concluding remarks are presented in Sect. 5.

2 Literature Review

Identifying the brain tumor is a challenging task for the radiologists. Recently, several deep learning based approaches are proposed to aid in faster diagnosis of the diseases. Segmentation of the infected region is most common and critical practice involved in the diagnosis. In addition, the segmented region can be provided with label (classification) to indicate what type of anomaly or infection is present in the image.

In contrast to the traditional approaches, Cheng et al. [7] proposed a brain tumor classification approach using augmented tumor region instead of original tumor region as RoI (region of interest). Authors utilized the bag of word (BOW) technique to segment and extract local features from RoI. Dictionary is used for encoding the extracted local features maps that are passed through SVM (support vector machine) classifier. The approach outperformed the traditional classification techniques with the accuracy of 91.28% but the performance is limited by the data availability. In similar work, Ismael et al. [14] proposed an approach of combining statistical features along with neural networks by using filter combination: discrete wavelet transform (DWT)(represented by wavelet coefficient) and Gabor filter (for texture representation). For classification of the tumor, three layered neural network classifier is developed using multilayer perceptron network that is trained with statistical features. In contrast to Cheng et al. [7], authors also achieved promising results on the limited data samples with an overall accuracy of 91.9%.

Recently, capsule network [12] has shown great performance in many fields especially in biomedical image processing. Afshar et al. [1] proposed basic capsnet with three capsules in last layer representing three tumor classes. However, due to varied behavior (background, intensity, structure, etc.) of MRI image, the proposed model failed to extract optimal features representing the tumor structure. The author achieved the tumor classification accuracy of 78% and 86.5% using raw MRI images and tumor segmented MRI images respectively. In another approach, Pashaei et al. [20] utilized CNN and kernel extreme learning machine that comprises one hidden layer with 100 neurons to increase the robustness of the model. With several experimental trials, the authors achieved an accuracy of 93.68% but detects only 1% of the positive pituitary tumor cases out of the total pituitary tumor case. Deepak et al. [9] proposed a transfer learning approach that uses pre-trained GoogleNet model to extract features (referred as deep CNN features) with softmax classifier in the output layer to classify three tumor classes. Furthermore, the authors combine the deep CNN features and SVM model to analyse the classification performance. The authors achieved 97.1% accuracy but resulted in poor performance by standalone GoogleNet model due to overfitting with limited training image dataset, and misclassifications in meningioma tumor.

In another approach, Pernas et al. [11] proposed to process images in three different spatial scales along with multi pathways feature scales for classification and segmentation of brain tumor. The images are pre-processed with elastic transform for preventing overfitting. The model analyses entire image and classifies pixel by pixel in one of four possible output labels (i.e. 0-healthy, 1-meningioma, 2-glioma, and 3-pituitary tumor). The proposed approach outperformed existing approaches with 97.3% classification accuracy, but with poor segmentation performance. Following this context, in this article multi-task attention guided network (MAG-Net) is proposed based on the U-Net architectural design [26] that uses parallel depthwise separable convolution layers for multi-level feature extraction along with an attention mechanism to better extract tumor features for brain tumor classification and generate the corresponding tumor mask.

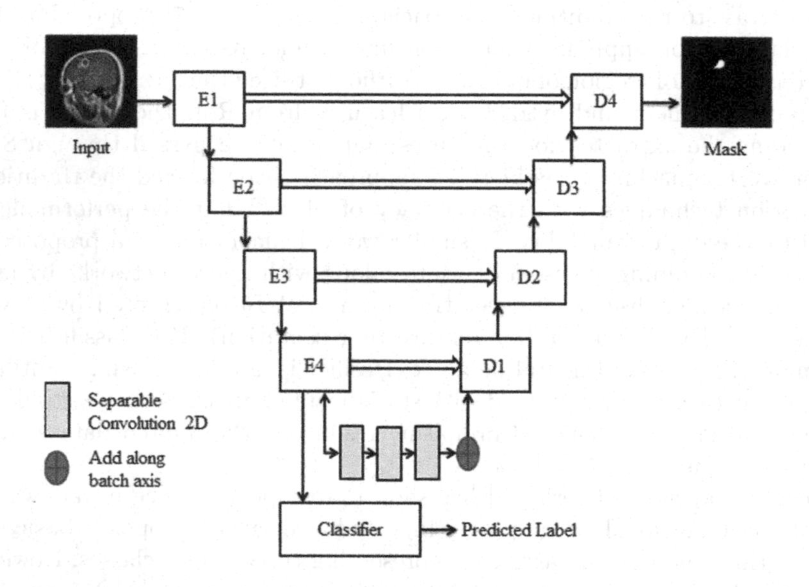

Fig. 1. Schematic representation of the architecture of MAG-Net model.

3 Proposed Work

The proposed mutli-task attention guided network (MAG-Net) model, as shown in Fig. 1, focuses on reducing overall computation, better feature extraction and optimizing the training parameters by reduction. The overall architectural design consists of an encoder, decoder, and classification module with 5.4M trainable parameters. The overall architectural design of the model is inspired by the U-Net encoder-decoder style [23]. Due to its state-of-the-art potential, this model is the most prominent choice among the researchers to perform biomedical image segmentation [17].

In MAG-Net to reduce the number of training parameters without the cost of performance, standard convolution operations are replaced with depthwise separable convolution. In addition, the skip connections are equipped with attention filters [19] to better extract the feature maps concerning the tumor regions. The attention approach filters the irrelevant feature maps in the skip connection by assigning weights to highlight its importance towards the tumor regions. Besides, the encoder block is equipped with parallel separable convolution filters of different sizes, where the extracted feature maps are concatenated for better feature learning. These features are then passed to the corresponding decoder blocks via attention enabled skip connections to aid in feature reconstruction with the help of upsampling operation. The bottleneck layer connects the feature extraction path to the feature reconstruction path. In this layer filters of different sizes are used along with the layer normalization. Furthermore, the classification is performed using the extracted feature maps obtained from the final encoder block.

3.1 Encoder

To detect the shape and size of varying image like brain tumor it is required to use separable convolution of different sizes. Inspired from the concept of inception neural network [22] the encoder segment is consist of separable convolutions of 1 $\times$ 1, 3 $\times$ 3, and 5 $\times$ 5 kernels. Each of separable convolutions are followed by layer normalization. The extracted feature maps are fused with add operation that are downsampled by max pooling operation. Figure 2, shows the proposed encoder architecture of MAG-Net model for some input feature map, $\mathcal{F}_i \in \mathcal{R}^{w \times h \times d}$, where w, h and d are the width, height and depth of the feature map.

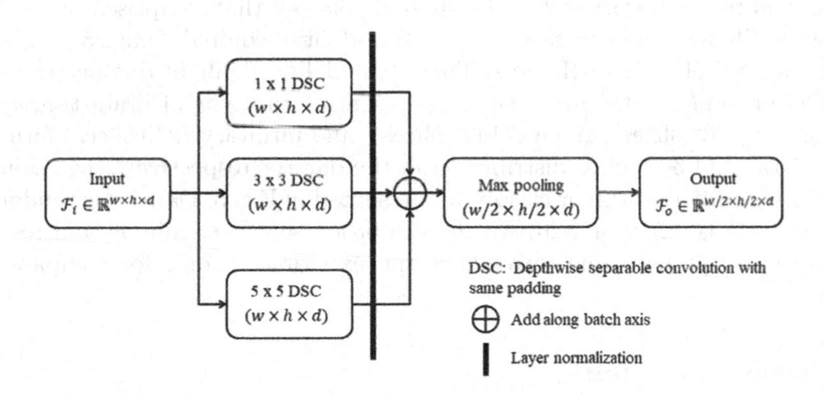

Fig. 2. Proposed MAG-Net 2D encoder module.

3.2 Decoder

The decoder component follows from the encoder block and that tend to reconstruct the spatial dimension to generate the output mask in same dimension as input. It consists of upsampling of the feature maps along with the concatenation with the attention maps followed by a separable convolution operation. Long skip connections [10] are used to propagate the attention feature maps from encoder to decoder to recover spatial information that was lost during downsampling in encoder. By using attention in the skip connection it helps the model to suppress the irrelevant features.

3.3 Classification

This module classifies the brain tumor MRI images into respective classes i.e. meningioma, glioma, and pituitary tumor by utilizing the features extracted from the encoder block. This encoder block act as backbone model for both classification and segmentation, thereby reducing the overall complexity of the model. In this classification block the feature maps of the last encoder block act as input that are later transformed into 1D tensor by using global average pooling. The pooled feature maps are then processed with multiple fully connected layers. The classification output is generated from the softmax activated layer that generates the probability distribution of the tumor classes for an image.

4 Experiment and Results

4.1 Dataset Setup

The present research work utilizes Figshare [6] dataset that comprises of 2D MRI scan with T1-weighted contrast-enhanced modality acquired from 233 patients to form a total of 3064 MRI scans. The T1 modality highlight distinct features of the brain tumor with three classes representing the type of brain tumor i.e. meningioma (708 slices), glioma (1426 slices), and pituitary (930 slices) forming 23%, 46.5%, and 30% class distribution in the dataset respectively. The sample MRI slices of different tumor classes are presented in Fig. 3. Dataset is randomly split into 80% training and 20% of the validation set. The training and testing composition kept the same throughout the experiment trails for comparative analysis.

4.2 Training and Testing

The MAG-Net model is trained and evaluated on the Figshare dataset. The training phase is accompanied with early-stopping [3] to tackle the overfitting problem, and Adam as a learning rate optimiser [27]. Cross entropy based loss functions are most popularly used for model training and validating segmentation and classification tasks. Following this, binary cross entropy and categorical cross entropy functions are employed for training the model for binary tumor mask

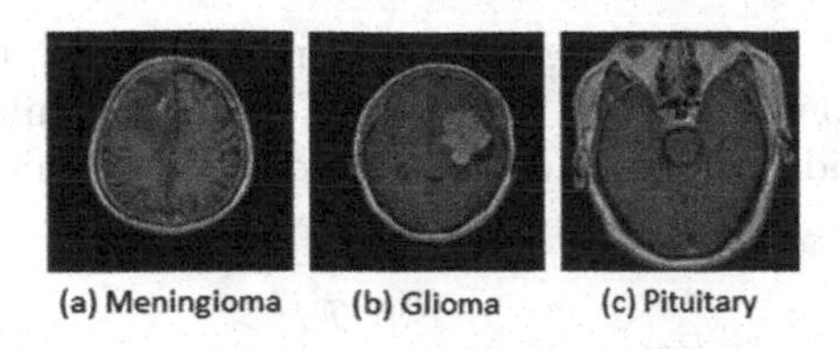

Fig. 3. A slice of MRI scan with T1 modality showing different tumor classes: meningioma, glioma, and pituitary

generation and classification respectively. Binary cross entropy (BCE, shown in Eq. 1) is a sigmoid activation [16] followed by cross entropy loss [29] that compares each of the predicted probabilities to actual output. Categorical cross entropy (CE, shown in Eq. 2) is a softmax activation function followed by cross-entropy-loss that compares the output probability over each tumor class for each MRI image.

$$\mathcal{L}_{BCE} = -\frac{1}{N}\sum_{i=1}^{N}(y_i.log(p(y_i)) + (1 - y_i).log(1 - P(y_i))) \tag{1}$$

where y represents actual tumor mask, $p(y)$ represents predicted tumor mask and N is the total number of images.

$$\mathcal{L}_{CE} = \sum_{i}^{C} t_i log(f(s_i)) \tag{2}$$

where C is the no. of class, $f(s_i)$ is the probability of occurrence of each class t_i represents 1 for true label and 0 for others.

For segmentation the most popular evaluation matrics are dice coefficient (shown in Eq. 3) and intersection-over-union (IoU/Jaccard index) (shown in Eq. 4), and hence are utilized to evaluate the trained MAG-Net model. TP defines correctly classified predictions FP defines wrongly classified, and FN defines missed objects of each voxel.

$$DiceCoefficient = \frac{2 * TP}{2 * TP + FP + FN} \tag{3}$$

$$IoU = \frac{TP}{TP + FP + FN} \tag{4}$$

To evaluate classification module of the MAG-Net, accuracy, precision, recall, f1-score and micro average metrics are considered for better quantification and visualization of the performance of the model. Precision of the class, as shown in Eq. 5, quantifies about the positive prediction accuracy of the model. Recall is the fraction of true positive which are classified correctly (shown in Eq. 6). F1-score quantifies the amount of correct predictions out of all the positive predictions (shown in Eq. 7). Support quantifies the true occurrence in the specified dataset of the respective class. Micro average (μ_{avg}) (shown in Eq. 8, Eq. 9 and Eq. 10) is calculated for precision, recall, and F1-score. To compute micro average (μ_{avg}), the test dataset is divided into two sub dataset, on each of which the true positive, false positive and false negative predictions are identified.

$$Precision = \frac{TP}{(TP + FP)} \tag{5}$$

$$Recall = \frac{TP}{(FN + FP)} \tag{6}$$

$$F1 - score = \frac{2 * Recall * Precision}{(Recall + Precision)} \tag{7}$$

$$\mu_{avg}(Precision) = \frac{TP_1 + TP_2}{(TP_1 + TP_2 + FP_1 + FP_2)} \tag{8}$$

$$\mu_{avg}(Recall) = \frac{TP_1 + TP_2}{(TP_1 + TP_2 + FN_1 + FN_2)} \tag{9}$$

$$\mu_{avg}(F1 - score) = HM(\mu_{avg}(Precision), mu_{avg}(Recall)) \tag{10}$$

where TP_1, FP_1, and FN_1 belong to the first set and TP_2, FP_2, and FN_2 belongs to the different sets. HM is the harmonic mean.

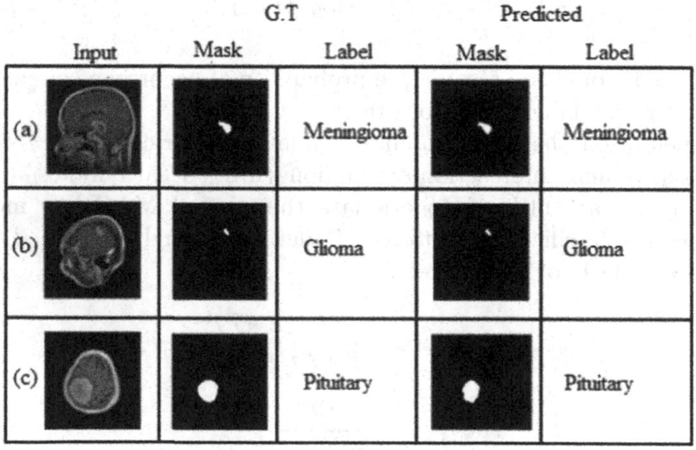

	Input	G.T Mask	G.T Label	Predicted Mask	Predicted Label
(a)			Meningioma		Meningioma
(b)			Glioma		Glioma
(c)			Pituitary		Pituitary

Fig. 4. Qualitative results of brain tumor segmentation and classification on MRI images (a, b, and c of different tumor classes) using MAG-Net model.

4.3 Results

The MAG-Net outputs the segmented mask of a given MRI image consisting of tumor region corresponding to meningioma, glioma, and pituitary as classified by the model. For randomly chosen MRI slices, Fig. 4 presents the segmentation and classification results of model. The visual representation confirms that the results are close to the ground truth of respective tumor classes.

Table 1. Comparative analysis of the MAG-Net with the existing segmentation models on test dataset.

Model	Accuracy	Loss	Dice coefficient	Jaccard index	Parameters
U-Net	99.5	0.024	0.70	0.55	31M
wU-Net	99.4	0.034	0.66	0.49	31M
Unet++	99.5	0.028	0.65	0.49	35M
MAG-Net	**99.52**	**0.021**	**0.74**	**0.60**	**5.4M**

* Bold quantities indicate the best results.

Table 1 represents the result of the proposed work for segmentation in the form of accuracy, loss, dice coefficient, Jaccard index, and trainable parameters along with comparative analysis with other popular approaches. The proposed framework outperforms the other approaches in segmenting tumor with the dice and IoU score of 0.74 and 0.60 respectively. In contrast to other models, MAG-Net achieved best results with minimal trainable parameters. The other popular approaches taken in comparative analysis for segmentation are U-Net [15], U-Net++ [18,31], and wU-Net [5].

Table 2. Comparative analysis of the MAG-Net with the existing classification models on test dataset using confusion matrix.

Model	Acc.	Loss		Meningioma	Glioma	Pituitary
VGG16	93.15	0.26	Meningioma	114	25	3
			Glioma	13	271	1
			Pituitary	0	1	185
VGG19	93.8	0.25	Meningioma	114	21	7
			Glioma	11	274	0
			Pituitary	0	1	185
ResNet50	94.2	0.31	Meningioma	123	12	7
			Glioma	16	266	3
			Pituitary	1	0	185
MAG-Net	**98.04**	**0.11**	Meningioma	**134**	**7**	**1**
			Glioma	**1**	**282**	**2**
			Pituitary	**1**	**0**	**185**

* Bold quantities indicate the best results.

Table 2 and Table 3 represent the results of the proposed work for classification in the form of accuracy, loss, confusion matrix, and classification report for meningioma, glioma, and pituitary tumor along with comparative analysis with other state-of-the-art approaches: VGG-16 [28], VGG-19 [28], and ResNet50 [25]. With exhaustive experimental trials it is observed that MAG-Net outperformed the existing approaches with significant margin in all the metrics.

Table 3. Comparative analysis of the MAG-Net with the existing classification models on test dataset considering classification report as evaluation parameter.

Model	Classes	Precision	Recall	F1-score	Support
VGG16	Meningioma	0.90	0.80	0.85	142
	Glioma	0.91	0.85	0.93	285
	Pituitary	0.98	0.99	0.99	186
	Micro avg.	0.93	0.93	0.93	613
VGG19	Meningioma	0.91	0.80	0.85	142
	Glioma	0.93	0.96	0.94	285
	Pituitary	0.96	0.99	0.98	186
	Micro avg.	0.93	0.93	0.93	613
ResNet-50	Meningioma	0.88	0.87	0.87	142
	Glioma	0.93	0.99	0.94	285
	Pituitary	0.95	0.99	0.97	186
	Micro avg.	0.94	0.94	0.94	613
MAG-Net	Meningioma	**0.99**	**0.94**	**0.96**	142
	Glioma	**0.98**	**0.99**	**0.98**	285
	Pituitary	**0.98**	**0.99**	**0.99**	186
	Micro avg.	**0.98**	**0.98**	**0.98**	613

* Bold quantities indicate the best results.

It is observed that unlike other state-of-the-art models, the MAG-Net model achieved promising results due to the reduction in the overall computation, better feature extraction and training parameters optimization. As shown in Table 1 raw U-Net displayed similar performance but at the cost of large number of trainable parameters. In the MAG-Net model, the encoder block is developed by replacing convolution layers with parallel depthwise separable convolution of various sizes connected in parallel which resulted in better multi-scale feature learning for varying shapes and sizes of the tumor. For reducing spatial loss during feature reconstruction, attention mechanism is used in skip connections for better feature reconstruction. To reduce the overall complexity of the model the feature extracted by encoder blocks are reused to classify the type of brain tumor.

5 Conclusion

In this paper, the complex task of brain tumor segmentation and classification is addressed using multi-task attention guided network (MAG-Net). This a U-Net based model that features reduction in the overall computation, better feature extraction and training parameters optimization. The proposed architecture achieved significant performance on the Figshare brain tumor dataset by exploiting the state-of-the-art advantages of U-Net, depthwise separable convolution and attention mechanism. The MAG-Net model recorded the best classification and segmentation results compared to the existing classification and segmentation approaches. It is believed that this work can also be extended to other domains involving classification and segmentation tasks.

Acknowledgment. We thank our institute, Indian Institute of Information Technology Allahabad (IIITA), India and Big Data Analytics (BDA) lab for allocating the centralised computing facility and other necessary resources to perform this research. We extend our thanks to our colleagues for their valuable guidance and suggestions.

References

1. Afshar, P., Mohammadi, A., Plataniotis, K.N.: Brain tumor type classification via capsule networks. In: 2018 25th IEEE International Conference on Image Processing (ICIP), pp. 3129–3133. IEEE (2018)
2. Albawi, S., Mohammed, T.A., Al-Zawi, S.: Understanding of a convolutional neural network. In: 2017 International Conference on Engineering and Technology (ICET), pp. 1–6. IEEE (2017)
3. Brownlee, J.: Use early stopping to halt the training of neural networks at the right time (2018). https://machinelearningmastery.com/how-to-stop-training-deep-neural-networks-at-the-right-time-using-early-stopping/. Accessed 17 Apr 2021
4. Cancer.Net: Brain tumor: diagnosis (2020). https://www.cancer.net/cancer-types/brain-tumor/diagnosis. Accessed 20 Mar 2021
5. CarryHJR: Nested UNet (2020). https://github.com/CarryHJR/Nested-UNet/blob/master/model.py. Accessed 11 Mar 2021
6. Cheng, J.: Brain tumor dataset (2017). https://figshare.com/articles/dataset/brain_tumor_dataset/1512427
7. Cheng, J., et al.: Enhanced performance of brain tumor classification via tumor region augmentation and partition. PLoS ONE **10**(10), e0140381 (2015)
8. Chollet, F.: Xception: deep learning with depthwise separable convolutions. In: Proceedings of the IEEE Conference on Computer Vision and Pattern Recognition, pp. 1251–1258 (2017)
9. Deepak, S., Ameer, P.: Brain tumor classification using deep CNN features via transfer learning. Comput. Biol. Med. **111**, 103345 (2019)
10. Drozdzal, M., Vorontsov, E., Chartrand, G., Kadoury, S., Pal, C.: The importance of skip connections in biomedical image segmentation. In: Carneiro, G., et al. (eds.) LABELS/DLMIA -2016. LNCS, vol. 10008, pp. 179–187. Springer, Cham (2016). https://doi.org/10.1007/978-3-319-46976-8_19

11. Díaz-Pernas, F.J., Martínez-Zarzuela, M., Antón-Rodríguez, M., González-Ortega, D.: A deep learning approach for brain tumor classification and segmentation using a multiscale convolutional neural network. Healthcare **9**(2), 153 (2021). https://doi.org/10.3390/healthcare9020153, https://app.dimensions.ai/details/publication/pub.1135094000 and www.mdpi.com/2227-9032/9/2/153/pdf
12. Hinton, G.E., Sabour, S., Frosst, N.: Matrix capsules with EM routing. In: International Conference on Learning Representations (2018)
13. Işın, A., Direkoğlu, C., Şah, M.: Review of MRI-based brain tumor image segmentation using deep learning methods. Procedia Comput. Sci. **102**, 317–324 (2016)
14. Ismael, M.R., Abdel-Qader, I.: Brain tumor classification via statistical features and back-propagation neural network. In: 2018 IEEE International Conference on Electro/information Technology (EIT), pp. 0252–0257. IEEE (2018)
15. Jain, A.: Brain tumor segmentation U-Net (2020). https://github.com/adityajn105/brain-tumor-segmentation-unet. Accessed 8 Jan 2021
16. Jamel, T.M., Khammas, B.M.: Implementation of a sigmoid activation function for neural network using FPGA. In: 13th Scientific Conference of Al-Ma'moon University College, vol. 13 (2012)
17. Minaee, S., Boykov, Y.Y., Porikli, F., Plaza, A.J., Kehtarnavaz, N., Terzopoulos, D.: Image segmentation using deep learning: a survey. IEEE Trans. Pattern Anal. Mach. Intell. (2021). https://doi.org/10.1109/TPAMI.2021.3059968
18. MrGiovanni: U-Net++ Keras (2020). https://github.com/MrGiovanni/UNetPlusPlus. Accessed 12 Mar 2021
19. Oktay, O., et al.: Attention U-Net: learning where to look for the pancreas (2018)
20. Pashaei, A., Sajedi, H., Jazayeri, N.: Brain tumor classification via convolutional neural network and extreme learning machines. In: 2018 8th International Conference on Computer and Knowledge Engineering (ICCKE), pp. 314–319. IEEE (2018)
21. Punn, N.S., Agarwal, S.: CHS-Net: a deep learning approach for hierarchical segmentation of COVID-19 infected CT images. arXiv preprint arXiv:2012.07079 (2020)
22. Punn, N.S., Agarwal, S.: Inception U-Net architecture for semantic segmentation to identify nuclei in microscopy cell images. ACM Trans. Multimedia Comput. Commun. Appl. (TOMM) **16**(1), 1–15 (2020)
23. Punn, N.S., Agarwal, S.: Multi-modality encoded fusion with 3D inception U-Net and decoder model for brain tumor segmentation. Multimedia Tools Appl. **80**(20), 30305–30320 (2020). https://doi.org/10.1007/s11042-020-09271-0
24. Punn, N.S., Agarwal, S.: Modality specific U-Net variants for biomedical image segmentation: a survey. arXiv preprint arXiv:2107.04537 (2021)
25. raghakot: Keras-ResNet (2017). https://github.com/raghakot/keras-resnet. Accessed 18 Mar 2021
26. Ronneberger, O., Fischer, P., Brox, T.: U-Net: convolutional networks for biomedical image segmentation. In: Navab, N., Hornegger, J., Wells, W.M., Frangi, A.F. (eds.) MICCAI 2015. LNCS, vol. 9351, pp. 234–241. Springer, Cham (2015). https://doi.org/10.1007/978-3-319-24574-4_28
27. Ruder, S.: An overview of gradient descent optimization algorithms (2017)
28. Thakur, R.: Step by step VGG16 implementation in Keras for beginners (2019). https://towardsdatascience.com/step-by-step-vgg16-implementation-in-keras-for-beginners-a833c686ae6c. Accessed 20 Mar 2021
29. Zhang, Z., Sabuncu, M.R.: Generalized cross entropy loss for training deep neural networks with noisy labels. arXiv preprint arXiv:1805.07836 (2018)

30. Zhou, T., Ruan, S., Canu, S.: A review: deep learning for medical image segmentation using multi-modality fusion. Array **3**, 100004 (2019)
31. Zhou, Z., Siddiquee, M., Tajbakhsh, N., Liang, J.U.: A nested U-Net architecture for medical image segmentation. arXiv preprint arXiv:1807.10165 (2018)

Smartphone Mammography for Breast Cancer Screening

Rohini Basu[1], Meghana Madarkal[2], and Asoke K. Talukder[3,4](✉)

[1] Cybernetic Care, Bangalore, India
rohini.basu@cyberneticcare.com
[2] University of Maryland, Maryland, USA
[3] National Institute of Technology Karnataka, Surathkal, India
[4] SRIT India Pvt Ltd, Bangalore, India
asoke.talukder@sritindia.com

Abstract. In 2020 alone approximately 2.3 million women were diagnosed with breast cancer which caused over 685,000 deaths worldwide. Breast cancer affects women in developing countries more severely than in developed country such that over 60% of deaths due to breast cancer occur in developing countries. Deaths due to breast cancer can be reduced significantly if it is diagnosed at an early stage. However, in developing countries cancer is often diagnosed when it is in the advanced stage due to limited medical resources available to women, lack of awareness, financial constraints as well as cultural stigma associated with traditional screening methods. Our paper aims to provide an alternative to women that is easily available to them, affordable, safe, non-invasive and can be self-administered. We propose the use of a smartphone's inbuilt camera and flashlight for breast cancer screening before any signs or symptoms begin to appear. This is a novel approach as there is presently no device that can be used by women themselves without any supervision from a medical professional and uses a smartphone without any additional external devices for breast cancer screening. The smartphone mammography brings the screening facility to the user such that it can be used at the comfort and privacy of their homes without the need to travel long distances to hospitals or diagnostic centers. The theory of the system is that when visible light penetrates through the skin into the breast tissue, it reflects back differently in normal breast tissue as compared to tissue with anomalies. A phantom breast model, which mimics real human breast tissue, is used to develop the modality. We make use of computer vision and image processing techniques to analyze the difference between an image taken of a normal breast and that of one with irregularities in order to detect lumps in the breast tissue and also make some diagnosis on its size, density and the location.

Keywords: Smartphone · Breast cancer screening · Early detection · Mammography

1 Introduction

Breast cancer is the most common cancer diagnosed in woman worldwide. It is also the second most common cancer overall [1]. Nearly one in nine women will develop breast

S. N. Srirama et al. (Eds.): BDA 2021, LNCS 13147, pp. 16–30, 2021.
https://doi.org/10.1007/978-3-030-93620-4_2

cancer at some point in their lives, among which 30% are terminal [2]. Early detection is vital in any cancer. However, in case of breast cancer more treatment options are available if detected early and it also increases the chances of survival by 40%.

In the United States (US) the mortality rates due to breast cancer reduced by 40% from 1989 to 2017. This decline is attributed to increased mammography screening rates starting from mid-1980's [3]. In contrast, breast cancer in LMIC (Low- and Middle-Income Countries) represents one-half of all cancer cases and 62% of the deaths globally [4]. Breast cancer disproportionately affects young women in LMICs, such that 23% of new breast cancer cases occur among women aged 15 to 49 years in LMICs versus 10% in high-income countries [5].

In India, the latest statistics suggest that in the near future breast cancer will surpass cervical cancer, which is currently the most common gynecological cancer among women in India [6]. This increase in incidence rate comes with an increased mortality rate which can be attributed to non-existent screening facilities, especially among the underserved population. Studies have shown that not only women coming from low-income rural backgrounds but also metropolitan cities, despite understanding the importance of early detection and having access to medical facilities, did not seek medical help when required [7]. Women, especially from rural background, have cited embarrassment, which can be due to the presence of male health workers or doctors, and cost as factors that dissuade them from screening. They are also often discouraged from screening by family members and community due to prevailing cultural norms, conservative mindset and social stigma associated with traditional screening methods [8].

One of the most commonly used breast cancer screening technique is mammography, which utilizes low dosage X-ray. However, it has a significant false positive rate of approximately 22% in women under 50 and is also not recommended for women under 30 years of age due to radiation exposure [9]. Magnetic resonance imaging (MRI) and Ultrasound are screening techniques that are sometimes used in addition to mammography for more accurate results, but they too have limitations such as limited specificity for MRI and low sensitivity for ultrasound. One of the major drawbacks of the above-mentioned techniques is that they are available only at diagnostic centers in cities. Some of these equipment's are available in mobile clinics and rarely available for the rural and the disadvantaged population. They are also expensive screening methods which discourages women from low-income background to use them.

Handful of research studies suggest the non-viability of mammography in India, considering the cost benefit analysis [10]. Alternatives are recommended for screening. Therefore, it is imperative that we develop an imaging system for early detection of breast cancer that can be used by women, especially the underserved and vulnerable group who do not have access to medical facilities. A system that can be used by the women themselves so the stigma and embarrassment associated with being in contact with male doctors or health workers can be avoided. It must be affordable, easy to use, accessible, safe and can be used privately.

Keeping all these in mind we have developed a novel smartphone mammogram system that uses only the camera and flashlight that are inbuilt in all smartphones. While there are a few papers which have also proposed the use of smartphone for cancer detection, they require an additional external device which makes the system expensive, bulky and difficult to use for general public by themselves. Our system is non-invasive and fully functional without any external attachments which makes it easy to use for women by themselves privately without any assistance from a medical professional. It is safe as it uses visible light from the flashlight, to do screening, which we are exposed to in our everyday life. Furthermore, the system is affordable and easily accessible as the average cost of a smartphone in India is $150 while the average cost of a single mammography is $106 [11]. Due to its affordability, smartphone penetration in rural India is over 59% as of February 2021 and in the next 3 years the number of smartphone users in rural areas will be higher than that of urban areas [12]. The smartphone mammography system is especially useful for the underserved and disadvantaged population who cannot afford and do not have easy access to medical facilities by bringing the medical facility at the doorstep of the beneficiary thereby eliminating the need to travel long distances to hospitals or diagnostic centers. Using the camera and flashlights in smartphones we detect anomalies in the breast tissue using visible spectrum of light that is portable, compact, and economically viable. The idea we worked on is that the absorption and reflection of visible lights will be different for normal tissues and lumps in the breast. For the purpose of experiment, we used Blue Phantom's breast model [13]. Image processing and computer vision techniques are used for analysis and detecting presence of lumps in the breast.

2 Related Work

The use of light as a way to take images of the breast is quickly emerging as a non-invasive approach for breast cancer screening. Several authors have proposed methods that use visible and near-infrared spectral window of light for non-invasive imaging. One paper presents Diffuse Optical Tomography (DOT) for breast imaging which uses near-infrared light to examine tissue optical properties [14]. In DOT the near-infrared light sequentially excites the breast tissue at several locations and then the scattered light is measured at several locations at the surface of the breast which is used to compute parameters such as the concentration of water, lipid, oxy-hemoglobin and deoxy-hemoglobin [15]. These parameters are used to calculate total oxygen saturation and hemoglobin concentration which is linked to angiogenesis, which in turn is critical for autonomous growth and the spread of breast cancer. Although it is being increasingly used as a breast cancer detection technique, DOT is bulky, has high cost and requires complete darkness [16].

Another study discusses Hyperspectral Imaging (HSI) for detection of breast cancer which uses a broad range of light which covers the ultraviolet, visible, and near-infrared regions of the electromagnetic spectrum. This technique captures hundreds of images in contiguous, narrow, and adjacent spectral bands to create a 3D hypercube that contains spatial as well as spectral data of the imaged scene. It measures diffusely reflected light after it has experienced several scattering and absorption events inside the tissue, creating an optical fingerprint of the tissue which displays the composition and morphology of the tissue, that can be utilized for tissue analysis [17]. HSI is however costly and complex which makes it difficult to use. Although DOT and HSI are non-invasive techniques, they are either expensive which makes it unaffordable for people with low incomes or they are bulky and complex which makes it difficult for people to use on themselves for self-test.

In one of the pioneering works, Peters et al. measured the scattering coefficients of optical absorption for specimens of normal and diseased human breast tissues. They measured the total attenuation coefficients for thin slices of tissue cut on a microtome using standard integrating sphere techniques [18]. Very recently Taroni did a pilot study on an optical mammograph that operates at 7 red-near infrared wavelengths, in the same geometry as x-ray mammography, but with milder breast compression. The instrument is mounted on wheels and is portable [19].

Several papers have proposed the use of a smartphone along with an external device for cancer screening. Joh et al. developed a chip, along with EpiView — a smartphone-based imaging platform. They developed an external handheld imaging device that could rapidly switch between brightfield and fluorescence modes without extensive re-assembly or recalibration. The switchable imaging units are attached to a common base unit that is mounted on a smartphone [20]. Another paper proposes the use of smartphone with an infrared camera for breast cancer detection using infrared thermography. The infrared camera takes thermal infrared images of the breast which are then sent to the smartphone for analysis [21]. Papers [22] and [23] discuss the use of AI and Machine Learning for disease prediction and classification. They used the Breast Cancer Wisconsin (Diagnostic) dataset for breast cancer prediction and detection.

3 System Description

Light reflects, scatters, and absorbs differently in healthy normal tissue and tissue with anomalies [18]. When light penetrates human tissue, it goes through multiple scattering and absorption events. The depth of light penetration into human tissue depends on how strongly the tissue absorbs light. Absorption spectra shows the oxygen saturation and concentration of hemoglobin which reveals angiogenesis which is critical for cancer growth. The vital chromophores-tissue components absorbing light for visible light are blood and melanin whose absorption coefficient decreases with increase in wavelength of light. When light falls on human tissue surface it is either reflected directly or scattered due to spatial variations in tissue density and then reemitted to the surface. Due to multiple scattering, the reemitted light is random in direction which is known as diffuse reflectance.

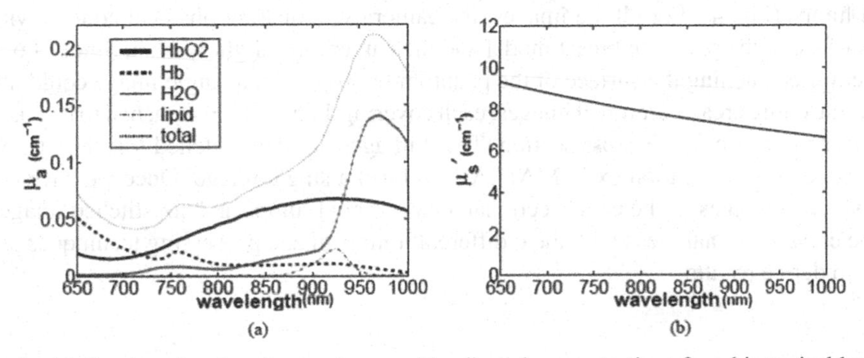

Fig. 1. (a) Spectra of major chromophores with adjusted concentrations found in typical breast tissue (b) Scattering spectra

The diffuse reflectance signal can be used to probe the microenvironment of the tissue non-invasively as it contains information about scattering and absorbing components deep within the tissue. Alterations in tissue morphology due to disease (breast cancer) progression can affect scattering signals which should lead to corresponding changes in the pattern of light reflected from the tissue [24]. We have conducted experiments to determine relation between light reflected and change in size of anomaly, light reflected and change in density of anomaly. Figure 1 (a) shows the spectrum of absorption and scattering in the near-infrared range of major chromophores found in breast tissue such as oxygenated hemoglobin, deoxygenated hemoglobin (Hb), water and lipid [8].

In order to perform experiments, we made use of a Blue Phantom tissue model [13]. The Blue Phantom simulated human tissue model is generally used for training clinicians in the psychomotor skills associated with breast ultrasound guided fine needle biopsy procedures. The model has been created with a highly realistic synthetic human tissues that is able to mimic the physicochemical, mechanical as well as the optical and thermal properties of live tissue. For the mammogram we used the mobile smartphone camera with the inbuilt flashlight.

To simulate the breast cancer tumor, we used lumps of different sizes and densities as shown in Fig. 2. The density of the Blue Phantom breast model is comparable to that of a normal human breast tissue and the lumps are of higher density than the phantom breast model. The breast model includes components such as skin, subcutaneous fat, bulk fat, a natural wear layer of dead skin at the surface and three discrete layers of epidermis, dermis and hypodermis which are present in an actual breast [10]. We used a total of 6 lumps which are categorized as small low-density lump (diameter = 1.2 cm), small high-density lump (diameter = 1 cm), medium low-density lump (diameter = 2.2 cm), medium high-density lump (diameter = 1.9 cm), large low-density lump (diameter = 2.8 cm) and large high-density lump (diameter = 3 cm). All lumps are assumed to be spherical in shape. The actual density of all the lumps is unknow.

To begin with, we choose two 3 cm by 3 cm cross sectional areas on the phantom breast model as shown in Fig. 3. We chose Area 1 for placing the small lumps and Area 2 for placing the medium and the large lumps. At both the areas the lumps are placed at the center of the cross-sectional area which is row 2 column 2 at a depth of 2 cm. First, images of the Area 1 and Area 2 are taken without any lumps and then we place the lumps one by one in their respective areas and take images of the phantom breast model with lumps (Fig. 4). For all the images, the camera was making physical contact with the surface of the phantom breast model and the camera flashlight was on. Since phone camera was touching the surface of the phantom breast model a single image could not cover the entire area, therefore 9 images, each covering 1 cm $\times$ 1 cm area, had to be taken to cover the entire 3 $\times$ 3 cross section. The 9 images are then stitched together, using the image processing toolbox in MATLAB, to form a single image. Once we have the two stitched images of the cross-sectional areas without lumps and six stitched images of the cross-sectional areas with the 6 different lumps, image processing techniques are used to derive results.

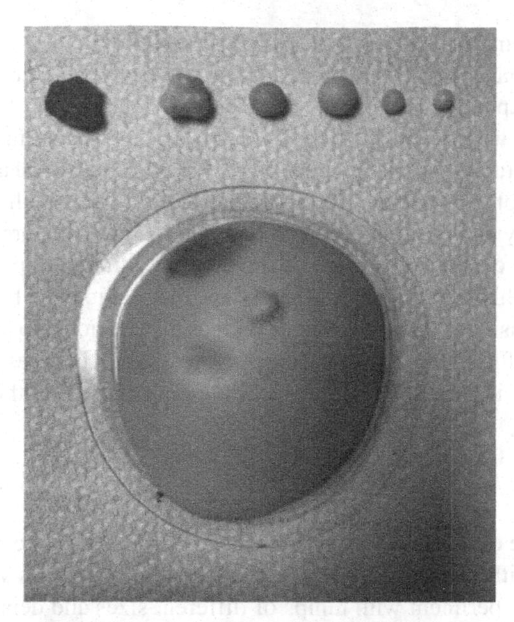

Fig. 2. Phantom breast model with 6 lumps

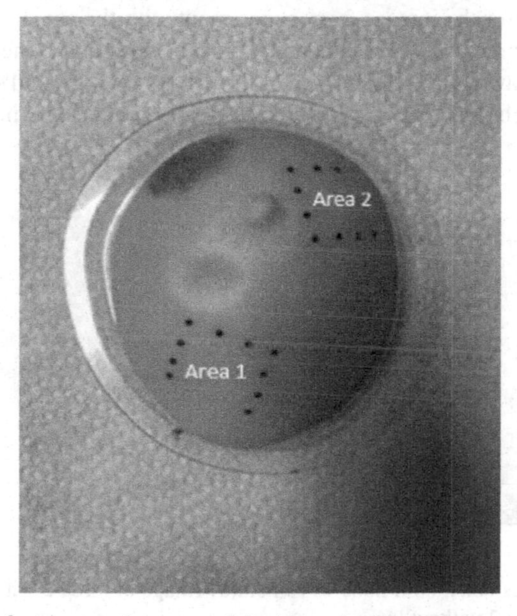

Fig. 3. Phantom breast model with marked 3 cm × 3 cm areas

This experiment is to determine the relationship between the density of the lump and the intensity of the light received by the smartphone camera, as well as the relationship between the size of lump and the intensity of light. This experiment had many variables such as the overlapping or exclusion of some parts in the 3 cm × 3 cm cross-section when taking the 9 images and a change in angle of image being taken.

Second set of experiments were also performed to eliminate the variables in the previous experiment where instead of the camera touching the surface of the phantom breast model, we captured a single image of the 3 × 3 cross-sectional area by placing the camera at a distance of approximately 2 cm with flashlight on. All other parameters such as depth, size, density and position of lumps being the same as the previous experiment. Here also images are captured of cross-sectional areas without lumps and then with the six types of lumps. This experiment is to determine the relationship between the density of the lump and intensity of light received as well as the relationship between the size of lump and intensity of light. Once again image processing techniques are used to derive some results which will be explained in the Simulations section of the paper.

4 Simulation

We performed all the experiments using iPhone 7s. All images were captured using the iPhone 7s camera with flashlight. All image processing techniques was carried out on MATLAB. For the experiment with lumps of different sizes and densities, background subtraction was performed. Each of the six images with lumps were background subtracted with the two images without lumps based on which lump was placed in which area. This was done for the images from experiment with the camera touching the surface of the phantom breast as well as the images from experiment where the camera was approximately 2 cm away from the surface of the breast. Background subtraction allows us to see how the light has reflected differently due to scattering when there is a lump present.

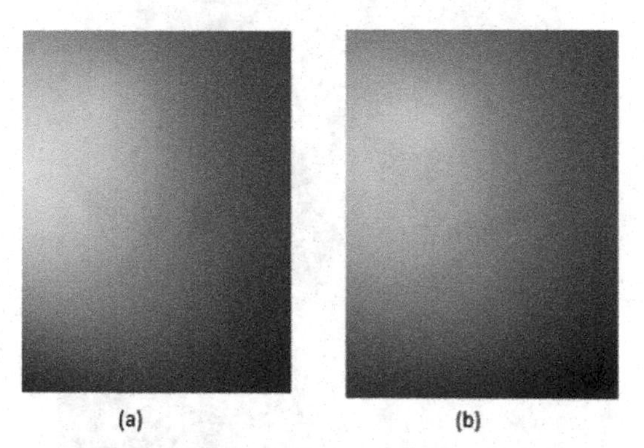

Fig. 4. Original images taken on iPhone 7s camera(a) Image of Area 2 with large high-density lump (b) Image of Area 2 without lump

The MATLAB function '`imsubtract`' is used to perform background subtraction. This function subtracts each element in one image with the corresponding element in the other image and when the subtraction goes into negative the function normalizes it to zero. In cases where the subtraction results in negative value, the '`double`' function in MATLAB is used. The 'double' function is used to first convert both images to double and then subtract them without using 'imsubtract'. This way the corresponding pixels of the two images are subtracted which gives a normalized RGB scale image that is easier to understand and visualize. The normalized images are then analyzed to see the relation between size and density of lump with light intensity. In order to do this, we first find the average RGB values of the image by adding the square of RGB values of each pixel and dividing by the total number of pixels and take the square root of the result. If $(n \times m)$ is the pixel dimension of the image then the below equations are used to calculate the average of RGB values.

$$R_{avg} = \sqrt{\frac{(R_{1\times1})^2 + (R_{1\times2})^2 + (R_{2\times1})^2 + (R_{1\times1})^2 + \cdots + (R_{n\times m})^2}{(n \times m)}}$$

$$G_{avg} = \sqrt{\frac{(G_{1\times1})^2 + (G_{1\times2})^2 + (G_{2\times1})^2 + (G_{1\times1})^2 + \cdots + (G_{n\times m})^2}{(n \times m)}}$$

$$B_{avg} = \sqrt{\frac{(B_{1\times1})^2 + (B_{1\times2})^2 + (B_{2\times1})^2 + (B_{1\times1})^2 + \cdots + (B_{n\times m})^2}{(n \times m)}}$$

Then the intensity is found by adding the RGB values and dividing it by three. The size of lump is then plotted against intensity values to get an insight on the relation between them. The below equation shows the intensity of an image.

$$Image\ Intensity = \frac{R_{avg} + G_{avg} + B_{avg}}{3}$$

5 Results

Figure 5 shows the result of subtracting large high-density image from no lump image for the experiment where the camera touches the phantom surface. The parts of the image that is black means that there is no difference in the light intensity between the images with and without lump.

When there is no difference in light intensity it means that light reflected and refracted from a surface of the same density. The part of the image which has the most color is where the light has scattered most when reflecting. As mentioned before, light reflects differently when there are anomalies in the tissue which results in change in light intensity. Therefore, according to Fig. 5 the lump is located in the middle cell of the right most row.

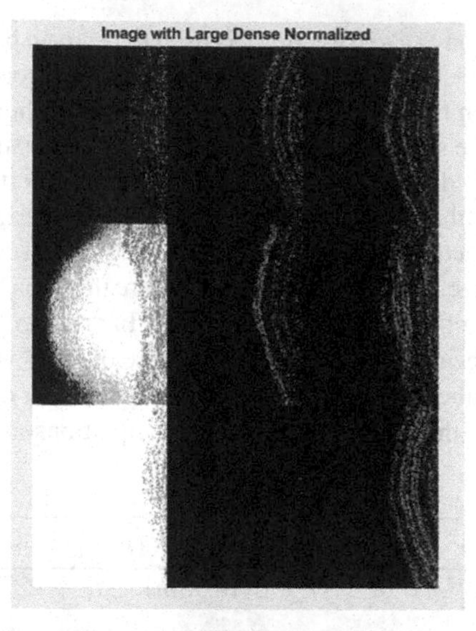

Fig. 5. Large high-density lump when camera in contact with phantom surface

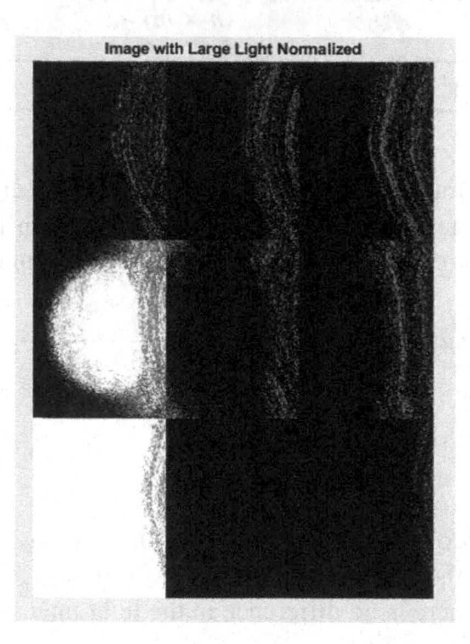

Fig. 6. Large low-density lump when camera in contact with phantom surface

Figure 6 shows the result of subtracting large low-density image from no lump image for the experiment where the camera touches the phantom surface. Similar to Fig. 5 we can see the lump at the middle cell of the rightmost row. When Figs. 5 and 6 are compared we see that large high-density lump image has lower intensity than the large low-density lump. This is because when the density is higher there is more scattering of light which reduces the intensity of light reflected.

Figures 7 and 8 show the result of subtracting medium high-density and medium-low density image from no lump image for the experiment where the camera is touching the phantom surface respectively. Similar to Figs. 5 and 6 we can see that most of the image is black except for the middle cell in rightmost row where the lump is located.

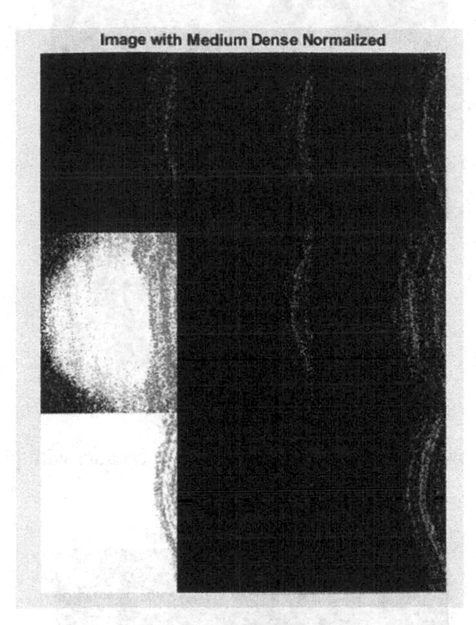

Fig. 7. Medium high-density lump when camera in contact with phantom surface

Figure 9 shows the result of subtracting large high-density lump from no lump image wherein both images are taken where the camera is 2 cm above the phantom surface. Here also we see that most of the image is black which means the breast tissue is normal and the areas that have color is where the lump is located.

In all of the images where the camera is in contact with the breast surface there is some scattered red color outside of the area that has been identified as where the lump is located. This is due to camera position and angle being different between the images taken without lump and with lump due to which the light has reflected with different intensity. However, in Fig. 9 where the camera was placed 2 cm above the phantom surface the red scattering is far lesser since only a single image has been taken here which eliminates the problem of changes in camera position and angles. In all of the above images the lumps have been detected in the same area which demonstrates that the position of the lump has been identified accurately regardless of lump size and density.

Figures 10 and 11 show graphs of diameter of lumps vs their intensity value when the camera is in contact with the phantom breast model surface. It can be inferred from the graph that the lumps with low density have a higher intensity value than the lumps with high density for all sizes. This is because high density lumps cause more scattering of light which reduces its intensity. We can deduce from the graph that larger the size of the lump, lower the intensity value. This is because when the size is large, light is

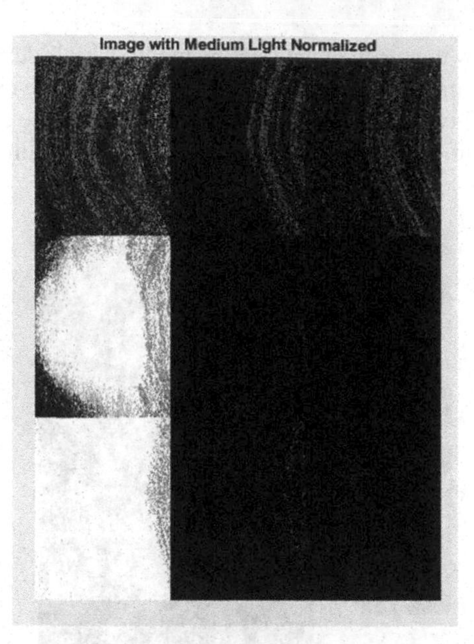

Fig. 8. Medium low-density lump when camera in contact with phantom surface

Fig. 9. Normalized large high-density lump when camera is 2 cm above phantom surface

reflected differently and scattered over a greater surface area which reduces the intensity of light.

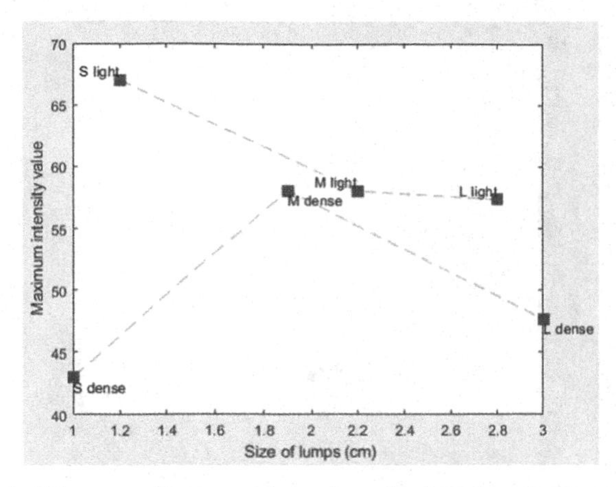

Fig. 10. Diameter of lumps vs intensity value based on density of lump

In Fig. 10 we can spot some irregularities in the light intensity values such as the small high-density lump with lower value than medium and large high-density lumps. The medium and large low-density lumps have almost the same value. These irregularities are due to changes in the angle or positioning of the camera or some parts of the image may be missing or overlapping when the 9 1 cm by 1 cm images taken are stitched as mentioned in Sect. 4. These irregularities can be reduced when the camera does not touch the surface of the phantom model. Table 1 shows the diameter of lumps versus intensity value when the camera is not in contact with the breast model. Here we can see that as the size of the lump increases the light intensity decreases and it also eliminates some irregularities seen when camera touches the surface of the phantom model.

Table 1. Size of lumps and its associated light intensity in no-contact case

Lump description	Lump size (cm)	Light intensity
Small low-density lump	1.2	73
Small high-density lump	1.0	68
Medium low-density lump	2.2	59
Medium high-density lump	1.9	57
Large low-density lump	2.8	53
Large high-density lump	3.0	48

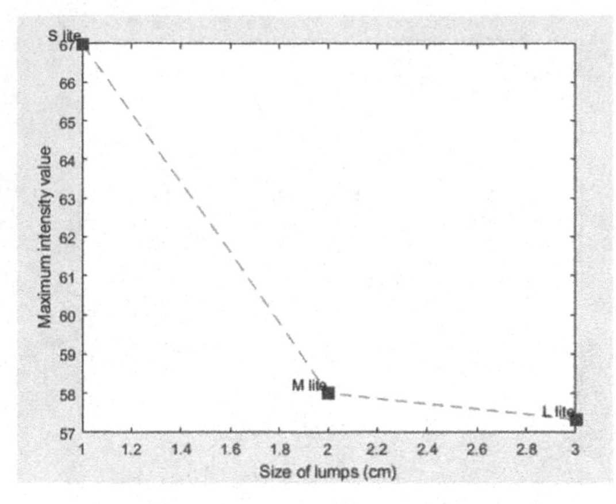

Fig. 11. Diameter of lumps vs Intensity Value based on size of lump

Therefore, we can not only detect the presence of anomaly in the breast tissue by performing image processing techniques we can also make some assumptions on the size and the density of the anomaly present in the breast tissue by analyzing the intensity values. This system has now been implemented using Flutter where the images are taken on an android smartphone and the image processing carried out at the backend in Python.

6 Conclusion and the Future Work

The basic goal of this work is to propose a technology that is easily accessible, affordable, and widely available for breast cancer screening. Inbuilt camera and flashlight of a smartphone serves that very well. We propose optical tomography to validate the viability of smartphone mammography. This is a unique method as it is the first time that only a smartphone, without any additional external device, is being used for breast cancer screening. Moreover, it also has the potential to overcome the socio-cultural barrier as a woman can herself use the smartphone without the help of a trained medical professional.

Our work identifies the anomaly in the tissue if any and provide an approximation of the size and position of this lump or anomaly based on the intensity of the reflected light. One of the many challenges we faced when performing the experiments is the precise positioning of the camera to enable decent background subtraction. This is important if we aim to deduce the position or the size of the lump. It is possible that only a portion of the lump is captured owing to camera positioning errors. Future experiments would aim to stabilize the light source and detector (camera) and only move the breast model around to accommodate for the positioning errors. As the flashlight on the camera is not very strong as a light source, it would be prudent to decouple the light source and detector system which in turn will also eliminate lighting and shadow issues due to ambient light.

Currently, we are working with a single source/detector system but we plan to leverage the use of multiple source/detector pairs placed alternatively and excited one after the other to achieve precise reconstruction of the tissue and for error correction. We also hope to do human trials on women with and without breast cancer to validate our system. Moreover, we are porting the MATLAB algorithms presented in this paper into Python3 and use the smartphone mammograph as an AIoT device. The AIoT smartphone mammograph will be connected to the oncologists in remote towns through the telemedicine system [25].

References

1. Labib, N.A., Ghobashi, M.M., Moneer, M.M., Helal, M.H., Abdalgaleel, S.A.: Evaluation of BreastLight as a tool for early detection of breast lesions among females attending National Cancer Institute, Cairo University. Asian Pac. J. Cancer Prev. **14**(8), 4647–4650 (2013). https://doi.org/10.7314/apjcp.2013.14.8.4647
2. American Cancer Society: Breast Cancer Facts & Figures 2019–2020. American Cancer Society, Inc., Atlanta (2019)
3. Narod, S.A., Iqbal, J., Miller, A.B.: Why have breast cancer mortality rates declined? J. Cancer Policy **5**, 8–17 (2015). https://doi.org/10.1016/j.jcpo.2015.03.002
4. Vieira, R.A., Biller, G., Uemura, G., Ruiz, C.A., Curado, M.P.: Breast cancer screening in developing countries. Clinics (Sao Paulo) **72**(4), 244–253 (2017). https://doi.org/10.6061/clinics/2017(04)09
5. Forouzanfar, M.H., Foreman, K., Lozano, R.: The Challenge Ahead: Progress and Setbacks in Breast and Cervical Cancer. Institute for Health Metrics and Evaluation (IHME) Seattle (2011)
6. Singh, S., Shrivastava, J.P., Dwivedi, A.: Breast cancer screening existence in India: a non-existing reality. Indian J. Med. Paediatr. Oncol. **36**(4), 207–209 (2015). https://doi.org/10.4103/0971-5851.171539
7. Kathrikolly, T., Shetty, R., Nair, S.: Opportunities and barriers to breast cancer screening in a rural community in Coastal Karnataka, India: a qualitative analysis. Asian Pac. J. Cancer Prev. **21**(9), 2569–2575 (2020). https://doi.org/10.31557/APJCP.2020.21.9.2569
8. Parsa, P., Kandiah, M., Rahman, H.A., Zulkefli, N.M.: Barriers for breast cancer screening among Asian women: a mini literature review. Asian Pac. J. Cancer Prev. **7**(4), 509–514 (2006)
9. Choe, R.: Diffuse Optical Tomography and Spectroscopy of Breast Cancer and Fetal Brain (2005)
10. Kumar, J.U., Sreekanth, V., Reddy, H.R., Sridhar, A.B., Kodali, N., Prabhu, A.: Screening mammography: a pilot study on its pertinence in Indian population by means of a camp. J. Clin. Diagn. Res. **11**(8), TC29–TC32 (2017)
11. The Hindu. https://www.thehindu.com/sci-tech/health/Breast-cancer-mdash-a-wake-up-call-for-Indian-women
12. Singh, J.: Growth and potential of wireless internet user in rural India. Psychol. Edu. An Interdis. J. **58**(2), 1010–1022 (2021)
13. Blue Phantom Homepage. https://www.bluephantom.com
14. Kim, M.J., Su, M.Y., Yu, H.J.: US-localized diffuse optical tomography in breast cancer: comparison with pharmacokinetic parameters of DCE-MRI and with pathologic biomarkers. BMC Cancer **16**(1) (2016). Article no. 50. https://doi.org/10.1186/s12885-016-2086-7
15. Hadjipanayis, C.G., Jiang, H., Roberts, D.W., Yang, L.: Current and future clinical applications for optical imaging of cancer: from intraoperative surgical guidance to cancer screening. Semin. Oncol. **38**(1), 109–118 (2011). https://doi.org/10.1053/j.seminoncol.2010.11.008
16. Lannaccone, S., Gicalone, M., Berettini, G., Potí, L.: An innovative approach to diffuse optical tomography using code division multiplexing. Europ. Fut. Technol. Conf. Exhib. **7**, 202–207 (2011). https://doi.org/10.1016/j.procs.2011.09.056

17. Lin, J.L., Ghassemi, P., Chen, Y., Pfefer, J.: Hyperspectral imaging with near infrared-enabled mobile phones for tissue oximetry. Opt. Biophotonics Low-Res. Settings IV **10485**, 895–910 (2018). https://doi.org/10.1117/12.2290870
18. Peters, V.G., Wyman, D.R., Patterson, M.S., Frank, G.L.: Optical properties of normal and diseased human breast tissues in the visible and near infrared. Phys. Med. Biol. **35**(9), 1317–1334 (1990). https://doi.org/10.1088/0031-9155/35/9/010
19. Taroni, P., Paganoni, A.M., Ieva, F., Pifferi, A.: Non-invasive optical estimate of tissue composition to differentiate malignant from benign breast lesions: a pilot study. Sci. Rep. **7** (2017). Article no. 40683. https://doi.org/10.1038/srep40683
20. Joh, D.Y., Heggestad, J.T., Zhang, S., Anderson, G.R. et al.: Cellphone enabled point-of-care assessment of breast tumor cytology and molecular HER2 expression from fine-needle aspirates. NPJ Breast Cancer **7** (85), (2021). https://doi.org/10.1038/s41523-021-00290-0
21. Ma, J., et al.: A portable breast cancer detection system based on smartphone with infrared camera. Vibroeng. PROCEDIA **26**, 57–63 (2019). https://doi.org/10.21595/vp.2019.20978
22. Hailemariam, Y., Yazdinejad, A., Parizi, R.M., Srivastava, G. et al.: An empirical evaluation of AI deep explainable tools. In: IEEE Globecom Workshops, 1–6 (2020). https://doi.org/10.1109/GCWkshps50303.2020.9367541
23. Koppu, S., Maddikunta, P.K.R., Srivastava, G.: Deep learning disease prediction model for use with intelligent robots. Comput. Electr. Eng. Int. J. **87** (2020). Article no. 106765. https://doi.org/10.1016/j.compeleceng.2020.106765
24. Lu, G., Fei, B.: Medical hyperspectral imaging: a review. J. Biomed. Opt. **19**(1) (2014). Article no. 010901. https://doi.org/10.1117/1.JBO.19.1.010901
25. Talukder, A.K., Haas, R.E.: AIoT: AI meets IoT and web in smart healthcare. In: 13th ACM Web Science Conference 2021 (WebSci '21 Companion), ACM, New York, NY, USA (2021). https://doi.org/10.1145/3462741.3466650

Bridging the Inferential Gaps in Healthcare

Asoke K. Talukder[1,2,3](✉)

[1] Computer Science and Engineering, National Institute of Technology Karnataka, Surathkal, India
asoke.talukder@sritindia.com
[2] SRIT India Pvt Ltd., Bangalore, India
[3] Cybernetic Care, Bangalore, India

Abstract. Inferential gaps are the combined effect of reading-to-cognition gaps as well as the knowledge-to-action gaps. Misdiagnoses, medical errors, prescription errors, surgical errors, under-treatments, over-treatments, unnecessary lab tests, etc. – are all caused by inferential gaps. Late diagnosis of cancer is also due to the inferential gaps at the primary care. Even the medical climate crisis caused by misuse, underuse, or overuse of antibiotics are the result of serious inferential gaps. Electronic health records (EHR) had some success in mitigating the wrong site, wrong side, wrong procedure, wrong person (WSWP) errors, and the general medical errors; however, these errors continue to be quite significant. In the last few decades the disease demography has changed from quick onset infectious diseases to slow onset non-communicable diseases (NCD). This changed the healthcare sector in terms of both training and practice. In 2020 the COVID-19 pandemic disrupted the entire healthcare system further with change in focus from NCD back to quick onset infectious disease. During COVID-19 pandemic misinformation in social media increased. In addition, COVID-19 made virtual healthcare a preferred mode of patient-physician encounter. Virtual healthcare requires higher level of audit, accuracy, and technology reliance. All these events in medical practice widened the inferential gaps further. In this position paper, we propose an architecture of digital health combined with artificial intelligence that can mitigate these challenges and increase patient safety in the post-COVID healthcare delivery. We propose this architecture in conjunction with *diseasomics*, *patholomics*, *resistomics*, *oncolomics*, *allergomics*, and *drugomics* machine interpretable knowledge graphs that will minimize the inferential gaps. Unless we pay our attention to this critical issue immediately, medical ecosystem crisis that includes medical errors, caregiver shortage, misinformation, and the inferential gaps will become the second, if not the first leading cause of death by 2050.

Keywords: Inferential gaps · Reading-to-cognition gap · Knowledge-to-action gap · Digital health · Knowledge graph · AI · IoT · Patient temporal digital twin · Patient spatial digital twin · Patient molecular digital twin · Physician digital twin · Digital triplet · Healthcare science · Medical ecosystem crisis

S. N. Srirama et al. (Eds.): BDA 2021, LNCS 13147, pp. 31–43, 2021.
https://doi.org/10.1007/978-3-030-93620-4_3

1 Introduction

The inferential gap is the gap between the accurate true knowledge and the consumed or used knowledge. Inferential gaps may be the reading-to-cognition gaps at the cognition level or the knowledge-to-action gaps at the decision making level. Reading-to-cognition gap is the distance between interpreted knowledge by a caregiver at the cognition level compared to the accurate error-free knowledge. Knowledge-to-action gap is the gap between the knowledge applied by a caregiver at a point-of-care and the accurate knowledge required to make an error-free decision.

In the context of healthcare, knowledge-to-action gap can be defined as the distance between knowledge required for error-free medical decision and erroneous knowledge used by a physician or a nurse during patient encounter. In contrast, reading-to-cognition gap refers to the gap between the true error-free knowledge and the way it is presented or understood by a caregiver. This can be attributed to underdeveloped or incomplete evidence, incomplete design of system and processes of care, as well as the inability to accommodate every patient's diverse demand and needs. Inferential gaps are measured by the quantification of medical errors, the adverse events, never events, patient injury, and medical ecosystem change.

Misdiagnoses, medical errors, prescription error, surgical errors, under-treatment, over-treatment, unnecessary pathological test or unnecessary radiology orders, etc. – are all caused by the inferential gaps during a patient-physician encounter. Even the late diagnosis of cancer is caused by inferential gaps at the primary care. Antibiotic resistance crisis is caused by inferential gaps as well.

A physician is required to make a perfect decision with imperfect information. During a clinical decision, a professional is required to "fill in" where they lack knowledge or evidence. This is also the case in empirical decision making under uncertainty and missing or unknown knowledge. The breadth of the inferential gap varies according to the experience of the physician, the availability of the medical knowledge and its relevance to clinical decision making. It also depends on the active memory and the recall capability of the caregiver. There is another dimension of inferential gap due to the time lag between clinical research outcome and its use at the point-of-care which is estimated to be 17 years [1].

A landmark report "To Err Is Human: Building a Safer Health System" released in November 1999 by the U.S. Institute of Medicine (IOM) resulted in increased awareness of U.S. medical errors [2]. The report was based upon two large analyses of multiple studies; one conducted in Colorado and Utah and the other in New York, by a variety of organizations. The report concluded that between 44,000 to 98,000 people die each year as a result of preventable medical errors. In Colorado and Utah hospitals, 6.6% of adverse events led to death, as compared to 13.6% in New York hospitals. In both of these studies, over half of these adverse events were preventable. For comparison, fewer than 50,000 people died of Alzheimer's disease and 17,000 died of illicit drug use in the same year. As a result of the report more emphasis was put on patient safety such that President Bill Clinton signed the Healthcare Research and Quality Act of 1999.

A 2016 study in the US placed the yearly death rate due to medical error in the U.S. alone at 251,454. This study found that medical error was the third leading cause of deaths in the US only after heart attack and cancer [3]. A 2019 meta-analysis identified 12,415

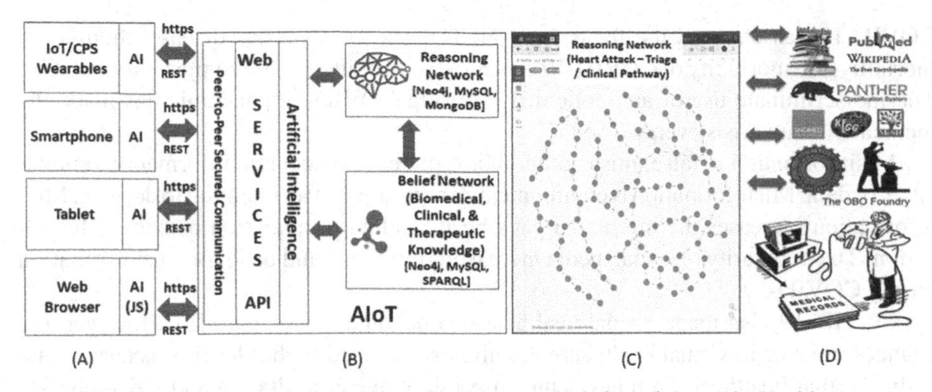

Fig. 1. No-error architecture that will eliminate the inferential gaps in Healthcare

scientific publications related to medical errors and outlined the impactful themes of errors related to drugs/medications, medicinal information technology, critical/intensive care units, children, and mental condition (e.g., burnout, depression) of caregiver. The study concluded that the high prevalence of medical errors revealed from the existing literature indicates the criticality of future work invested in preventive approaches [4]. Though all these reports were related to US healthcare delivery -- it does not imply that inferential gaps or medical errors are absent outside of the US.

In the last few decades, the disease demography changed across the world from infectious disease to non-communicable diseases (NCD). NCDs kill 41 million people each year, equivalent to 71% of all deaths globally, out of which 77% deaths are in low- and middle-income countries. Each year, more than 15 million people die from NCD between the ages of 30 and 69 years; 85% of these "premature" deaths occur in low- and middle-income countries. Cardiovascular diseases account for most NCD deaths which is 17.9 million people annually, followed by cancers (9.3 million), respiratory diseases (4.1 million), and diabetes (1.5 million). These four groups of diseases account for over 80% of all premature NCD deaths [5].

Infectious diseases are quick onset diseases whereas NCD are slow onset disease with high interdependence of conditions. Low and quick onset infectious diseases need different processes of medical management compared to NCD. The shift from infectious to NCD was mainly due to mass vaccination and the discovery of the miracle drug penicillin and other antibiotics that reduced morbidity and mortality due to viral and bacterial infections respectively. However, indiscriminate abuse of antibiotics caused antibiotic resistance in the bacteria and made most of the antibiotics useless. Antibiotic abuse is an example of serious inferential gap, which is causing antibiotic resistance and likely to overpass the mortality of cancer by 2050 making inferential gap the second leading cause of deaths globally. Antibiotic resistance is called medical climate crisis that is estimated to cost the world up to 100 trillion US Dollars by 2050 [6].

The COVID-19 pandemic has been a wakeup-call for the healthcare sector across the globe. It has disrupted the whole medical ecosystem and the entire healthcare system starting from patientcare to medical education. Healthcare systems, be it in a high-income country or a low-income country is in a critical state and needs urgent attention.

COVID-19 has changed the focus from NCD back to infectious disease, though the mortality and morbidity of COVID-19 was mostly attributed to the comorbidity of NCD. The indiscriminate use of antibiotic drugs during COVID-19 pandemic has made the medical climate crisis even worse [7].

Misinformation is false, misleading, incomplete, or inaccurate information, knowledge, or data. Misinformation is communicated regardless of an intention to deceive. Misinformation increases the inferential gaps be it reading-to-cognition or knowledge-to-action. The popularity of social media increased the spread and belief on misinformation during COVID-19.

COVID-19 also made virtual healthcare to become mainstream [8]. However, the chances of error in virtual healthcare are higher and need higher level of accuracy and audit. Virtual healthcare in many setups use telephone consultation and video consultation without any medical records. Virtual healthcare also makes the medical data vulnerable to ransomware and security attacks.

Cancer disease progression can be divided into three phases, namely, (1) pre-cancerous or premalignancy, (2) malignancy or early-stage cancer, and (3) metastasis. The pre-cancerous and the early-stage cancers show some signs and symptoms that are often misdiagnosed at the primary care due to inferential gaps. This reading-to-cognition and knowledge-to-action gap prevents the timely referral to the specialized care. Cancer mortality and morbidity can be reduced substantially if phase-1 and phase-2 diagnosis are made efficient with oncology knowledge available at the primary care.

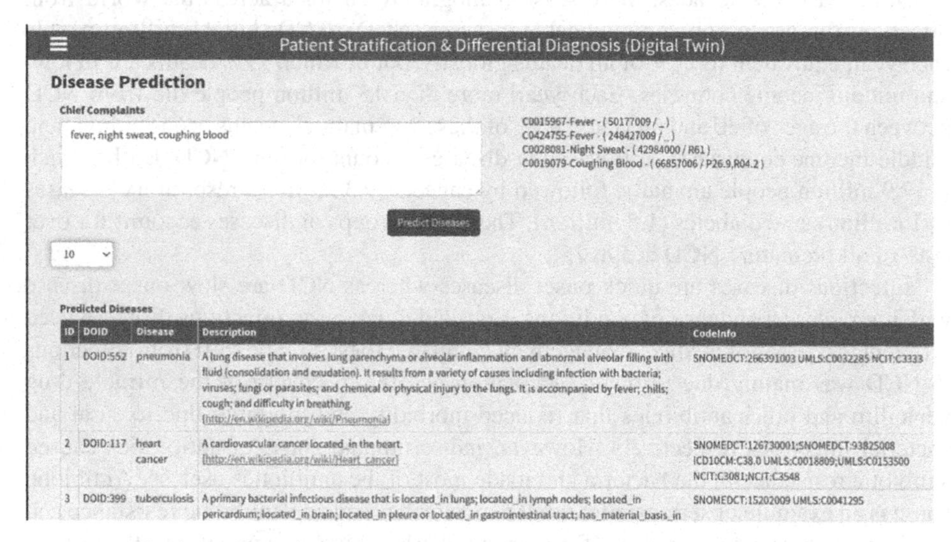

Fig. 2. Differential diagnosis at the point-of-care by No-error technology

In this position paper we discuss the modalities of how the inferential gaps in healthcare can be eliminated in the post-COVID era and in virtual healthcare. Patient safety will be ensured with the use of No-error architecture as shown in Fig. 1. In this figure, (A) shows various user interfaces (smartphone, Web) for nurses, physicians, surgeons, and wearables, IoT, CPS (Cyber Physical Systems) devices for caregivers and patients, etc., (B) shows the basic architecture of secured peer-to-peer communication with IoT, AI and knowledge graphs, (C) shows the rule-based reasoning engine, and (D) shows the knowledge sources that include actionable and machine interpretable knowledge networks. To realize No-error here we propose healthcare science that will include digital health, digital twins, digital triplets, machine interpretable actionable knowledge, and the use of artificial intelligence (AI).

2 Digital Health

Digital health is a multi-disciplinary domain involving many stakeholders with a wide range of expertise in healthcare, engineering, biology, medicine, chemistry, social sciences, public health, health economics, and regulatory affairs. Digital health includes software, hardware, telemedicine, wearable devices, augmented reality, and virtual reality. Digital health relates to using data science, information science, telecommunication, and information technology to remove all hurdles of healthcare.

EHR (Electronic Health Records) is at the center of digital health transformation. EHR systems are made up of the electronic patient chart and typically include functionality for computerized provider order entry (CPOE), medical notes, laboratory, pharmacy, imaging, reporting, and medical device interfaces. Ideally, the system creates a seamless, legible, comprehensive, and enduring record of a patient's medical history and treatment or a digital twin of the patient. EHRs have been widely adopted for both inpatient and outpatient settings that reduced the medical errors and increased the patient safety. The EHR is currently underutilized – it is used only as a repository of medical encounter records. However, it is required to mature as a knowledge source.

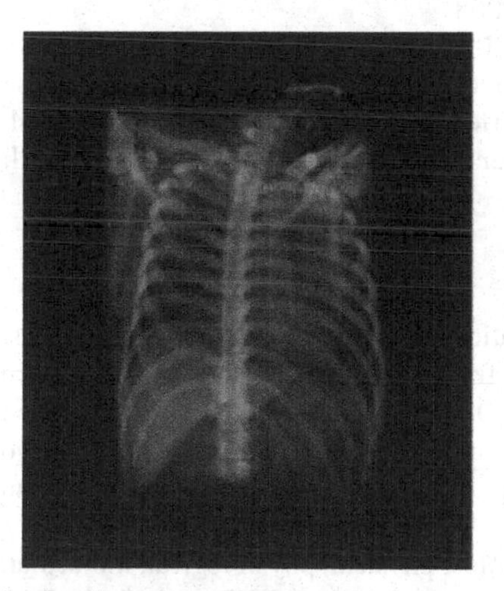
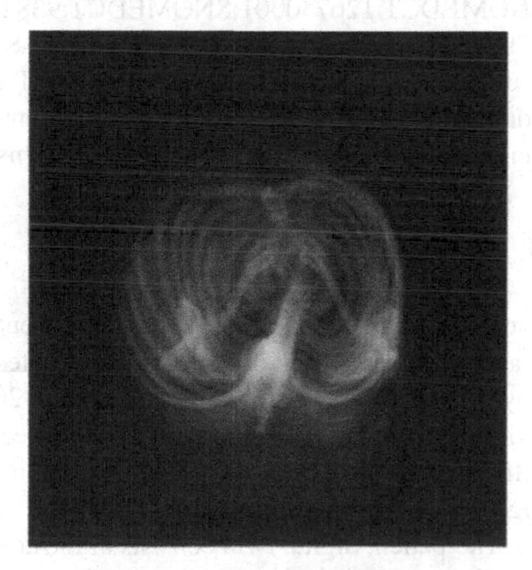

Fig. 3. The volume rendering of chest CT image using Computer Vision and WebGL as part of healthcare science.

A study conducted from 209 primary care practices between 2006–2010 showed that, within all domains, EHR settings showed significantly higher rates of having workflows, policies and practices that promote patient safety compered to paper record settings. While these results were expected in the area of medication management, EHR use was also associated with reduction of inferential gaps in areas that had no prior expectation of association [9]. Moreover, EHR increased the efficiency in insurance claims settlements. In a review of EHR safety and usability, researchers found that the switch from paper records to EHRs led to decreases in medication errors, improved guideline adherence, and enhanced safety attitudes and job satisfaction among physicians. There are indications to suggest that WSWP errors are reduced through the use of EHR as well. However, the transition to this new way of recording and communicating medical information has also introduced new possibilities of errors and other unanticipated consequences that can present safety risks [10]. The scale down version of EHR is personal health record (PHR) which is suitable for medical records storage and retrievals in small clinics or individual family doctors who cannot offer investments in large EHR.

Figure 2 shows a knowledge-driven digital health example. This use case uses knowledge graph and AI as presented in Fig. 1. The upper part of this Fig. 2 shows elimination of reading-to-cognition gaps, whereas the lower half shows elimination of knowledge-to-action gaps. In the upper half, the clinician enters the signs and symptoms as described by the patient in the 'Chief Complaints' section of the screen. In this example, symptoms are "fever", "night sweat", "coughing blood". The human understandable symptoms are converted into machine interpretable UMLS, ICD10, and SNOMED codes using NLP and MeteMap2020 [11]. The machine interpretable codes are C0015967-Fever - (50177009/_), C0424755-Fever - (248427009/_), C0028081-Night Sweat - (42984000/R61), C0019079-Coughing Blood - (66857006/P26.9,R04.2). These three symptoms co-occur in three diseases namely, DOID:552 (pneumonia: SNOMEDCT:266391003 UMLS:C0032285 NCIT:C3333), DOID:117 (heart cancer: SNOMEDCT:126730001;SNOMEDCT:93825008 ICD10CM:C38.0, UMLS:C0018809;UMLS:C0153500, NCIT:C3081;NCIT:C3548), and DOID:399 (tuberculosis: SNOMEDCT:15202009, UMLS:C0041295). It may be noted that both input and output have been converted from human understandable English language to machine interpretable codes. This ensures no-error documentation as well.

3 Digital Twin

A digital twin is the accurate digital representation of an object in computers. The first practical definition of digital twin originated from NASA in an attempt to improve physical model simulation of spacecraft in 2010. One of the basic use cases of digital twin is simulation of various states of the object. In the context of healthcare there are two different types of digital twins, namely, (1) the patient digital twin, and (2) the physician digital twin.

The patient digital twin consists of the complete physical, physiological, molecular, and disease lifecycle data of a patient (or person) constructed from EHR, pathological test data, radiology images, medication history, genetic test, behavior, and lifestyle related data. The patient digital twin can be further divided into patient spatial digital twins, patient temporal digital twin, and patient molecular digital twin. The physician digital twin is the physicians' brain or mind that houses the actionable medical knowledge.

The physician digital twin will be constructed from the literature, textbooks, biomedical ontologies, clinical studies, research outcomes, protocols, and the EHR (Fig. 1).

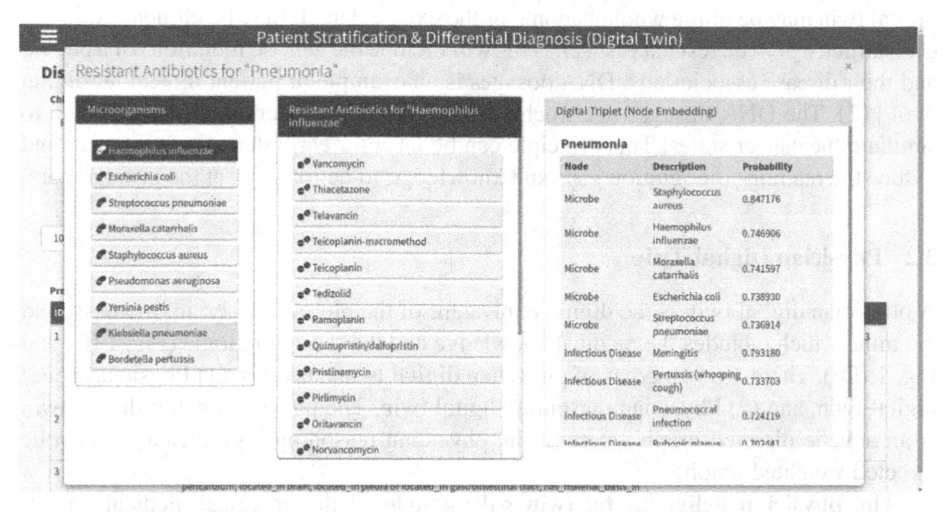

Fig. 4. The 2AI&7D use case to counter the Medical Climate Crisis.

3.1 Patient Digital Twin

The patient digital twin is the digital equivalent of the patient or a patient population. Patient digital twin is further broken down into (1) Patient spatial digital twin, (2) Patient temporal digital twin, and (3) Patient molecular digital twin. These will have all the spatial, temporal, and molecular (genomic) records and history of the patient's life combined with the environmental factors related to the individual's health.

Patient spatial digital twins are constructed from patient's medical records, diagnostic records, medical images, and medication records at any point-in-time. A point-in-time will include a single episode of a disease incidence. A single episode of disease incidence can combine multiple patient-physician interactions over a short time duration. Also, the lifestyle data like exercise, smoking, alcohol, regular health indicators captured by wearables are added as the environmental data. The challenge with this data is that these data may contain unstructured data like medical notes in text, handwritten notes, voice as well as images of ECG, X-rays, or CT/MRI etc. These data are human understandable and cannot be used for simulation. However, EHR combined with AI/ML/DL will convert the human understandable data into machine interpretable data as shown in Fig. 2, that can be used for simulation. Figure 3 shows the digital twin of a chest CT scan of a patient. This is the volume rendering of the 169 CT frames using WebGL. Some impressions are visible in this digital twin which were not visible in the naked eye and the radiologist missed when a normal DICOM Viewer was used.

Patient temporal digital twin is constructed by combining multiple episodes of disease incidence of a patient over a period-of-time. This will include the trajectory of different diseases for a person over a long period of time. This digital twin helps understand the disease and health patterns as a person grows older.

Patient molecular digital twin is the genomic data of a patient or a person. This digital twin may be of the whole genome or the exome data. It may be clinical exome or even targeted genetic test data as well. This will include the genetic mutations of a patient and their disease associations. DNA dosages is an example of patient molecular digital twin [12]. The DNA dosages use stochiometric matrix over genomic (exome) data to simulate the cancer states. This principle can be used for early detection of cancer and reduce the reading-to-cognition gaps and knowledge-to-action gaps at the primary care.

3.2 Physician Digital Twin

A physician digital twin is the digital equivalent of the physician's brain, memory, and the mind which includes the medical knowledge and the reasoning logic (Fig. 1 (C) and Fig. 1 (D)). There are two types of physician digital twins, namely (1) Physician belief digital twin, and (2) Physician reasoning digital twin. The physician belief digital twin is an acyclic directed graph whereas, the physician reasoning digital twin is a cyclic directed weighted graph,

The physician belief digital twin will include all the universal medical knowledge that is essential at any point-of-care – be it primary, secondary, tertiary, or specialized care. Universal medical knowledge includes biomedical ontologies combined with supplementary knowledge from PubMed, textbooks, Wikipedia, clinical research, biomedical networks, and other medical literature.

An ontology is a formal description of knowledge as a set of concepts within a domain and the relationships between them. Open Biological and Biomedical Ontologies at the OBO Foundry contains about 200 ontologies (http://www.obofoundry.org/). These ontologies are manually curated by experts with the best possible knowledge available as of date. The greatest advantage of ontologies is that they are peer-reviewed and eliminate the reading-to-cognition gaps altogether.

Biological and biomedical ontologies are unipartite directed graphs like DOID (Disease Ontology) or GO (Gene Ontology). However, there are few bipartite networks as well like the DisGeNET (Disease gene Network) etc. There are some multipartite networks as well like NCIt, or SNOMED CT. These biomedical ontologies or networks will become knowledge when multiple ontologies and networks are integrated semantically and thematically into a multipartite properties graph. When this multipartite graph is stored in a properties graph database and accessible through application programming interface (API) over Internet, this experts' knowledge becomes knowledge graph. Figure 4 shows the physician digital twin that includes right disease-causing agent, and right drug for antibiotic stewardship.

The physician reasoning digital twin is the reasoning logic used by the physician to arrive at a medical or surgical decision at the point-of-care. It depends on factors that is not always deterministic and cannot be defined simply by if-then-else logic.

4 Digital Triplet

Digital triplet is the third sibling of the digital twins with added intelligence. Digital triplets are constructed from the knowledge graphs using vector embedding. Vector embedding helps construct a lower dimension knowledge space where the structure of the original knowledge is retained. This means that the distance between objects in the embedded space retains the similarity of the original space. Vector embedding of knowledge graph can be done through graph neural network (GNN) or node2vec. Digital triplets help predict missing links and identify labels. The digital triplet will address part of the challenges related to missing and unknown knowledge [13]. Figure 4 shows the digital triplet of pneumonia.

5 Artificial Intelligence and Related Technologies

The patient digital twin will grow as and when more data is generated. EHR data is converted into machine interpretable data through AI. Lifestyle data and health data captured by wearables and sensors will be filtered through AI models. In healthcare all components of AI will be used for the elimination of knowledge-to-action gaps and ensure patient safety. Following are the AI components used to reduce inferential gaps:

1. **Speech Recognition:** During the patient encounter, speech recognition will help capture the conversation and store the interaction in machine readable text during virtual patient-physician encounter.
2. **Speech Synthesis:** Speech synthesis is used for text to speech conversion. In virtual healthcare and home-care this will play a significant role.
3. **Optical Character Recognition (OCR):** Some prescriptions or test reports will be available as scanned image; OCR will be used to extract the content and convert in machine readable format.
4. **Natural Language Processing:** Natural language processing will be used to extract medical terms from human understandable text. Medical notes are human understandable; however, for virtual healthcare the human understandable medical notes will be converted into machine interpretable UMLS, ICD, NCI, and SNOMED codes through the use of NLP (Fig. 2).
5. **Deep Learning:** Deep learning is used for cognitive function of images and other unstructured medical contents. Medical image classification and computer assisted diagnosis (CAD) will increase the radiologists' efficiency.
6. **Image Segmentation:** Image segmentation is very useful for histopathology and medical images. Virtual reality and augmented reality will benefit from these techniques.
7. **Computer Vision/WebGL:** This is used for medical images. This is used for preprocessing of medical images like X-ray, ECG, CT etc. This will also be used for 3D image rendering as shown in Fig. 3.
8. **Generative Adversarial Network:** Generative adversarial networks (GANs) consist of a generative network and a discriminative network. These are very useful for construction of high quality synthetic medical images. Generative networks will be very useful in medical training.

9. **Bluetooth and IoT:** Bluetooth Low Energy (BLE) protocol is used for medical device or IoT devices integration to smartphone/edge computer communication [14].
10. **Smartphone Sensors:** Realtime data from inbuilt sensors such as accelerometer, gyroscope, GPS can be used to detect occurrence of fall due to stroke and send location to emergency contacts. Smartphone sensors are used for anemia [15] and breast cancer screening [16].
11. **Face & Facial Gesture Recognition:** Useful in mental healthcare sector for emotion analysis. Facial landmarks and cues such as lip biting or eye flipping can help to interpret patients metal condition for instance stress or anxiety.
12. **WebRTC and WoT:** Web of Things (WoT) and WebRTC will be used for secured realtime data exchange between two endpoints [14].

AI driven digital health care is essential for a smart healthcare, smart hospital, and precision health that will optimize the clinical process and the time taken by the physician to provide the best possible patient care. Digital health care combined with AI will reduce disease burden and increase the health equity by offering the right care at the right time at the right price for everyone from anywhere at any point-of-care.

6 Knowledge Graphs

Human knowledge allows people to think productively in various domains. The ways experts and novices acquire knowledge are different. This is due to the reading-to-cognition function of human mind. Existing knowledge allows experts to think and generate new knowledge. Experts have acquired extensive knowledge that affects their cognitive skills – what they notice (read) and how they organize, represent, and interpret information in their environment. This, in turn, affects their abilities to remember, reason, and solve problems and infer [17].

In AI we save knowledge in knowledge graphs. Knowledge graph is a knowledge base that uses a graph-structured topology like synapses and neurons to represent knowledge in computers as concepts and their interrelationships. Knowledge graphs can be stored in any database; however, properties graph databases are most suitable for this function.

Knowledge graphs equipped with the latest and accurate medical knowledge accessible by machines can eliminate both reading-to-cognition gaps and knowledge-to-actions gaps. Following are examples of various machine interpretable actionable knowledge.

1. **Diseasomics** – Diseasomics contains knowledge of diseases and their associations with symptoms, genetic associations, and genetic variations (mutations). Diseasomics knowledge graph is constructed through the semantic integration of disease ontology, symptoms of diseases, ICD10, SNOMED, DisGeNET, PharmGKB. Diseasomics also includes the spatial and temporal comorbidity knowledge of diseases. The spatial and temporal knowledge of a disease is extracted from millions of EHR data and thematically integrated with the ontology knowledge [18, 19]. Figure 2 shows how the diseasomics knowledge helps perform the differential diagnosis.

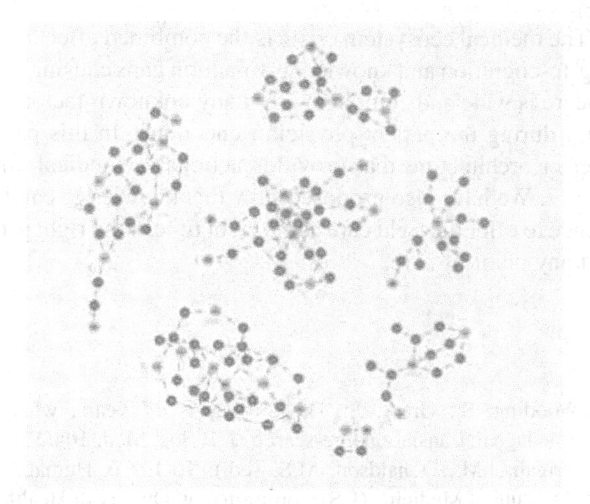

Fig. 5. Integration of the patient temporal digital twin and the physician digital twin in Oncolomics knowledge graph stored in Neo4j properties graph database.

2. **Patholomics** – Patholomics includes the pathology knowledge and its association with diseases [18, 19]. This is constructed from the EHR pathological test data. The biomarkers outside the normal range of a test result namely the Hyper and Hypo markers are used for the construction of this knowledge graph. From a hyper or hypo marker we can determine the disease association.
3. **Resistomics** – Resistomics includes all knowledge related to antibiotic resistance and antibiotic stewardship. It is constructed from EHR data, and knowledge obtained from EUCAST and WHO, and WHOCC [7]. To stop the inferential gaps in antibiotic usage, antibiotic stewardship is proposed. A use case of eliminating the inferential gaps in antibiotic abuse is the 2AI&7D model [7]. 2AI is Artificial Intelligence and Augmented Intelligence. 7D is right Diagnosis, right Disease-causing agent, right Drug, right Dose, right Duration, right Documentation, and De-escalation. Figure 4 sows some interface to the knowledge extracted from 2AI&7d digital twins. The third column in Fig. 4 in the pop-up screen shows the digital triplet for pneumonia.
4. **Oncolomics** – For the early detection of cancer, specialized oncology knowledge must be available at the primary care. Oncolomics includes machine interpretable actionable cancer related knowledge that can be used by any caregiver agnostic to their level of expertise [13]. Figure 5 shows the oncolomics on the Neo4j browser. This knowledge graph uses the computer interpretable ICD-O codes.
5. **Drugomics** – Drugomics includes knowledge about drugs. It includes the drug ontology, drug-bank data, and the drug-drug interactions [20].
6. **Allergomics** -- Allergomics is the knowledge of allergies and their relationships with the allergy causing agents.

7 Conclusion

The Post-COVID healthcare is heading for a crisis. We call this crisis as the Medical Ecosystem Crisis caused by medical errors, caregiver shortage, misinformation, and the inferential gaps. The medical ecosystem crisis is the combined effect of inferential gaps caused by reading-to-cognition and knowledge-to-action gaps causing patient harm. The domain of healthcare is wide and complex with many unknown factors that a physician needs to deal with during the patient-physician encounter. In this position paper, we presented a No-error architecture that provides actionable medical knowledge at the point-of-care 24 × 7. We have also proposed how this knowledge can be used through artificial intelligence to offer the right care at the right time at the right price for everyone from anywhere at any point-of-care.

References

1. Morris, Z.S., Wooding, S., Grant, J.: The answer is 17 years, what is the question: understanding time lags in translational re-search. J. R Soc. Med. **104**(12), 510–520 (2011)
2. Kohn, L.T., Corrigan, J.M., Donaldson, M.S. (ed.): To Err is Human: Building a Safer Health System. Institute of Medicine (US) Committee on Quality of Health Care in America Washington (DC), National Academies Press (US), (2000)
3. Makary, M.A., Daniel, M.: Medical error—the third leading cause of death in the US. BMJ 2016, 353. https://www.bmj.com/content/353/bmj.i2139 (2016)
4. Atanasov, A.G., et al.: First, do no harm (gone wrong): total-scale analysis of medical errors scientific literature. Front. Public Health. **16**(8), 558913 (2020). https://doi.org/10.3389/fpubh.2020.558913
5. Non-Communicable Diseases. https://www.who.int/news-room/fact-sheets/detail/noncommunicable-diseases
6. O'Neill, J.: Antimicrobial Resistance: Tackling a crisis for the health and wealth of nations. The Review on Antimicrobial Resistance (2014)
7. Talukder, A. K., Chakrabarti, P., Chaudhuri, B. N., Sethi, T., Lodha, R., Haas, R. E.: 2AI&7D Model of Resis-tomics to Counter the Accelerating Antibiotic Resistance and the Medical Climate Crisis. BDA2021, Springer LNCS (2021)
8. Webster, P.: Virtual health care in the era of COVID-19. The Lancet. WORLD REPORT| 395(10231), P1180–P1181 (2020)
9. Tanner, C., Gans, C., White, J., Nath, R., Poh, J.: Electronic health records and patient safety: co-occurrence of early ehr implementation with patient safety practices in primary care settings. Appl. Clin. Inform. **6**(1), 136–147 (2015). https://doi.org/10.4338/ACI-2014-11-RA-0099
10. Electronic Health Records. Patient safety Network. https://psnet.ahrq.gov/primer/electronic-health-records
11. MetaMap2020. https://lhncbc.nlm.nih.gov/ii/tools/MetaMap.html
12. Talukder, A.K., Majumdar, T., Heckemann, R.A.: Mechanistic metabolic model of cancer using DNA dosages (02/13/2019 06:15:43). SSRN: https://ssrn.com/abstract=3335080. https://doi.org/10.2139/ssrn.3335080 (2019)
13. Talukder, A.K., Haas, R.E.: Oncolomics: digital twins & digital triplets in cancer care. In: Accepted for presentation at Computational Approaches for Cancer Workshop (CAFCW21), in conjunction with Super Computing Conference (SC21) (2021)
14. Talukder A.K., Haas, R.E.: AIoT: AI meets IoT and web in smart healthcare. In: 13th ACM Web Science Conference 2021 (WebSci '21 Companion), June 21–25, 2021, Virtual Event, United Kingdom. ACM, New York, NY, USA, 7p (2021). https://doi.org/10.1145/3462741.3466650

15. Ghosal, S., Das, D., Udutalapally, V., Talukder, A.K., Misra, S.: sHEMO: Smartphone spectroscopy for blood hemoglobin level monitoring in smart anemia-care. IEEE Sens. J. **21**(6), 8520–8529 (2021). https://doi.org/10.1109/JSEN.2020.3044386
16. Basu, R., Madarkal, M., Talukder, A.K.: Smartphone Mammography for Breast Cancer Screening of Disadvantaged & Conservative Population. BDA2021, Springer LNCS, (2021)
17. Bransford, J.D., Brown, A.I., Cocking, R.R. (eds.): How People Learn: Brain, Mind, Experience, and School. National Academy Press, Washington, D.C. (2000)
18. Talukder, A.K., Sanz, J.B., Samajpati, J.: 'Precision health': balancing reactive care and proactive care through the evidence based knowledge graph constructed from real-world electronic health records, disease trajectories, diseasome, and patholome. In: Bellatreche, L., Goyal, V., Fujita, H., Mondal, A., Reddy, P.K. (eds.) BDA 2020. LNCS, vol. 12581, pp. 113–133. Springer, Cham (2020). https://doi.org/10.1007/978-3-030-66665-1_9
19. Talukder, A.K., Schriml, L., Ghosh, A., Biswas, R., Chakrabarti, P., Haas, R.E.: Diseasomics: Actionable Machine Inter-pretable Disease Knowledge at the Point-of-Care. Under review (2021)
20. Drug-drug interaction Netwoirk. https://snap.stanford.edu/biodata/datasets/10001/10001-ChCh-Miner.html

2AI&7D Model of Resistomics to Counter the Accelerating Antibiotic Resistance and the Medical Climate Crisis

Asoke K. Talukder[1,2](✉), Prantar Chakrabarti[3], Bhaskar Narayan Chaudhuri[4], Tavpritesh Sethi[5], Rakesh Lodha[6], and Roland E. Haas[7]

[1] National Institute of Technology Karnataka, Surathkal, India
[2] SRIT India Pvt Ltd., Bangalore, India
asoke.talukder@sritindia.com
[3] Vivekananda Institute of Medical Sciences, Kolkata, India
prantar@cyberneticcare.com
[4] Peerless Hospitex Hospital and Research Center Ltd., Kolkata, India
[5] Indraprastha Institute of Information Technology Delhi, Delhi, India
tavpriteshsethi@iiitd.ac.in
[6] All India Institute of Medical Sciences, Delhi, India
[7] The International Institute of Information Technology, Bangalore, India
roland.haas@iiitb.ac.in

Abstract. The antimicrobial resistance (AMR) crisis is referred to as 'Medical Climate Crisis'. Inappropriate use of antimicrobial drugs is driving the resistance evolution in pathogenic microorganisms. In 2014 it was estimated that by 2050 more people will die due to *antimicrobial resistance* compared to cancer. It will cause a reduction of 2% to 3.5% in Gross Domestic Product (GDP) and cost the world up to 100 trillion USD. The indiscriminate use of antibiotics for COVID-19 patients has accelerated the resistance rate. COVID-19 reduced the window of opportunity for the fight against AMR. This man-made crisis can only be averted through accurate actionable antibiotic knowledge, usage, and a knowledge driven *Resistomics*. In this paper, we present the 2AI (Artificial Intelligence and Augmented Intelligence) and 7D (right Diagnosis, right Disease-causing-agent, right Drug, right Dose, right Duration, right Documentation, and De-escalation) model of antibiotic stewardship. The resistance related integrated knowledge of resistomics is stored as a *knowledge graph* in a Neo4j properties graph database for 24 × 7 access. This actionable knowledge is made available through smartphones and the Web as a Progressive Web Applications (PWA). The 2AI&7D Model delivers the right knowledge at the right time to the specialists and non-specialist alike at the point-of-action (Stewardship committee, Smart Clinic, and Smart Hospital) and then delivers the actionable accurate knowledge to the healthcare provider at the point-of-care in realtime.

Keywords: AMR · Diseasomics · Resistomics · Antimicrobial resistance · Antibiotic resistance decay · ASP · Antimicrobial stewardship program · Probabilistic knowledge graph · Deterministic knowledge graph · Reasoning network · Belief network · Bayesian learning · CDS · Clinical decision support ·

S. N. Srirama et al. (Eds.): BDA 2021, LNCS 13147, pp. 44–53, 2021.
https://doi.org/10.1007/978-3-030-93620-4_4

Knowledge networks · 2AI · 7D · Patient stratification · Differential diagnosis · Predictive diagnosis · Vector embedding · node2vec

1 Introduction

In the last few decades, *antimicrobial resistance* (AMR) has turned into a crisis. In the US approximately 50% of antimicrobial usage in hospitals and up to 75% of antibiotic usage in long-term care facilities may be inappropriate or unnecessary [1]. Antibiotic-resistant bacteria and resistant genes are also transferred into the human body through food. This crisis is man-made. Misuse and injudicious use of antibiotics have reduced its life-saving potential. This crisis is referred to as a 'Medical Climate Crisis'. Research on *anti-microbial resistance* (AMR) in 2014 showed continued rise in resistance. By 2050, it could lead to 10 million deaths every year – more than cancer deaths and result into a reduction of 2% to 3.5% in Gross Domestic Product (GDP) costing the world up to 100 trillion USD [2].

Indiscriminate use of antibiotics during the COVID-19 pandemic shrunk the window of opportunity to fight against the medical climate crisis. In the US, 96% of hospitalized patients diagnosed with COVID-19 were prescribed antibiotics within the first 48 h of hospitalization. However only 29% of the patients were confirmed to have bacterial co-infection along with Coronavirus viral infection [3].

To fight this medical climate crisis, we have created the *Resistomics* knowledge graph that integrates data from microbiology, sensitivity tests, and hospital ICU (Intensive Care Unit). We have integrated this Resistomics knowledge with Diseasomics knowledge that was created in our previous works [4, 5]. We extended the 4D model proposed by Joseph and Rodvold [6] to offer a 7D model. Our 7D model includes (1) the right Diagnosis, (2) the right Disease-causing-agent, (3) the right Drug, (4) the right Dose, (5) the right Duration, (6) the right Documentation, and (7) the De-escalation. The 7D actionable knowledge is stored in a *Neo4j properties graph database* as an actionable machine interpretable knowledge graph.

We used a 2AI (Artificial Intelligence and Augmented Intelligence) model to construct the Diseasomics and Resistomics knowledge graphs. We performed node2vec node embedding on our knowledge graph to discover unknown knowledge as part of augmented knowledge. This knowledge system is integrated with the hospital EHR (Electronic health Record) system such that automated instant *triggers, alerts, escalations,* and *de-escalations* are delivered at the *point-of-action* through smartphones to stop antimicrobial abuse at the *point-of-care*. The model presented in this paper will not only reduce antibiotic resistance but will also help promoting antibiotic resistance decay to increase the life of existing antibiotics. This is the first time a comprehensive system is built that manages the entire lifecycle of antibiotic resistance that protects both patients and antibiotic drugs and finally counters the medical climate crisis.

2 Related Work

Antibiotic stewardship is an effort to improve and measure how antibiotics are prescribed by clinicians and used by patients. The first systematic assessment of antibiotic use was published in Canada in 1966 [7]. Kuper et al. suggested use of IT (Information Technology) into the antimicrobial stewardship program [8].

In recent times, *artificial intelligence* (AI) is being used to support the Antimicrobial Stewardship Program (ASP) [9]. These AI techniques are fundamentally *predictive AI that* relies on correlation, regression, and association. There are some commercial systems that address the 4D part of the resistance as suggested by Joseph and Rodvold [6]. To best of our knowledge this is the first time a machine interpretable comprehensive resistomics knowledge as presented here addressing all areas of antibiotic resistance vulnerability has been constructed that covers the entire lifecycle from diagnosis to therapeutics.

3 The Root Cause of Antibiotic Resistance

Antibiotic resistance occurs due to antibiotic misuse, antibiotic overuse, or antibiotic underuse. Misuse is caused by incorrect diagnosis of the disease or the wrong empirical judgement of the causative disease-causing bacteria. For example, pneumonia and tuberculosis are respiratory tract infections with similar symptoms. Potentially 26.5 million tuberculosis courses/year are prescribed globally for 5.3 million/year non-tuberculosis patients [10].

In case of overdose or underdose, the diagnosis of the disease, the causative bacteria, and the curative agents (drug) are correct with incorrect dosages of the drug. Not completing an antibiotic course falls into the underuse category and is one of the main reasons of personal antibiotic resistance. It may be due to the use of a broad-spectrum drug where a narrow spectrum is sufficient. It may also be caused by incorrect route of administration like using the drug through intravenous route where an oral route may be sufficient or vice versa. Many studies showed that switch from intravenous to oral in hospitals reduce hospitalization time and the cost of care with lesser side effects [11].

3.1 Solving the Antibiotic Misuse Crisis

The symptom of an infectious disease depends on the disease manifestation and not on the causative bacteria. For example, symptoms of pneumonia be it bacterial pneumonia or viral pneumonia are overlapping. Even pneumonia and tuberculosis share common symptoms, though tuberculosis is caused by a different type of bacterial agent. The diagnosis of an infectious disease is mostly done using 'trial-of-antibiotics'-empirical antibiotic treatment, as a 'rule-out' diagnostic process, which causes resistance [10]. In China a study found that out of a total of 74,648 antibiotics prescriptions in rural primary care – only 8.7% of the antibiotic prescriptions were appropriate [12]. The root cause of antibiotic misuse is the knowledge gap. The National Academy of Medicine, USA found that it takes 17 years for a research result to reach the point-of-care [13]. This lag also leads to misuse.

3.2 Antibiotic Overuse and Underuse

To mitigate the overuse/underuse incidents, we have added the DDD (Defined Daily Dose) defined by WHO (World Health Organization) Collaborating Centre for Drug

Statistics Methodology [14]. We also added the ATC (Anatomical Therapeutic Chemical) code to remove chances of errors. WHO does not define the days of therapy (DOT). We, therefore, looked at few million patient data in the EHR and computed the average DOT for a particular diagnosis and an antibiotic used for similar patient profiles. The spectrum of antibiotic drug prescribed may be inaccurate as well. To provide the spectrum information of an antibiotic we used AWaRe and WHO data. The route of administration information is also integrated.

4 The Solution to Contain Antibiotic Resistance

Joseph and Rodvold proposed the 4D model of optimal antimicrobial therapy [6]. We enhanced this model by adding three additional Ds of the right Diagnosis, the right Disease-causing agent, and the right Documentation to upgrade it to the 7D model of Antibiotic therapy. To achieve 7D, we constructed four knowledge graphs; namely, (1) The *Diseasomics knowledge graph*, (2) Categorical *Belief knowledge graph* that includes static knowledge of inherent antibiotic resistance and (3) Probabilistic *Belief knowledge graph* constructed from EHR data through Bayesian learning combined with antimicrobial and microorganism knowledge, and (4) the rule-based *Reasoning knowledge graph*. Conceptually the belief network in our model can be considered equivalent of the *human semantic memory*; whereas the reasoning network combined with hospital EHR is equivalent to the *episodic memory* as defined in neuropsychology. This knowledge base is integrated with the hospital EHR system.

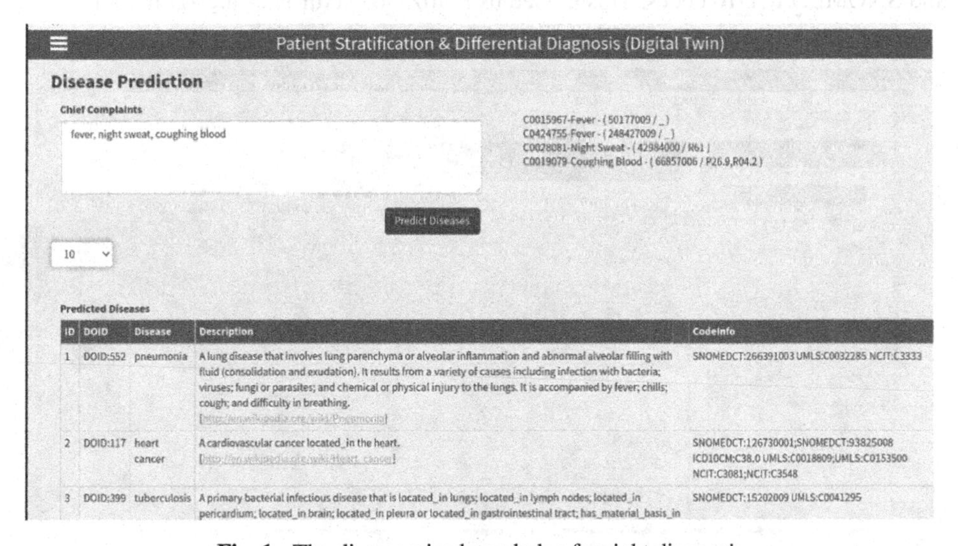

Fig. 1. The diseasomics knowledge for right diagnosis

A Probabilistic Belief network is a weighted graph with edges having probabilistic weights constructed by Bayesian learning [15] form the antibiogram culture-sensitivity data that includes healthcare specific knowledge of resistance. Antibiogram tests are used by hospitals to track resistance patterns within a hospital and *hospital acquired infections* (HAI). We used Bayesian networks on antibiogram sensitivity data generated at a tertiary care hospital pediatric ICU at AIIMS Delhi [16].

4.1 Diseasomics Knowledge Graph

In the diseasomics knowledge graph [4] we integrated Symptoms, Diseases, Spatial disease comorbidity, Temporal disease comorbidity, Disease-Gene network, Pharmacogenomics, UMLS, ICD10, and SNOMED data. Figure 1 shows the results of patient stratification and differential diagnosis from the diseasomics knowledge graph. This addresses the right Diagnosis function of the 7D model.

In Fig. 1 the care provider entered three symptoms in human understandable English natural language. These three symptoms are 'fever', 'night sweat', and 'coughing blood' in the "Chief Complaints" field. It may be noted that a different caregiver may use 'hemoptysis' medical term instead of 'coughing blood'. For this, we use an NLP (Natural Language Processing) engine to convert these human understandable English words into machine interpretable biomedical concepts. We used the UMLS, SNOMED, and ICD10 codes shown on the right side of the user input. The symptom of 'coughing blood' is associated with 36 diseases. Fever is associated with 474 diseases. Night sweat is associated with 44 diseases. We combined all three associated diseases and used the set intersection of the result-sets. Finally, we got three diseases namely 'pneumonia', 'heart cancer', and 'tuberculosis' in the 'Predicted Diseases' section as shown in Fig. 1. The output also proves the DOID (Disease Ontology ID), Disease Name, Disease Description, and SNOMED/ICD10 codes. This helped us realize the sixth-D of the 7D model.

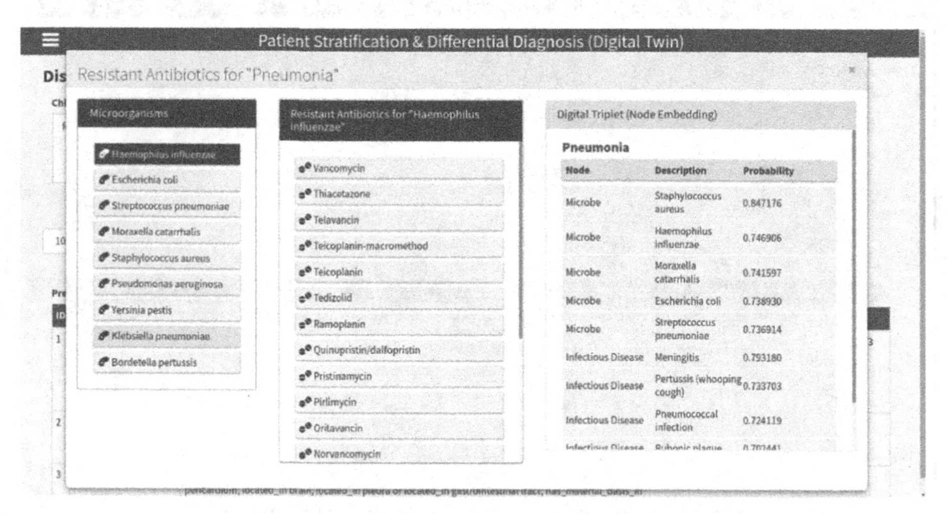

Fig. 2. The Resistomics knowledge for the right disease-causing-agent and the right drug

4.2 Categorical Belief Knowledge Graph

For the deterministic belief graph we used the 'Intrinsic Resistance' data of antibiotics based on EUCAST v3.2, 2020 updated on 24 September 2020 and downloaded from [17]. The intrinsic resistance data is available in 93,892 rows of Microsoft excel data that can be interpreted by human experts only.

Our basic motivation is to convert this human understandable knowledge into machine interpretable knowledge, so that a computer algorithm can read and interpret the knowledge. Following data transformation, there were 8,904 nodes of microorganisms and 141 antibiotic nodes. We converted this data into a (V,E) Vertex-Edge graph structure and loaded into the Neo4j properties graph database. This is evidence based accurate antibiotic resistance knowledge interpretable by computer algorithms agnostic to the domain knowledge of the caregiver.

These nodes (vertices) in the intrinsic resistance graph are connected through 93,892 directed relationships (edges) with relationship type as 'Intrinsic'. Figure 2 shows the second and third Ds namely 'right Disease-causing agent' and the 'right Drug' from this intrinsic resistance. In the first column of the pop-up window, we can see the pathogenic microorganisms causing pneumonia. Here we have the second-D – the right Disease-causing-agent or the causative bacteria for the disease. In the second column we present all antibiotics that the pathogen has become resistant to. In other words, the pathogen is susceptible to all antibiotics outside of this list. This helps us to get the third-D – the right Drug.

4.3 Vector Embedding Through Node2Vec

The Knowledge graph transformed tacit knowledge of specialists into explicit knowledge interpretable by machines and renders it to individuals who are not domain experts. Vector Embedding on the other hand takes the knowledge and helps discover unknown or missing knowledge to offer augmented intelligence. In this work we converted the Neo4j knowledge graph into a NetworkX graph and used the Python node2vec library [18]. Vector embedding offers a technique to reduce the dimension of the data with retention of the original structure. The result of vector embedding is provided with the disease-causing-agent in the third and rightmost column of the pop-up-window in Fig. 2. In case of pneumonia, we can see that Meningitis and Pneumonia are semantically similar with probability 0.793180. This knowledge is very valuable and contributes to spatial and temporal comorbidity and useful as a predictive diagnosis.

4.4 Probabilistic Belief Knowledge Graph

For this part of the knowledge graph, we have used antibiogram data downloaded from https://osf.io/57y98/ [16]. The original data is presented in human understandable Microsoft excel format. The original data was generated from a total of 5,061 bacterial cultures during December 2014 to February 2018 at AIIMS Hospital in Delhi, India.

We used the '`bnlearn`' R package for this task [19]. Because our data volume is not high, we chose the score-based hill-climbing method for the structure learning algorithm [20]. An edge and its direction were accepted when both were present in at least 50% of

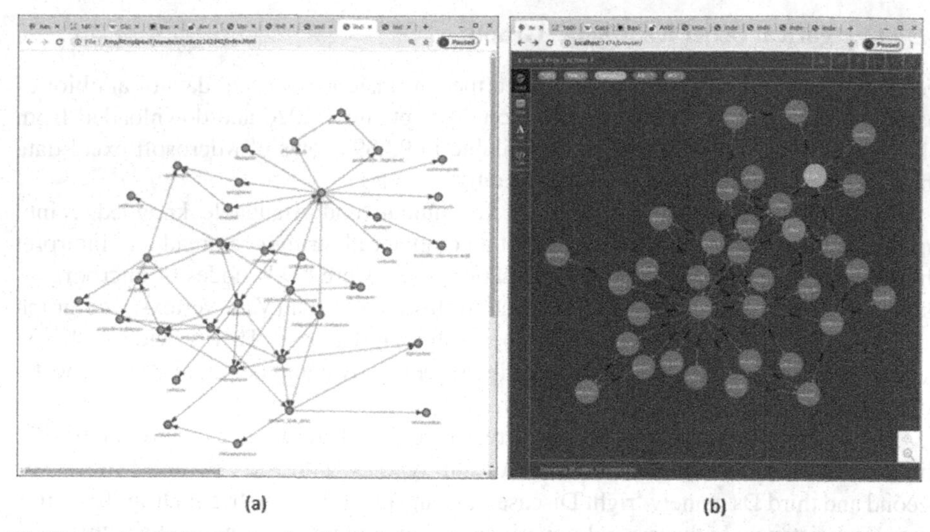

Fig. 3. Probabilistic Reasoning knowledge network (Bayesian learning network). (a) Bayesian network in GraphML. (b) Bayesian network converted to the Neo4j graph database shown in the Neo4j browser

the learned structures. The stability of the structures was assessed through AIC (Akaike Information Criteria) agreement. The probabilistic Bayesian graph constructed from the antibiogram data has 32 antibiotics (variables) that were used to test the susceptibility of different antibiotics on 17 organisms. The Bayesian network thus constructed has totally 116 relationships. The GraphML Bayesian network is shown in Fig. 3(a) that is used for interactive decision making. Figure 3(b) shows the graph in Neo4j browser that will be accessed through API. The relationship between an organism and an antibiotic in the Neo4j graph is named as 'Bayes'.

We used the WHO AWaRe database (https://adoptaware.org/) as provenance. We also added the DDD (Defined Daily Dose) from chemical and drugs database managed by WHO Collaborating Centre (https://www.whocc.no/) as a node attribute. Moreover, we computed the joint probability of conditional probabilities of resistance/susceptibility of an antibiotic for a pathogenic microorganism. This provides us the intervention or do-calculus as proposed in Counterfactual theory [21].

4.5 De-escalation (Site-Specific and Patient-Specific Resistance)

Intrinsic resistance is managed through the belief knowledge network and hospital specific resistance is managed through the Bayesian knowledge network. However, there is another type of antibiotic resistance, which is patient or person specific. It has been shown that development of resistance and the resistance decay for a patient depends on the time interval between antibiotic therapies [22]. This information is available only in the EHR. Therefore, whenever an antibiotic is prescribed for a patient at the point-of-care, the EMR records for the patient are searched to identify the complete antibiotic therapy history. We normalized the brand name to the generic antibiotic name along with the antibiotic groups (class) using the biomedical data available with the AMR package in R [23]. We used MetaMap [24] and NLP (Natural Language Processing) to extract the ICD10 code from clinical notes using UMLS biomedical databases as proposed in [25]. When we observe a change of antibiotic prescription from one group (class) to another for the same ICD10 diagnosis code within 10 to 20 days, we consider this to be a potential case of patient specific antibiotic resistance and a case for *escalation.* If an antibiotic belonging to 'Watch' or 'Reserve' category, an escalation is done irrespective. We have developed a mobile application using progressive web application (PWA) cross-platform development platform Flutter for this function (Fig. 4).

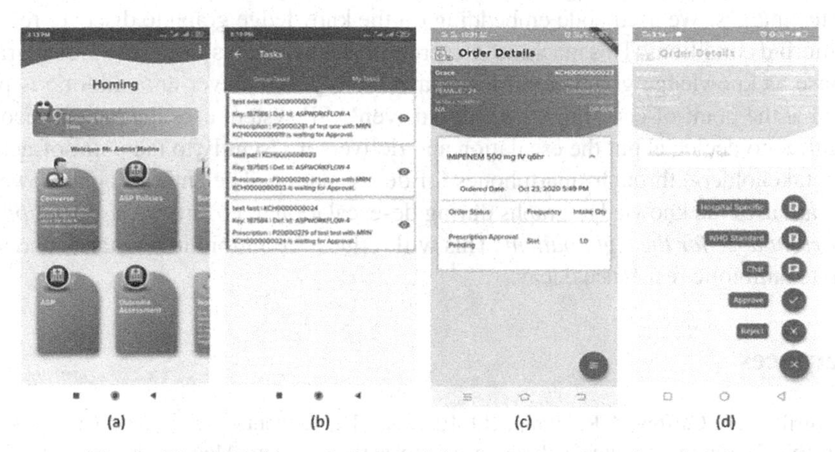

Fig. 4. The ASP de-escalation decision support at the point-of-action

Figure 4(a) shows some services available in the CDSS at the *point-of-action.* The ASP icon in Fig. 4(a) is used for ASP *de-escalation.* Antimicrobial de-escalation is a process to stop antibiotic abuse. In Fig. 4(b) three prescriptions are awaiting de-escalation from the ASP stakeholders. Figure 4(c) shows the order details of the prescribed antibiotic. Figure 4(d) allows the approver to approve, reject, cancel, and refer to site specific data.

4.6 The Right Automated Documentation

In evidence-based medicine machine interpretable documentation plays a critical role. Recordings of every episode and every event within an episode will help patients, caregivers, providers, and the public health department to understand the infectious disease epidemiology, its manifestations, and resistance patterns. In our system, everything, starting from the symptom to the therapeutics are automatically recorded. These records contain human understandable multilingual natural language along with the machine interpretable codes like ICD10, SNOMED, UMLS, ATC etc. This addresses the sixth-D (the right Documentation).

5 Conclusion

In this research paper we have presented the first 2AI and 7D techniques to counter the medical climate crisis. We used human understandable real-world medical (culture-sensitivity) data from hospitals and converted them into machine interpretable deterministic and probabilistic knowledge. We collated biomedical knowledge from different sources and converted it into ontology-based *machine interpretable actionable knowledge*. We made EHR data machine-interpretable using NLP to extract diagnosis and antibiotic usage history. We linked all this knowledge for antibiotic stewardship and clinical decision support using machine learning, ontologies, and knowledge management techniques. We used node embedding on the knowledge graph to discover related or comorbid conditions. This machine interpretable knowledge is stored in a Neo4j graph database as knowledge graphs for 24×7 availability. Whenever an antibiotic is prescribed at the point-of-care, the knowledge-driven ASP system uses the patient records at realtime to decide about the escalation and delivers it instantly to the point-of-action (ASP stakeholders) through smartphones for de-escalation. The antimicrobial stewardship team uses the knowledge graphs during de-escalation to ensure *the right antibiotic at the right dose for the right patient*. This will reduce the antibiotic resistance and will promote antibiotic resistance decay.

References

1. Morrill, H.J., Caffrey, A.R., Jump, R.L.P., Dosa, D., LaPlante, K.L.: Antimicrobial stewardship in long-term care facilities: a call to action. J. Am. Med. Direct. Assoc. **17**(2), 183.e1–183.e16 (2016). https://doi.org/10.1016/j.jamda.2015.11.013
2. O'Neill, J.: Antimicrobial resistance: tackling a crisis for the health and wealth of nations. https://amr-review.org/sites/default/files/AMR%20Review%20Paper%20-%20Tackling%20a%20crisis%20for%20the%20health%20and%20wealth%20of%20nations_1.pdf. (2014)
3. Could Efforts to Fight the Coronavirus Lead to Overuse of Antibiotics? (2021) https://www.pewtrusts.org/en/research-and-analysis/issue-briefs/2021/03/could-efforts-to-fight-the-coronavirus-lead-to-overuse-of-antibiotics
4. Talukder, A.K., Schriml, L., Ghosh, A., Biswas, R., Chakrabarti, P., Haas, R.E.: Diseasomics: Actionable Machine Interpretable Disease Knowledge at the Point-of-Care, Submitted (2021)
5. Talukder, A.K., Haas, R.E.: AIoT: AI meets IoT and web in smart healthcare. In: 13th ACM Web Science Conference 2021 (WebSci '21 Companion), June 21–25, 2021, Virtual Event, United Kingdom (2021)
6. Joseph, J., Rodvold, K.A.: The role of carbapenems in the treatment of severe nosocomial respiratory tract infections. Expert Opin. Pharmacother. **9**(4), 561–575 (2008). https://doi.org/10.1517/14656566.9.4.561.PMID:18312158
7. Wikipedia. https://en.wikipedia.org/wiki/Antimicrobial_stewardship
8. Kuper, K.M., Nagel, J.L., Kile, J.W., May, L.S., Lee, F.M.: The role of electronic health record and "add-on" clinical decision support systems to enhance antimicrobial stewardship programs. Infect. Control Hosp. Epidemiol. **40**(5), 501–511 (2019). https://doi.org/10.1017/ice.2019.51. Epub 2019 Apr 25. PMID: 31020944
9. Dengb, J.L.S., Zhang, L.: A review of artificial intelligence applications for antimicrobial resistance. Biosafety and Health (Available online 11 August 2020) (2020)

10. Divala, T.H., et al.: Accuracy and consequences of using trial-of-antibiotics for TB diagnosis (ACT-TB study): protocol for a randomised controlled clinical trial. BMJ Open **10**(3), e033999 (2020). https://doi.org/10.1136/bmjopen-2019-033999.PMID:32217561;PMCID: PMC7170647,(2020)
11. Cyriac, J.M., James, E.: Switch over from intravenous to oral therapy: a concise overview. J. Pharmacol. Pharmacother. **5**(2), 83–87 (2014). https://doi.org/10.4103/0976-500X.130042
12. Chang, Y., et al.: Clinical pattern of antibiotic overuse and misuse in primary healthcare hospitals in the southwest of China. PLoS ONE **14**(6), e0214779 (2019). https://doi.org/10.1371/journal.pone.0214779
13. Institute of Medicine: Crossing the Quality Chasm: A New Health System for the 21st Century. National Academy Press, Washington, D.C. (2001)
14. WHO Collaborating Centre. https://www.whocc.no/atc_ddd_index/
15. Timo, J.T. Koski, J.N.: A review of bayesian networks and structure learning. 40(1), 51–103 (2012)
16. Sethi, T., Maheshwari, S., Nagori, A., Lodha, R.: Stewarding antibiotic stewardship in intensive care units with Bayesian artificial intelligence [version 1; referees: awaiting peer review], Welcome Open Research 2018, 3:73 Last updated: 18 JUN 2018 (2018)
17. Antibiotic Resiatance dataset. https://msberends.github.io/AMR/articles/datasets.html
18. Grover, A., Leskovec, J.: node2vec: Scalable feature learning for networks. 2016. Knowledge Discovery and Data Mining (2016)
19. Scutari, M.: Learning Bayesian networks with the bnlearn R package. J. Stat. Softw. **35**(3), 1–22 (2010)
20. Su, C., Andrew, A., Karagas M.R., Borsuk, M.E.: Using Bayesian networks to discover relations between genes, environment, and disease. BioData Mining **6**, 6. (2013)
21. Pearl, J.: The Do-Calculus revisited. In: de Freitas, N., Murphy, K. (eds.), Proceedings of the Twenty-Eighth Conference on Uncertainty in Artificial Intelligence, Corvallis, OR, AUAI Press, 4–11 (2012)
22. Bakhit, M., Hoffmann, T., Scott, A.M., et al.: Resistance decay in individuals after antibiotic exposure in primary care: a systematic review and meta-analysis. BMC Med **16**, 126 (2018)
23. Berends, M.S., Luz., C.F, Friedrich, A.W., Sinha, B.N.M., Albers, C.J., Glasner, C.: AMR - An R package for working with antimicrobial resistance data. bioRxiv (2019). https://doi.org/10.1101/810622
24. MetaMap. https://metamap.nlm.nih.gov/
25. Talukder, A.K., Sanz, J.B., Samajpati, J.: 'Precision health': balancing reactive care and proactive care through the evidence based knowledge graph constructed from real-world electronic health records, disease trajectories, diseasome, and patholome. In: Bellatreche, L., Goyal, V., Fujita, H., Mondal, A., Reddy, P.K. (eds.) BDA 2020. LNCS, vol. 12581, pp. 113–133. Springer, Cham (2020). https://doi.org/10.1007/978-3-030-66665-1_9

Tooth Detection from Panoramic Radiographs Using Deep Learning

Shweta Shirsat[1(✉)] and Siby Abraham[2]

[1] University Department of Computer Science, University of Mumbai, Mumbai 400098, India
[2] Center of Excellence in Analytics and Data Science, NMIMS Deemed to be University, Mumbai 400056, India
siby.abraham@nmims.edu

Abstract. The proposed work aims at implementing a Deep Convolutional Neural Network algorithm specialized in object detection. It was trained to perform tooth detection, segmentation, classification and labelling on panoramic dental radiographs. A dataset of dental panoramic radiographs was annotated according to the FDI tooth numbering system. Mask R-CNN Inception ResNet V2 object detection algorithm was able to give excellent results in terms of tooth segmentation and numbering. The experimental results were validated using standard performance metrics. The method could not only give comparable results to that of similar works but could detect even missing teeth, unlike similar works.

Keywords: Tooth detection · Deep Learning · Neural network · Transfer learning · Radiographs

1 Introduction

Deep Learning has become a buzzword in the technological community in recent years. It is a branch of Machine Learning. It is influenced by the functioning of the human brain in designing patterns and processing data for decision making. Deep neural networks are suitable for learning from unlabelled or unstructured data. Some of the key advantages of using deep neural networks are their ability to deliver high-quality results, eliminating the need for feature engineering and optimum utilization of unstructured data [1]. These benefits of deep learning have given a huge boost to the rapidly developing field of computer vision. Various applications of deep learning in computer vision are image classification, object detection, face recognition and image segmentation. An area that has achieved the most progress is object detection.

The goal of object detection is to determine which category each object belongs to and where these objects are located. The four main tasks in object detection include classification, labelling, detection and segmentation. Fast Convolutional Neural Networks (Fast-RCNN), Faster Convolutional Neural Network (Faster-RCNN) and Region-based Convolutional Neural Networks(R-CNN) are the most widely used deep learning-based object detection algorithms in computer vision. Mask R-CNN, an extension to faster-RCNN [2] is far superior to others in terms of detecting objects and generating high-quality masks. Mask-RCNN architecture is represented in Fig. 1

S. N. Srirama et al. (Eds.): BDA 2021, LNCS 13147, pp. 54–63, 2021.
https://doi.org/10.1007/978-3-030-93620-4_5

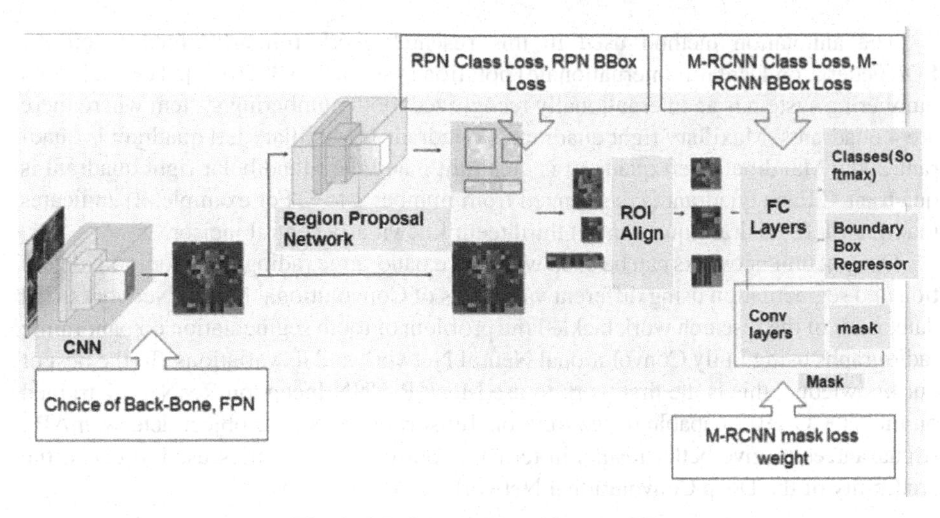

Fig. 1. Mask R-CNN architecture [3]

Mask-RCNN is also used for medical image diagnosis in locating tumours, measuring tissue volumes, studying anatomical structures, lesion detection, planting surgery, etc. Mask-RCNN uses the concept of transfer learning to drastically reduce the training time of a model and lower generalization results. Transfer learning is a method where a neural network model is trained on a problem that is similar to the problem being solved. These layers of the trained model are then used in a new model to train on the problem of interest. Several high-performing models can be used for image recognition and other similar tasks in computer vision. Some of the pre-trained transfer learning models include VGG, Inception and MobileNet [4].

Deep Learning is steadily finding its way to offer innovative solutions in the radiographic analysis of dental X-ray images. In dental radiology, the tooth numbering system is the format used by the dentist for recognizing and specifying information linked with a particular tooth. A tooth numbering system helps dental radiologists identify and classify the condition associated with a concerned tooth. The most frequently used tooth numbering methods are the Universal Numbering System, Zsigmondy-Palmer system, and the FDI numbering system [4] (Fig. 2).

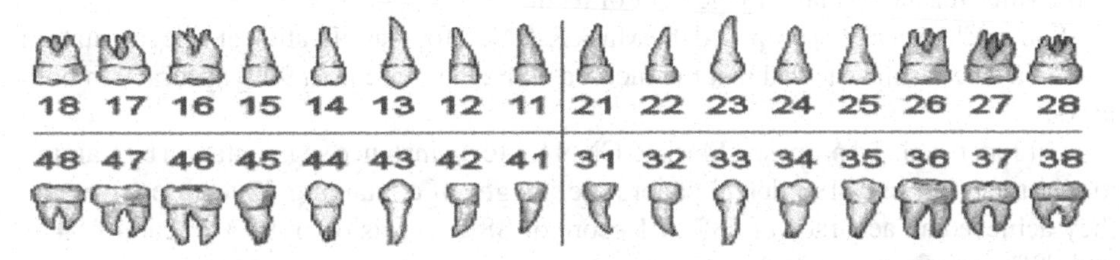

Fig. 2. FDI tooth numbering system chart [10]

The annotation method used in this research work primarily focuses on the FDI(Federation Dentaire Internationale) notation system(ISO 3950) [9]. The FDI tooth numbering system is an internationally recognized tooth numbering system where there are 4 quadrants. Maxillary right quadrant is quadrant 1, Maxillary left quadrant is quadrant 2, the Mandibular left quadrant is quadrant 3 and the Mandibular right quadrant is quadrant 4. Each quadrant is recognized from number 1 to 8. For example, 21 indicates maxillary left quadrant(quadrant 2) third teeth known as a central incisor.

Deep neural networks can be used with these panoramic radiographs for tooth detection and segmentation using different variations of Convolutional Neural Networks. Till date, most of the research work tackled the problem of tooth segmentation on panoramic radiographs using Fully Convolutional Neural Network and its variations. To the best of our knowledge, this is the first work to use Mask R-CNN Inception ResNet V2 trained on the COCO 2017 capable of working on Tensorflow version 2 object detection API. It guarantees to give better results in terms of performance metrics used to check the credibility of the Deep Convolutional Network algorithm used.

2 Related Works

There were a few attempts to apply deep learning techniques for teeth detection and segmentation.

Thorbjorn Louring Koch et al. [1] implemented a Deep Convolutional Neural Network (CNN) developed on the U-Net architecture for segmentation of individual teeth from dental panoramic X-rays. Here CNN reached the dice score of 0.934(AP). 1201 radiographic images were used for training, forming an ensemble that increased the score to 0.936.

Minyoung Chung et al. [3] demonstrated a CNN-based individual tooth identification and detection algorithm using direct regression of object points. The proposed method was able to recognize each tooth by labelling all 32 possible regions of the teeth including missing ones. The experimental results illustrated that the proposed algorithm was best among the state-of-the-art approaches by 15.71% in the precision parameter of teeth detection.

Dmitry V Tuzoff et al. [2] used the state-of-the-art Faster R-CNN architecture. The FDI tooth numbering system was used for teeth detection and localization. A classical VGG-16 CNN along with a heuristic algorithm was used to improve results according to the rules for the spatial arrangement of teeth.

Shuxu Zhao et al. [4] proposed the Mask R-CNN, for classification and segmentation of tooth. The results showed that the method achieved more than 90% accuracy in both tasks.

Gil Jader et al. [7] proposed Mask R-CNN for teeth instance segmentation by training the system with only 193 dental panoramic images of containing 32 teeth on average, they achieved an accuracy of 98%, F1-score of 88%, precision of 94%, recall of 84%, and 99% specificity.

Gil Jader et al. [8] performed a study of tooth segmentation and numbering on panoramic radiographs using an end-to-end deep neural network. The proposed work used Mask R-CNN for teeth localization on radiographic images. The calculated accuracy was 98% out of which F1-score was 88%, precision was 94%, recall was 84% and specificity of 99% of over 1224 radiographs.

Gil Jader et al. [9] also proposed a segmentation system based on the Mask R-CNN and transfer learning to perform an instance segmentation on dental radiographs. The system was trained with 193 dental radiographs having a maximum of 32 teeth. Accuracy achieved was 98%.

3 Methodology

Figure 3 represents Mask R-CNN applied on a set of dental radiographs to perform tooth identification and numbering.

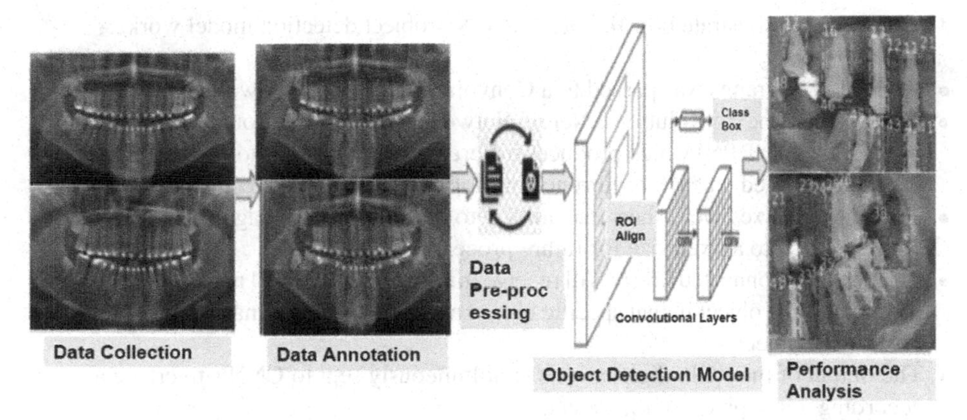

Fig. 3. Mask R-CNN architecture for tooth segmentation and numbering

3.1 Data Collection

To train a robust model we needed a lot of images that should vary as much as possible. The dataset of panoramic dental radiographs was collected from Ivison dental labs(UFBA_UESC_DENTAL_IMAGES_DEEP) [10]. The height and width of each panoramic dental radiograph ranged from 1014–1504 pixels and 2094 to 3432 pixels respectively. These radiographs were then resized to a fixed resolution of 800 * 600. The suitable format to store the radiographs was JPG.

3.2 Data Annotation

After collecting the required data these radiographs had to be annotated as per the FDI tooth numbering system. Rather than only annotating existing teeth in a radiograph, we annotated all 32 teeth including missing teeth.A JSON/XML file was created for each radiograph representing manually defined bounding boxes, and a ground truth label set for each bounding box. Though there were a variety of annotation tools available such as the VGG Image annotation tool, labelme, and Pixel Annotation Tool [8]. The proposed work uses labelme software because of its efficiency and simplicity. All annotations were verified by a clinical expert in the field.

3.3 Data Preprocessing

In addition to the labelled radiographs a TFRecord file needed to be created that could be used as input for training the model. Before creating TFRecord files, we had to convert the labelme labels into COCO format, as we had used the same as the pre-trained model. Once the data is in COCO format it was easy to create TFRecord files [9].

3.4 Object Detection Model

The below steps illustrate how the Mask R-CNN object detection model works:-

- A set of radiographs was passed to a Convolutional Neural Network.
- The results of the Convolutional Neural Network were passed through to a Region Proposal Network (RPN) which produces different anchor boxes known as ROI(Regions of Interest) based on each occurrence of tooth objects being detected.
- The anchor boxes were then transmitted to the ROI Align stage. It is essential to convert ROI's to a fixed size for future processing.
- A set of fully connected layers will receive this output which will result in the generating class of the object in that specific region and defining coordinates of the bounding box for the object.
- The output of the ROI Align stage is simultaneously sent to CNN's to create a mask according to the pixels of the object.

Hyper Parameter Tuning: The training was performed on dental radiographic images having 32 different objects that were identified and localized. The hyperparameter values of the object detection models were: Number of classes = 32; image_resolution = 512 * 512; mask_height * width = 33 * 33; standard_deviation = 0.01; IOU_threshold = 0.5; Score_converter = Softmax;batch_size = 8; No. of steps = 50,000; learning_rate_base = 0.008. The parameters like standard deviation, score_converter, batch_size, no.of epochs and fine_tune_checkpoint_type were optimised.

3.5 Performance Analysis

Test results after training the model for 50K epochs are shown in Fig. 4. To measure the performance accuracy of object detection models some predefined metrics such as Precision, Recall and Intersection Over Union(IoU) are required [8].

Precision: Precision is the capability of a model to identify only the relevant objects. It is the percentage of correct positive predictions and is given by Precision = TP/(TP + FP) [9].

Recall: Recall is the capability of a model to find all the relevant cases (all ground truth bounding boxes). It is the percentage of true positives detected among all relevant ground truths and is given by: Recall = TP/(TP + FN) [10].

Where,

TP = **true positive** is observed when a prediction-target mask pair has an IoU score that exceeds some predefined threshold;

FP = **false positive** indicates a predicted object mask has no associated ground truth object mask.

FN = **false negative** indicates a ground truth object mask has no associated predicted object mask.

IoU: Intersection over Union is an evaluation metric used to measure the accuracy of an object detector on a specific dataset [10].

IoU = Area of Overlap/Area of Union

4 Experimental Results

The experiment was executed on a GPU (1 × Tesla K80), with 1664 CUDA cores and 16 GB memory. The algorithm was running on TensorFlow version 2.4.1 having python version 3.7.3. In general the tooth detection numbering module demonstrated results for detecting each tooth from dental radiographs. Then it was also able to provide tooth numbers for each detected tooth as per the FDI tooth numbering system. The sample results are shown in Fig. 4. When the training process was successfully completed, a precision of 0.98 and recall of 0.97 was recorded as seen in Table 1.

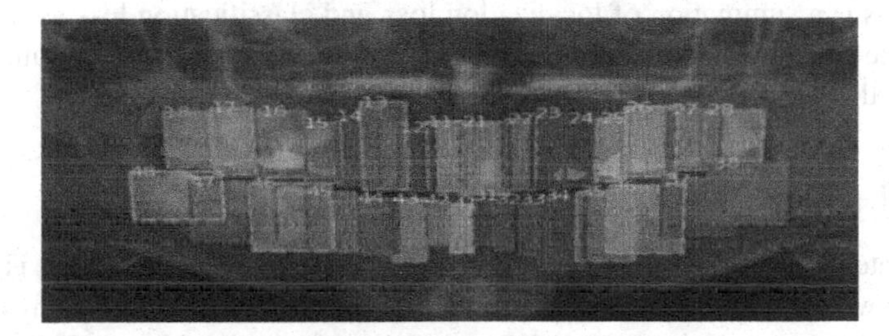

Fig. 4. Dental radiograph with results

Table 1. Evaluation metrics

Evaluation metric	Value
Precision	0.98
Recall	0.97
IoU	0.5

Along with evaluation metrics, a graphical representation of various loss functions presented on the tensorboard is given in Figs. 5, 6, 7 and 8.

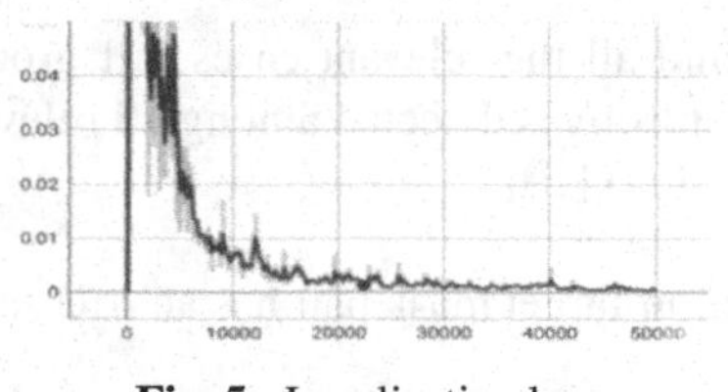

Fig. 5. Localization loss

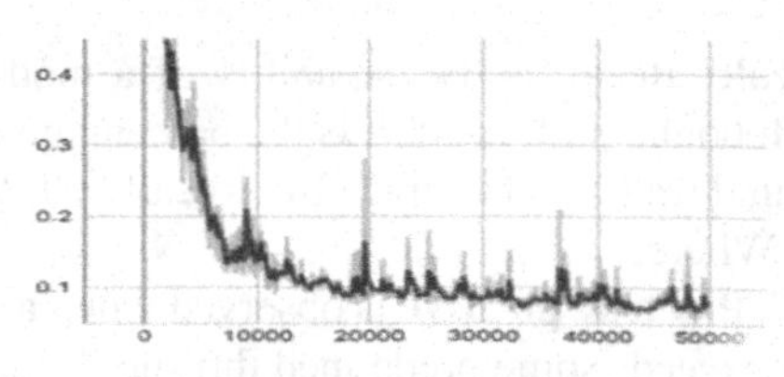

Fig. 6. Total loss

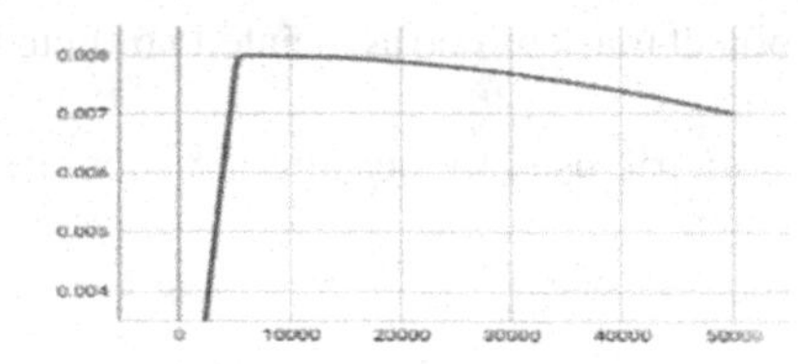

Fig. 7. Learning rate

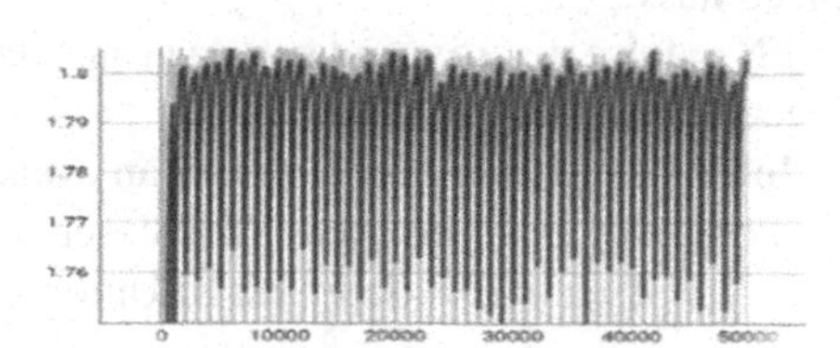

Fig. 8. Steps_per_epoch

4.1 Localization Loss

Localization loss is used to demonstrate loss between a predicted bounding box and ground truth [10]. As the training progresses, the localization loss decreases gradually and then remains stable as illustrated in Fig. 5.

4.2 Total Loss

The total loss is a summation of localization loss and classification loss as represented in Fig. 6. The optimisation model reduces these loss values until the loss sum reaches a point where the network can be considered as fully trained.

4.3 Learning Rate

Learning Rate is the most important hyperparameter for this model which is shown in Fig. 7. Here we can see there is a gradual increase in the learning rate after every batch recording the loss at every increment. When entering the optimal learning rate zone it is observed that there is a sudden drop in the loss function.

4.4 Steps Per Epoch

This hyperparameter is useful if there is a huge dataset with considerable batches of samples to train. It defines how many batches of samples are used in one epoch. In our tooth detection model, the total number of epochs was 50,000 with an interval of 100 as represented in Fig. 8.

5 Comparative Study

5.1 Comparison with Clinical Experts

The results provided by the proposed model were asked to be verified by a clinical expert for detected and undetected bounding boxes, confusion with similar teeth or missing tooth labels, failure in complicated cases and objects detected more than ground tooth.

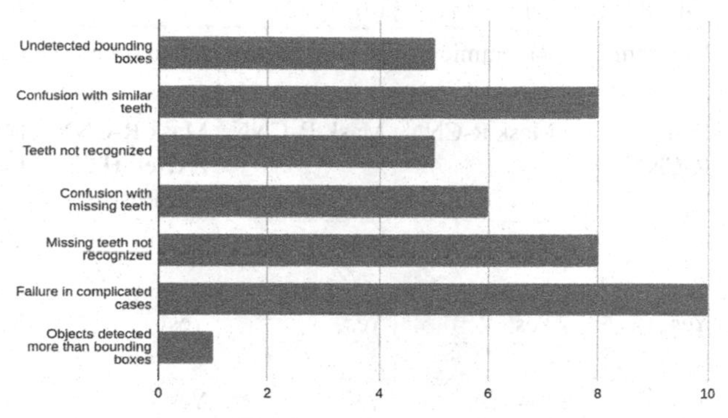

Fig. 9. Expert analysis

Figure 9 demonstrates the percentage difference between the expert reviews and predictions from the system.

Out of the total sample data of 25 radiographs there exists 8% confusion with similar kinds of teeth in a single radiograph. 5% of teeths were not correctly recognized by the algorithm. This algorithm was also able to identify tooth numbers for missing teeth but there exists a confusion of 6%. Under some complicated scenarios around 8% of missing teeth were not recognized. There exists around 10% failure in complicated cases such as tooth decay, impacted teeth, cavities or because of partial or full dental implantation. As per the graphical representation inaccuracies, though small in number, are attributed to a large extent because of the poor quality of the data and not the performance deficiency of the model.

5.2 Comparison with Other Works

Table 2 provides a comparative study of the effectiveness of the proposed model with similar works. Comparison has been done based on seven criterias as shown in Table 2.The proposed work is different from others as it demonstrates the implementation of the Mask R-CNN Inception ResNet V2 object detection model. Only one Mask R-CNN Model is supported with TensorFlow 2 object detection API at the time of writing of this paper [10]. The transfer learning technique was successfully implemented using a pre-trained COCO dataset. Along with tooth detection, tooth segmentation, tooth numbering we were also able to predict missing teeth. These missing teeth are known as edentulous spaces from maxilla or mandible. M-RCNN was able to correctly recognize edentulous spaces on a radiograph. This study demonstrates that the proposed method is far superior to the other state of the art models pertaining to tooth detection and localization. Also, there is a clear understanding of poor detection, wherever it occurred though small in number, and its verification is done by a clinical expert.

Table 2. Comparative study

Author	Minyoung Chung [3]	Shuxu Zhao [4]	Guohua Zhu [5]	Gil Jader [7]	Proposed work
Type of radiograph	Panoramic	Panoramic	Digital X-rays	Panoramic	Panoramic
Deep learning method	Fast R-CNN	Mask R-CNN	Mask R-CNN	Mask R-CNN, PANet, HTC and ResNeSt	Mask R-CNN Inception V2
Transfer learning	No	Yes	Yes	Yes	Yes
Tooth detection	Yes	Yes	Yes	Yes	Yes
Tooth segmentation	Yes	Yes	Yes	Yes	Yes
Tooth numbering	Yes	Yes	Yes	Yes	Yes
Missing tooth detection	No	Yes	No	No	Yes
Comparison with experts	No	No	No	Yes	Yes

6 Conclusion

The Mask R-CNN Inception ResNet V2 object detection model was used to train a dataset of dental radiographs for tooth detection and numbering. It was observed that the training results were exceptionally good especially in tooth identification and numbering with high IOU, precision and recall. The visualization results were considerably better than Fast R-CNN. The performance of our selected model was very close to the level of the clinical expert who was selected as a referee in this study. In future studies, we will consider working with more advanced models for periodontal bone loss, early caries diagnosis, and various periapical diseases.

References

1. Koch, T.L., Perslev, M., Igel, C., Brandt, S.S.: Accurate Segmentation of Dental Panoramic Radiographs with U-NETS. In: Published in Proceedings of 16th IEEE International Symposium on Biomedical Imaging, Venice, Italy (2019)

2. Tuzoff, D.V., et al.: Tooth detection and numbering in panoramic radiographs using convolutional neural networks. In: Published in Conference of Dental Maxillofacial Radiology in medicine, Russia, March-2019 (2019)
3. Chung, M.,et al.: Individual Tooth Detection and Identification from Dental Panoramic X-Ray Images via Point-wise Localization and Distance Regularization. Elsevier, Artificial Intelligence in medicine, South Korea (2020)
4. Zhao, S., Luo, Q., Liu, C.: Automatic Tooth Segmentation and Classification in Dental Panoramic X-ray Images. In: IEEE International Conference on Pattern Analysis and Machine Intelligence, China (2020)
5. Zhu, G., Piao, Z., Kim, S.C.: Tooth detection and segmentation with mask R-CNN. In: Published in International Conference on Artificial Intelligence in Information and Communication (2020)
6. Jader, G., Oliveira, L., Pithon, M.: Automatic segmentation teeth in X-ray images: Trends, a novel data set, benchmarking and future perspectives. In: Elsevier, Expert systems with applications volume 10, March-2018 (2018)
7. Gil Jader et al.: Deep instance segmentation of teeth in panoramic x-ray images. In: SIBGRAPI 31st International Conference on Graphics, Patterns and Images (SIBGRAPI), Brazil, May 2018 (2018)
8. Silva, B., Pinheiro, L., Oliveira, L., Pithon, M.: A study on tooth segmentation and numbering using end-to-end deep neural networks. In: Published in 33rd SIBGRAPI Conference on Graphics, Patterns and Images (SIBGRAPI), Brazil (2020)
9. Jader, G. et al.: Deep instance segmentation of teeth in panoramic X-ray images (2018). http://sibgrapi.sid.inpe.br/col/sid.inpe.br/sibgrapi/2018/08.29.19.07/doc/tooth_segmentation.pdf
10. Freny, R.K.: Essentials of Oral and Maxillofacial Radiology. Jaypee Brothers Medical Publishers(P) Ltd., New Delhi. MDS (Mumbai University)

Machine/Deep Learning

Hate Speech Detection Using Static BERT Embeddings

Gaurav Rajput, Narinder Singh Punn[(✉)], Sanjay Kumar Sonbhadra, and Sonali Agarwal

Indian Institute of Information Technology, Allahabad, Prayagraj 211015, Uttar Pradesh, India
{pse2017002,rsi2017502,sonali}@iiita.ac.in

Abstract. With increasing popularity of social media platforms hate speech is emerging as a major concern, where it expresses abusive speech that targets specific group characteristics, such as gender, religion or ethnicity to spread violence. Earlier people use to verbally deliver hate speeches but now with the expansion of technology, some people are deliberately using social media platforms to spread hate by posting, sharing, commenting, etc. Whether it is Christchurch mosque shootings or hate crimes against Asians in west, it has been observed that the convicts are very much influenced from hate text present online. Even though AI systems are in place to flag such text but one of the key challenges is to reduce the false positive rate (marking non hate as hate), so that these systems can detect hate speech without undermining the freedom of expression. In this paper, we use ETHOS hate speech detection dataset and analyze the performance of hate speech detection classifier by replacing or integrating the word embeddings (fastText (FT), GloVe (GV) or FT + GV) with static BERT embeddings (BE). With the extensive experimental trails it is observed that the neural network performed better with static BE compared to using FT, GV or FT + GV as word embeddings. In comparison to fine-tuned BERT, one metric that significantly improved is specificity.

Keywords: Hate speech detection · BERT embeddings · Word embeddings · BERT

1 Introduction

With growing access to internet and many people joining the social media platforms, people tend to post online as per their desire and tag it as freedom of speech. It is one of the major problems on social media that tend to degrade the overall user's experience. Facebook defines hate speech as a direct attack against people on the basis of protected characteristics: race, ethnicity, national origin, disability, religious affiliation, caste, sexual orientation, sex, gender identity and

G. Rajput, N. S. Punn, S. K. Sonbhadra, and S. Agarwal—Equal contribution.

S. N. Srirama et al. (Eds.): BDA 2021, LNCS 13147, pp. 67–77, 2021.
https://doi.org/10.1007/978-3-030-93620-4_6

serious disease [3], while for Twitter hateful conduct includes language that dehumanizes others on the basis of religion or caste [6]. In March 2020, Twitter expanded the rule to include languages that dehumanizes on the basis of age, disability, or disease. Furthermore, the hateful conduct policy was expanded to also include race, ethnicity, or national origin [6]. Following this context, hate speech can be defined as an abusive speech that targets specific group characteristics, such as gender, religion, or ethnicity.

Considering the massive amount of text what people post on social media, it is impossible to manually flag them as hate speech and remove them. Hence, it is required to have automated ways using artificial intelligence (AI) to flag and remove such content in real-time. While such automated AI systems are in place on social media platform, but one of the key challenges is the separation of hate speech from other instances of offensive language and other being higher false positive (marking non-hate as hate) rates of such system. Higher false positive rate means system will tag more non-hate content as hate content which can undermine the right to speak freely.

Hate speech can be detected using state-of-the-art machine learning classifiers such as logistic regression, SVM, decision trees, random forests, etc. However, deep neural networks (DNNs) such as convolutional neural networks (CNNs), long short-term memory networks (LSTMs) [15], bidirectional long short-term memory networks (BiLSTMs) [26], etc. have outperformed the former mentioned classifiers for hate speech detection [19]. Former classifiers do not require any word embedding [7] to work with while the latter ones i.e. DNNs requires word embeddings such as GloVe (GV) [20], fastText (FT) [16], Word2Vec [18], etc. Following this context, the present research work focuses on the scope of improvement of the existing state-of-the-art deep learning based classifiers by using static BERT [13] embeddings (BE) with CNNs, BiLSTMs, LSTMs and gated recurrent unit (GRU) [11].

1.1 BERT

Bidirectional encoder representations from transformers (BERT) was developed by Devlin et al. [13] in 2018. BERT is a transformer-based ML technique pre-trained from unlabeled data that is taken from Wikipedia (language: English) and BookCorpus. Transformer [28] model has two main parts: encoder and decoder. BERT is created by stacking the encoders. Two major strategies that BERT uses for training are masked language modelling (MLM) and next sentence prediction (NSP). The MLM strategy and fine-tuning of BERT is pictorially depicted in Fig. 1. In MLM technique 15% of the words in a sentence are selected randomly and masked. Based on the context of the other words (which are not masked) the model tries to predict the masked word.

In NSP technique model is given pairs of sentences as input. The model learns to predict if the second sentence in a selected pair is the subsequent sentence in original document. During the training phase half of the inputs are a pair in which second sentence is subsequent sentence to the first in the original document while the rest half of the input pairs has a randomly selected sentence as second sentence.

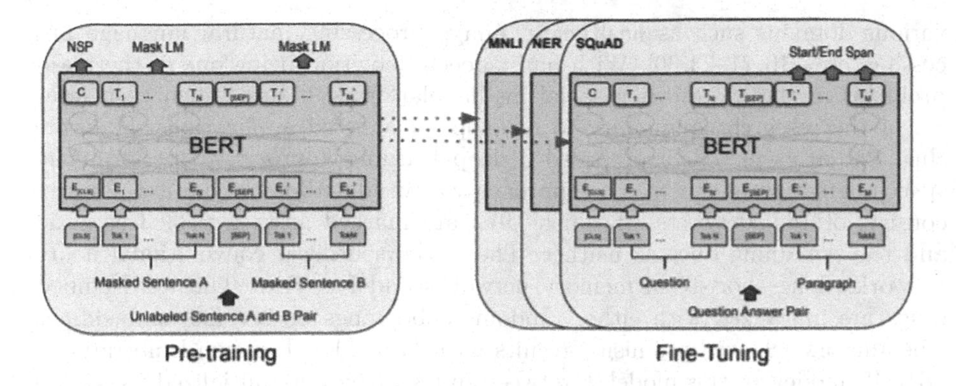

Fig. 1. Pre-training and fine-tuning of BERT.

1.2 Attention in Neural Networks

While processing a sentence in natural language for any NLP task, all words are not of equal importance, hence it is necessary to put more attention to the important words of the sentence. The importance of words depends on the context and is learned through training data. Bahdanau et al. [9] proposed the attention mechanism for improving machine translation that uses seq-to-seq model. It is done by aligning the decoder with the relevant input sentences and implementing attention. Steps for applying attention mechanism are as follows:

1 Produce the encoder hidden states.
2 Calculate alignment scores.
3 Soft-max the alignment scores.
4 Calculate the context vector.
5 Decode the output.
6 At each time step, based on the current state of decoder and input received by decoder, an element of decoder's output sequence is generated by decoder. Besides that decoder also updates its own state for next time step. Steps 2–5 repeats itself for each time step of the decoder until the output length exceeds a specified maximum length or end of sentence token is produced.

The rest of the paper is organised as follows: literature review in Sect. 2 which focuses on the related work and recent developments in this field, Sect. 3 describes the proposed methodology. Section 4 covers the exhaustive experimental trials followed by improved results in Sect. 5, and lastly in Sect. 6 the concluding remarks are presented.

2 Related Work

The advancements in deep learning technology have widen spectrum of its application tasks involving classification, segmentation, object detection, etc., across various domains such as healthcare, image processing, natural language processing, etc. [10,21–23,30]. With hate speech detection being one of the major problems in the evergrowing social media platforms, it has drawn keen interest of the research community to develop AI assisted applications. Following this, Badjatiya et al. [8] proposed a deep learning approach to perform hate speech detection in tweets. The approach was validated on the dataset [29] that consists of 16,000 tweets, of which 1972 are marked as racist, 3383 as sexist and the remaining ones as neither. The authors utilized convolutional neural networks, long short-term memory networks and FastText. The word embeddings are initialized with either random embeddings or GV [20] embeddings. The authors achieved promising results with "LSTM + Random Embedding + GBDT" model. In this model, the tweet embeddings were initialized to random vectors, LSTM was trained using back-propagation, and then learned embeddings were used to train a GBDT classifier.

Rizos et al. [25] experimented by using short-text data augmentation technique in deep learning for hate speech classification. For short-text data augmentation they used substitution based augmentation (ThreshAug), word position augmentation (PosAug) and neural generative augmentation (GenAug). For performing experiments they used the dataset [12] which consists of around 24k samples, of which 5.77% samples are marked as hate, 77.43% samples are marked as offensive and 16.80% samples as neither. The authors experimented with multiple DNNs such as CNN, LSTM and GRU. In addition, fastText, GloVe and Word2Vec were used as word embeddings. They achieved their best results by using GloVe + CNN + LSTM + BestAug, where BestAug is combination of PosAug and ThreshAug. Faris et al. [14] proposed a deep learning approach to detect hate speech in Arabic language context. They created their dataset by scraping tweets from twitter using an application programming interface (API) [1] and performed standard dataset cleaning methods. The obtained dataset have 3696 samples of which 843 samples are labelled as hate and 791 samples as normal while rest of the samples were labelled as neutral. Word2Vec and AraVec [27] were used for feature representation and embedding dimension was kept to 100. The authors achieved promising results using combination of CNN and LSTM with AraVec.

Ranasinghe et al. [24] in hate speech and offensive content identification in Indo-European languages (HASOC) shared task 2019 experimented with multiple DNNs such as pooled GRU, stacked LSTM + attention, LSTM + GRU + attention, GRU + capsule using fastText as word embedding on the dataset having posts written in 3 languages: German, English and code-mixed Hindi. Furthermore, they also experimented with fine-tuned BERT [13] which outperformed every above mentioned DNN for all 3 languages. In another work, Mollas et al. [19] proposed ETHOS dataset to develop AI based hate speech detection framework that have used FT [16], GV [20] and FT + GV as word embeddings with CNNs, BiLSTMs and LSTMs. In contrast to other datasets which are based on tweets scraped from Twitter, this new dataset is based on YouTube and Reddit comments. A binary version of ETHOS dataset has 433 sentences containing hate text and 565 sentences containing non-hate text. Besides, transfer learning was used to fine-tune BERT model on the proposed dataset that outperformed the above mentioned deep neural networks. The results of the aforementioned experiments are shown in Table 1, where bold values represent the highest value of metrics among all models [19].

Table 1. Performance of BERT (fine-tuned on binary ETHOS dataset) with various neural networks using FT, GV or FT + GV as word embedding [19]

Model	F1 score	Accuracy	Precision	Recall	Specificity
CNN + Attention + FT + GV	74.41	75.15	74.92	74.35	**80.35**
CNN + LSTM + GV	72.13	72.94	73.47	72.4	76.65
LSTM + FT + GV	72.85	73.43	73.37	72.97	76.44
BiLSTM + FT + GV	76.85	**77.45**	77.99	77.10	79.66
BiLSTM + Attention + FT	76.80	77.34	77.76	77.00	79.63
BERT	**78.83**	76.64	**79.17**	**78.43**	74.31

Ever since the researchers started using BERT [13] for natural language processing tasks, it has been observed that a fine-tuned BERT usually outperforms other state-of-the-art deep neural networks in same natural language processing task. The same has been observed in the results of experiment carried out by Mollas et al. [19]. Motivated from this, the experiments carried out in this paper aims to analyse the performance of fine-tuned BERT with other deep learning models.

3 Proposed Methodology

Following the state-of-the-art deep learning classification models, in the proposed approach the impact of BERT based embeddings is analyzed. The hate speech detection framework is designed by combining DNNs (CNN, LSTM, BiLSTM and GRU) with static BERT embeddings to better extract the contextual information. Initially, the static BERT embedding matrix is generated from large corpus of dataset, representing embedding for each word and later, this matrix is processed using DNN classifiers to identify the presence of hate. The schematic representation of the proposed model is shown in Fig. 2.

3.1 Static BERT Embedding Matrix

The embedding matrix contains the word embeddings for each word in dataset. Each row in the embedding matrix contains word embedding for a unique word and they are passed to the DNNs (by converting natural language sentences to vectors) that accepts input in fixed dimensions, therefore the word embeddings have to be static. Since BERT [13] gives contexualised embedding of each word according to the usage of the word in sentence, thereby same word will have different embeddings depending on the usage context unlike in other static word embeddings where each word has unique static embedding irrespective of the context in which it is used.

Initially, the raw BERT embeddings are generated using bert-embedding library [2] to provide contextualized word embedding. An embedding dictionary (key-value pair) is developed where key is the unique word and value is an array containing contextualized embeddings of that unique word. Since same word can be used in different context in different sentences, hence it will have different word embeddings depending on the context. Every contextualized embedding for a word are stored in the dictionary [5] by pushing the embedding into the vector corresponding to the unique word. Furthermore, the static BE of a word is obtained by taking mean of the vector containing the contexualized BERT embeddings of that word. For example, a word 'W' occurs 4 times in the dataset, then there will be 4 contexualized embeddings of 'W', let it be E_1, E_2, E_3, E_4. These 4 embeddings each of dimension (768,) are stored in the array corresponding to the key 'W' in the dictionary. Later, mean of E_1, E_2, E_3, E_4 is computed that represents the static BERT [13] embedding of 'W'. For words which are not in vocabulary, BERT [13] splits them into subwords and generate their embeddings, then take the average of embeddings of subwords to generate the embedding of the word which was not in vocabulary. Finally, by using keras Tokenizer [17] and static BERT embeddings we create the embedding matrix.

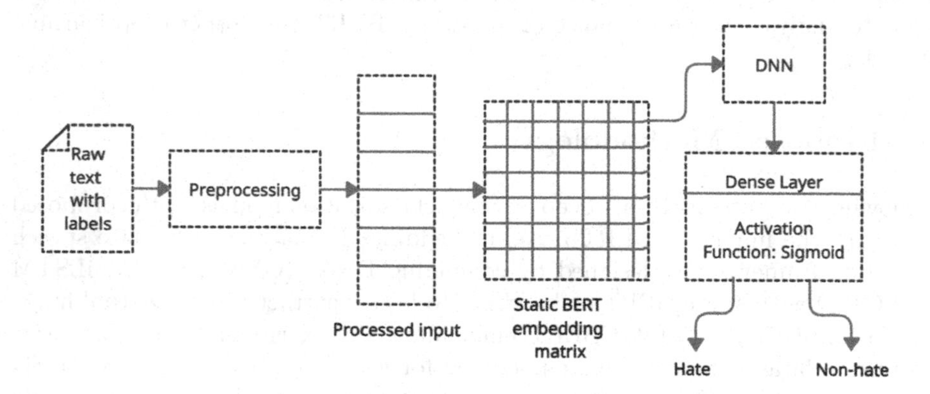

Fig. 2. Block diagram of proposed model.

4 Experiments

4.1 Choice of Dataset

Even though there are multiple datasets that are publicly available for hate speech detection but we chose to use binary version of ETHOS dataset [19] because it is the most recent dataset and the two classes (hate and non-hate) present in it are almost balanced as compared to other datasets. For example, Davidson dataset having around 24k samples (Hate speech: 5.77%, Offensive: 77.43% and 16.80% as Neither) [12] is highly imbalanced. ETHOS dataset address all such issues of available datasets.

The Shannon entropy can be used as a measure of balance for datasets. On a dataset of n instances, if we have k classes of size c_i we can compute entropy as follows:

$$H = -\sum_{i=1}^{k} \frac{c_i}{n} \log \frac{c_i}{n} \tag{1}$$

It is equal to zero if there is a single class. In other words, it tends to 0 when the dataset is very unbalanced and $log(k)$ when all the classes are balanced and of the same size n/k. Therefore, we use the following measure of balance (shown in Eq. 2) for a dataset [4]:

$$Balance = \frac{H}{\log\ k} = \frac{-\sum_{i=1}^{k} \frac{c_i}{n} \log \frac{c_i}{n}}{\log\ k} \tag{2}$$

Binary version of ETHOS dataset has 433 samples containing hate text and 565 samples containing non-hate text. For binary version of ETHOS dataset, $Balance = 0.986$ which is nearly equal to 1, indicating balance between classes.

4.2 Neural Network Architectures and Testing Environment

The proposed approach is trained and validated on the binary version of ETHOS dataset. For the purpose of comparison, the neural network architectures are kept exactly same as described by the Mollas et al. [19]. From the number of units in a neural network to the arrangement of layers in a neural network everything is kept same so as to create the same training and testing environment but change the word embeddings to static BERT embeddings.

To establish robust results, stratified k-fold validation technique with value of $k = 10$ is utilized. Furthermore, the training phase is assisted with callbacks such as early stopping (stop the training if performance doesn't improve) to avoid the overfitting problem and model-checkpointing (saving the best model). Finally, the trained model is evaluated using standard classification performance metrics i.e. accuracy, precision, recall (sensitivity), F1-score and specificity.

$$Accuracy = \frac{TP + TN}{TP + TN + FP + FN} \tag{3}$$

$$Precision = \frac{TP}{TP + FP} \tag{4}$$

$$Recall = \frac{TP}{TP + FN} \tag{5}$$

$$F1 - score = 2 \times \frac{Precision \times Recall}{Precision + Recall} \tag{6}$$

$$Specificity = \frac{TN}{TN + FP} \tag{7}$$

Where, *TP*: True Positive, *TN*: True Negative, *FP*: False Positive, *FN*: False Negative

5 Results and Discussion

Table 1 represents the results of experiment carried out by Mollas et al. [19] on binary version of ETHOS dataset, hence the DNNs uses only FT, GV or FT + GV as word embeddings. It is evident from the Table 1 that BERT (fine-tuned on binary ETHOS dataset) outperformed other models in all metrics except accuracy and specificity, its specificity stands at 74.31% which indicates high false positive hate speech classification.

Table 2. Comparative analysis of the performance of various DNNs with and without static BERT embeddings (BE).

Model	F1-score	Accuracy	Precision	Recall	Specificity
CNN + Attention + FT + GV	74.41	75.15	74.92	74.35	80.35
CNN + Attention + static BE	77.52	77.96	77.89	77.69	79.62
CNN + LSTM + GV	72.13	72.94	73.47	72.4	76.65
CNN + LSTM + static BE	76.04	76.66	77.20	76.18	79.43
LSTM + FT + GV	72.85	73.43	73.37	72.97	76.44
LSTM + static BE	79.08	79.36	79.38	79.37	79.49
BiLSTM + FT + GV	76.85	77.45	77.99	77.10	79.66
BiLSTM + static BE	**79.71**	**80.15**	**80.37**	**79.76**	**83.03**
BiLSTM + Attention + FT	76.80	77.34	77.76	77.00	79.63
BiLSTM + Attention + static BE	78.52	79.16	79.67	78.58	83.00
GRU + static BE	77.91	78.36	78.59	78.18	79.47
BERT	78.83	76.64	79.17	78.43	74.31

Bold model names represent static BERT embedding variants of the models
Bold values represent the highest value of any metric among all models

The Table 2 presents the obtained results on various DNNs, where bold model names represent BERT variant of a DNN model and bold quantities represent the highest values. It is observed that a deep neural network with static BERT embeddings outperforms the same deep neural network which is using word embedding as fastText, GloVe or fastText + GloVe in all metrics. For DNNs like CNN using attention, LSTM, CNN + LSTM, BiLSTM and BiLSTM using attention, the average (avg) increase in F1-score is 3.56%, accuracy is 3.39%, precision is 3.40%, recall is 3.55% and sensitivity is 2.37%. Hence, it is evident that static BERT embeddings tend to provide better feature representation as compared to fastText, GloVe or fastText + GloVe.

Furthermore, BiLSTM using static BERT embeddings (BiLSTM + static BE) performs better in all metrics as compared to other DNNs under consideration. In the results of experiments done by Mollas et al. [19], fine-tuned BERT outperformed other models in every metric with specificity as 74.31%, which increases to 83.03% using static BERT embeddings (BiLSTM + static BE), thereby attaining a significant increase of 8.72%.

6 Conclusion

In this article, the impact of performance in deep learning based hate speech detection using static BERT embeddings is analysed. With exhaustive experimental trials on various deep neural networks it is observed that using neural networks with static BERT embeddings can significantly increase the performance of the hate speech detection models especially in terms of specificity, indicating that model is excellent at correctly identifying non hate speeches. Therefore, it flags lesser non hate speech as hate speech, thereby protecting the freedom of speech. With such promising improvements in the results, the same concept of integrating static BERT embeddings with state-of-the-art models can further be extended to other natural language processing based applications.

References

1. Application programming interface. https://en.wikipedia.org/wiki/API. Accessed June 24 2021
2. BERT-embedding. https://pypi.org/project/bert-embedding/. Accessed June 10 2021
3. Community standards. https://www.facebook.com/communitystandards/hate_speech. Accessed 10 June 2021
4. A general measure of data-set imbalance. https://stats.stackexchange.com/questions/239973/a-general-measure-of-data-set-imbalance. Accessed 10 June 2021
5. Python dictionary. https://www.programiz.com/python-programming/dictionary. Accessed 24 June 2021
6. Updating our rules against hateful conduct. https://blog.twitter.com/en_us/topics/company/2019/hatefulconductupdate.html. Accessed 10 June 2021
7. Word embedding. https://en.wikipedia.org/wiki/Word_embedding. Accessed 24 June 2021
8. Badjatiya, P., Gupta, S., Gupta, M., Varma, V.: Deep learning for hate speech detection in tweets. In: Proceedings of the 26th International Conference on World Wide Web Companion, pp. 759–760 (2017)
9. Bahdanau, D., Cho, K., Bengio, Y.: Neural machine translation by jointly learning to align and translate. arXiv preprint arXiv:1409.0473 (2014)

10. Batra, H., Punn, N.S., Sonbhadra, S.K., Agarwal, S.: BERT based sentiment analysis: a software engineering perspective. arXiv preprint arXiv:2106.02581 (2021)
11. Cho, K., et al.: Learning phrase representations using RNN encoder-decoder for statistical machine translation. arXiv preprint arXiv:1406.1078 (2014)
12. Davidson, T., Warmsley, D., Macy, M., Weber, I.: Automated hate speech detection and the problem of offensive language. In: Proceedings of the International AAAI Conference on Web and Social Media, vol. 11 (2017)
13. Devlin, J., Chang, M.W., Lee, K., Toutanova, K.: BERT: pre-training of deep bidirectional transformers for language understanding. arXiv preprint arXiv:1810.04805 (2018)
14. Faris, H., Aljarah, I., Habib, M., Castillo, P.A.: Hate speech detection using word embedding and deep learning in the Arabic language context. In: ICPRAM, pp. 453–460 (2020)
15. Hochreiter, S., Schmidhuber, J.: Long short-term memory. Neural Comput. **9**(8), 1735–1780 (1997)
16. Joulin, A., Grave, E., Bojanowski, P., Douze, M., Jégou, H., Mikolov, T.: FastText.zip: compressing text classification models. arXiv preprint arXiv:1612.03651 (2016)
17. Keras-Team: Keras-team/keras. https://github.com/keras-team/keras. Accessed 10 June 2021
18. Mikolov, T., Sutskever, I., Chen, K., Corrado, G., Dean, J.: Distributed representations of words and phrases and their compositionality. arXiv preprint arXiv:1310.4546 (2013)
19. Mollas, I., Chrysopoulou, Z., Karlos, S., Tsoumakas, G.: ETHOS: an online hate speech detection dataset. arXiv preprint arXiv:2006.08328 (2020)
20. Pennington, J., Socher, R., Manning, C.D.: GloVe: global vectors for word representation. In: Proceedings of the 2014 Conference on Empirical Methods in Natural Language Processing (EMNLP), pp. 1532–1543 (2014)
21. Punn, N.S., Agarwal, S.: Inception U-Net architecture for semantic segmentation to identify nuclei in microscopy cell images. ACM Trans. Multimedia Comput. Commun. Appl. (TOMM) **16**(1), 1–15 (2020)
22. Punn, N.S., Agarwal, S.: Multi-modality encoded fusion with 3D inception U-Net and decoder model for brain tumor segmentation. Multimedia Tools Appl. **80**(20), 30305–30320 (2020). https://doi.org/10.1007/s11042-020-09271-0
23. Punn, N.S., Agarwal, S.: Automated diagnosis of COVID-19 with limited posteroanterior chest X-ray images using fine-tuned deep neural networks. Appl. Intell. **51**(5), 2689–2702 (2021)
24. Ranasinghe, T., Zampieri, M., Hettiarachchi, H.: BRUMS at HASOC 2019: deep learning models for multilingual hate speech and offensive language identification. In: FIRE (Working Notes), pp. 199–207 (2019)
25. Rizos, G., Hemker, K., Schuller, B.: Augment to prevent: short-text data augmentation in deep learning for hate-speech classification. In: Proceedings of the 28th ACM International Conference on Information and Knowledge Management, pp. 991–1000 (2019)
26. Schuster, M., Paliwal, K.: Bidirectional recurrent neural networks. IEEE Trans. Sig. Process. **45**, 2673–2681 (1997). https://doi.org/10.1109/78.650093
27. Soliman, A.B., Eissa, K., El-Beltagy, S.R.: AraVec: a set of Arabic word embedding models for use in Arabic NLP. Procedia Comput. Sci. **117**, 256–265 (2017)
28. Vaswani, A., et al.: Attention is all you need. arXiv preprint arXiv:1706.03762 (2017)

29. Waseem, Z., Hovy, D.: Hateful symbols or hateful people? Predictive features for hate speech detection on twitter. In: Proceedings of the NAACL Student Research Workshop, pp. 88–93 (2016)
30. Zhang, T., Gao, C., Ma, L., Lyu, M., Kim, M.: An empirical study of common challenges in developing deep learning applications. In: 2019 IEEE 30th International Symposium on Software Reliability Engineering (ISSRE), pp. 104–115. IEEE (2019)

Fog Enabled Distributed Training Architecture for Federated Learning

Aditya Kumar and Satish Narayana Srirama(✉)

School of Computer and Information Sciences, University of Hyderabad, Telangana 500046, India
{20mcpc03, satish.srirama}@uohyd.ac.in

Abstract. The amount of data being produced at every epoch of second is increasing every moment. Various sensors, cameras and smart gadgets produce continuous data throughout its installation. Processing and analyzing raw data at a cloud server faces several challenges such as bandwidth, congestion, latency, privacy and security. Fog computing brings computational resources closer to IoT that addresses some of these issues. These IoT devices have low computational capability, which is insufficient to train machine learning. Mining hidden patterns and inferential rules from continuously growing data is crucial for various applications. Due to growing privacy concerns, privacy preserving machine learning is another aspect that needs to be inculcated. In this paper, we have proposed a fog enabled distributed training architecture for machine learning tasks using resources constrained devices. The proposed architecture trains machine learning model on rapidly changing data using online learning. The network is inlined with privacy preserving federated learning training. Further, the learning capability of architecture is tested on a real world IIoT use case. We trained a neural network model for human position detection in IIoT setup on rapidly changing data.

Keywords: Internet of Things · Decentralized learning · Fog computing

1 Introduction

With advances in digital technology, Internet of Things (IoT) [6] devices are prevailing everywhere. Multiple sensors, cameras, mobiles, and smart gadgets are installed to provide support in decision making. As technology progresses, the reliance on such devices is increasing day by day. Deployment of various IoT devices has increased exponentially nowadays. The devices include simple sensors to very sophisticated industrial tools that exchange data/information through the internet. In the past few years, the number of IoT devices has increased rapidly. Currently, there are more than 10 billion IoT devices available worldwide, which is expected to be around 17 billion in 2025 and 26 billion by 2030 [8]. Every standalone device produces data which is shared with other devices for further processing. The IoT devices placed at the edge layer are generally resource

S. N. Srirama et al. (Eds.): BDA 2021, LNCS 13147, pp. 78–92, 2021.
https://doi.org/10.1007/978-3-030-93620-4_7

constrained. However, devices such as cameras or sensors generate continuous data by sensing the environment. These devices have very crucial information to mine that can be used to achieve a business goal. As the number of devices increases, the velocity and volume of data increases significantly over time. Processing and analysis of continuously generated data from resource constrained remote devices is a challenging task. Training a machine learning(ML)model on such large distributed data can be used to solve real world computational problems.

In the conventional machine learning paradigm, the training is done at the central server/cloud. The data is generated at the edge device, which is sent to the cloud. The cloud stores all data and performs training. The cloud-IoT architecture consumes large bandwidth while transferring raw data to the cloud that also creates network congestion [4]. The high latency is an issue in the cloud-IoT model that limits it for continuous learning. Additionally, data privacy concern is another major challenge in the data collection at the server. Fog computing [2,3] brings computational resources closer to the edge nodes that can efficiently process the raw data. Various data generating devices closely communicate with the nearest fog node for local computation. A fog node has enough computational capacity to process periodically collected data. The fog node can manoeuvre the associated IoT devices and directly communicates to the central server for further knowledge discovery. The distributed machine learning can be used to learn the hidden patterns from raw data efficiently. Federated learning trains a machine learning model from distributed data without sharing raw data to the server.

Distributed machine learning is a multi-node training paradigm where a participating node trains its model and collaborates with each other or the server for optimization. Federated learning [15] training proposed by McMahan et al. is a decentralized machine learning technique that can train an artificial neural network (ANN) without sending and storing raw data at the server. The algorithm trains a global model in collaboration with various devices without sharing data. Every participating device trains a model locally on their local data. The central curator coordinates with all the devices to create a global model. The locally trained model parameters such as weights and biases of ANN are shared with the central server rather than raw data. The server further aggregates model parameters from participating devices and creates a global model. This iterative process continues till the convergence of the model. In the entire training, the raw data is never shared with anyone that makes the system overall privacy preserving. Federated learning is applied to various tasks such as smart city, autonomous driving cars, industrial automation, etc. The data generating IoT devices are resources constrained that cannot train a machine learning model on the edge. Whereas, a fog computing paradigm brings computational efficiency near to IoT devices that can directly participate in federated learning. A fog enabled cloud-IoT model has the potential to quickly process continuous data.

In this paper, we have proposed an IoT-fog-cloud architecture to train a neural network on continuously generated distributed data. The paper tries to combine fog computing and federated learning for model creation. This addresses online training of continuously generating data using resource constrained devices. The proposed architecture is shown in Fig. 2. The IoT devices at the edge layer generate continuous data and share it with the fog node. The fog node is capable of online training that trains models in collaboration with the central cloud. The central server applies federated learning with the fog layer. The fog nodes capture periodic data from their associated IoT devices. It performs local model training and communicates with the cloud for global model optimization. The raw data generated at IoT devices are only shared with the local/nearest fog node. The fog layer is equipped with finite computational and storage resources that can process the raw data. The continuous data gets accumulated at fog layer/backup storage, whereas the training is done on a periodic/recent dataset only. Once the data is used for single shot training, it is not used for further training. The data does not leave the premises of fog architecture that makes the system more privacy sensitive. However, the stored data can be used for future references or it can be sent to the server if required by the application in performing long-term big data analytics. We have simulated the proposed architecture to train a neural network for safe position detection in real world Industrial Internet of Things (IIoT) setup using docker. The contributions of the paper are summarized in the following:

- Proposed a fog enabled distributed training architecture for machine learning tasks. A hybrid of Fog computing and Federated learning paradigm is used for the model training.
- Online continuous training is done with rapidly changing/growing datasets. The training includes only recent periodic data for modelling. The system assures privacy by restricting the raw data to the fog level. In addition, by not sharing raw data directly to the server, the system optimizes network bandwidth and congestion.
- We simulated the proposed architecture with Docker container. To test the learning capability of the model, we used radar data to train safe position classification in Human Robot (HR) workspace.

Rest of the paper is organized as follows. Section 2 discusses existing related articles. The proposed architecture and decentralized machine learning are discussed in 3. Experimental setup and numerical results are shown in Sect. 4. Section 5 concludes the paper with scope for future work.

2 Related Work

Data processing and machine learning need huge computational and storage resources to execute a specific task. Multiple IoT devices have generated and continue to generate voluminous data. One of the efficient ways to achieve such a task is to rent a cloud computing facility. With virtually infinite resources, the cloud executes complex model training on big data. The cloud-IoT faces various challenges such as Bandwidth, Latency, Uninterrupted, Resources-constraint and security [4]. The IoT devices are resource constraint which are connected through a wireless network to the cloud. These shortcomings obstruct smooth execution of the various tasks, specifically the real time processing. Fog computing proposes an alternative to the cloud that brings resources closer to IoT devices. It ensures low latency, network congestion, efficiency, agility and security [5]. This enabled efficient data processing at the fog layer and opens door to various applications such as smart cars, traffic control, smart buildings, real time security surveillance, smart grid, and many more. The fog layer has sufficiently enough resources to store and process raw data. This gives an advantage over a cloud to processing and decision making locally.

Bonomi et al. [2] have discussed about the role of fog computing in IoT and its applications. Fog computing provides localization that acted as a milestone in delay sensitive and real time applications. Data analytics on real time data have various applications based on context. Some of them, such as detection or controlling, need quick response typically in milliseconds or sub seconds, whereas other applications like report generation, global data mining tasks are long term tasks. Fog computing and cloud computing can performs interplay operations to achieve big data solutions. Fog responds to the real time processing task, which can be geographically distributed, and cloud computing proceed with big data analysis or knowledge consolidation. Due to proximity, fog computing is beneficial for delay sensitive applications, but it may lose importance when it gets congested. The number of jobs received at a particular fog node at a specific time may be higher, which cannot be processed quickly due to limited resources. Al-khafajiy et al. [1] have proposed a fog load balancing algorithm to request offloading that can potentially improve network efficiency and minimize latency of the services. The fog nodes communicate with each other to share their load optimally, which improves the quality of services of the network. Similarly, Srirama et al. [20], have studied utilizing fog nodes efficiently with distributed execution frameworks such as Akka, based on Actor programming model. In this context, it is also worth mentioning that scheduling of applications/tasks in cloud and fog has been studied extensively in the last few years, which is summarized in the related work of Hazra et al. [7].

The Cloud-fog computing paradigm is more efficient, scalable and privacy preserving for machine learning model training. So edge centric computing framework is needed to solve various real time operations. Munusamy et al. [16] have designed a blockchain-enabled edge centric framework to analyze the real time data in Maritime transportation systems. The framework ensures security and privacy of the network and exhibits low latency and power consumption. Distributed machine learning training offers parallel data processing over the edge of the network. Kamath et al. [9] propose a decentralized stochastic gradient descent method to learn linear regression on the edge of the network. The work utilizes distributed environment to train regression model using SGD. The method process data at device level and avoids sending it to the cloud. Federated learning is another decentralized ML technique that trains models using very large set of low resourced participating devices [10,15]. The proposed federated method collaborates with various participating devices to create a global ML model on decentralized data. Federated learning is done on low constraint devices that need efficient training strategies for uplink and down ink communication. Konečný et al. [11], talks about efficient communication between cloud and devices. The authors have suggested sketched and structured updates for server communication that reduce the amount of data sent to the server.

The fog-cloud architecture is well suited for distributed machine learning training. Li et al. [12] have used cloud-fog architecture for secure and privacy preserving distributed deep learning training. The local training is given at the fog layer then it coordinates with the cloud server for aggregation. Additionally, it uses encrypted parameters and authentication of valid fog node to ensure legit updates. The central node works like a master node for information consolidation. It synchronizes the training from various devices. Due to stragglers or mobility of devices such as vehicles, drone the synchronous update creates difficulty in training. Lu et al. [13] proposes asynchronous federated training for mobile edge computing. The training is done similar to federated learning, but global model aggregation is done asynchronously. To ensure the privacy and security of the shared model, it adds noise to the parameters before sending it to the server. Luo et al. [14] have proposed a hierarchical federated edge learning framework to train low latency and energy-efficient federated learning. The framework introduces a middle layer that partially offloads cloud computational work. The proposed 3-layered framework aggregates model parameters at both fog layer and cloud layer while training is done at the remote device. Fog enabled federated learning can facilitate distributed learning for delay-sensitive applications. Saha et al. [17] proposed fog assisted federated learning architecture for delay-sensitive applications. The federated learning is done between edge and fog layer, then the central node heuristically steps in for global model aggregation. This training is done on geographically distributed network that optimizes communication latency by 92% and energy consumption by 85%. Most of the research assumes that data generating IoT devices contain enough computational resources to train ML models. Additionally, these IoT devices participate in distributed training with the complete dataset. Our proposed architecture is a three layered network for machine learning training. The edge layer generates raw data only, and the cloud layer consolidates the global model. Whereas, the fog layer participates in decentralized machine learning training with the central server. The federated training is done on continuously changing dataset generated by the edge layer.

3 Decentralized Federated Learning

Decentralized federated learning aggregates locally trained models on the central server. A global model is created by combining multiple independently trained models. The conventional machine learning approaches collect possible data D to the server; then it learns a machine learning model M using a sophisticated algorithm. In contrast, federated learning trains its model without collecting all possible data to the central server. It is a collaborative learning paradigm where k number of participating devices trains a local model on their data D_i. Each participating device i contains their personal data D_i. The total data set is $D = \sum_{i=1}^{k} D_i$. Here, D_i is a collection of input data samples $(x_j, y_j)_{j=1}^{n}$ for supervised learning. Where $x_j \in \mathbb{R}^d$ is a d dimensional input data and $y_j \in \mathbb{R}$ is the associated label for input x_j. The device data D_i are remotely generated or centrally distributed.

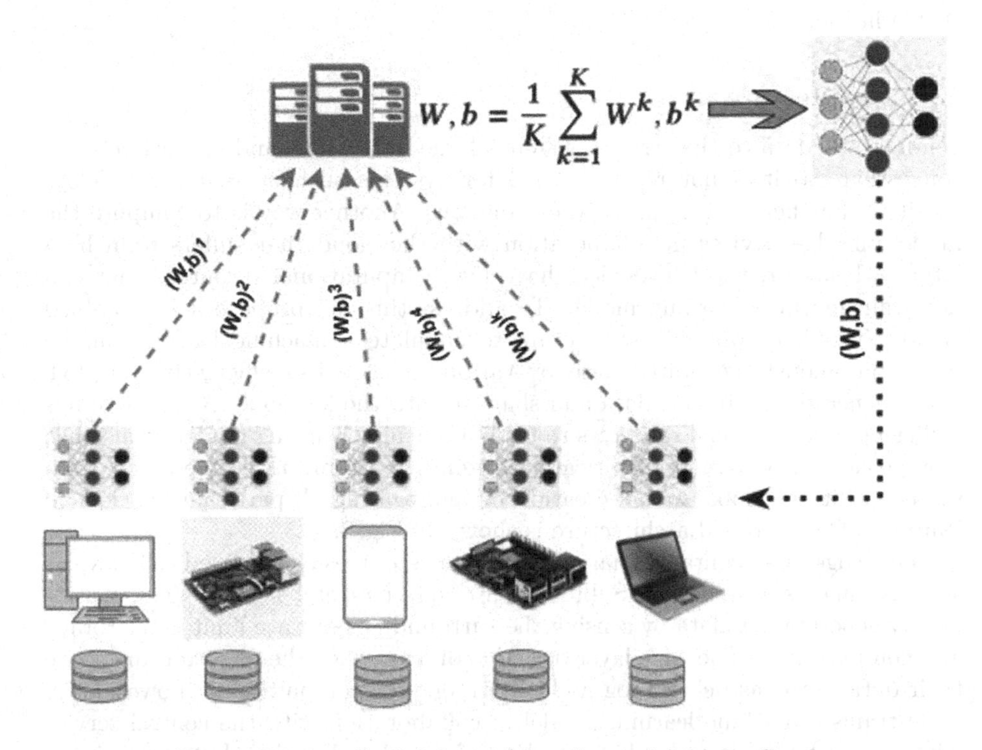

Fig. 1. Decentralized federated learning training paradigm

For every participating device, it is a machine learning task where it trains local model parameters using its data D_i. It takes input data $(x_j, y_j)_{j=1}^{n}$ to compute local parameters i.e. weights and biases. The loss function of every devices i for dataset D_i is $F_i(w) = \frac{1}{|D_i|} \sum_{j=1}^{n} f(h(w, x_j), y_j)$. Where $f(h(w, x_j), y_j)$ is the loss for j^{th} sample from D_i. The participating devices optimize the loss using an optimizer to find optimal parameters. In the subsequent step, the locally trained parameters (weights and biases) are shared with the central server for global model creation. The central curator receives all k locally trained models parameters and performs aggregation operations on them. The weights and biases (W, b) of respective layers of every model are aggregated. Thus, the aggregated model contains representation from all models, which is further fine-tuned in the next iteration. The global model training paradigm is done in collaboration with central serve as shown in Fig. 1. The aggregated/updated global parameters (W, b) are pushed back to local devices for the next round of training. It is an iterative process that optimizes global parameters using local model updates. The global loss function for the system is $F(w) = \frac{1}{|D|}\sum_{i=1}^{k}|D_i| \times F_i(w)$. The goal of federated learning is to learn a global model by combining all local models. This training cycle continues until convergence without accessing raw data as shown in Fig. 1.

3.1 Architecture

Sending raw data to the server consumes large bandwidth and creates network congestion. So it is not recommended for systems such as real time, delay-sensitive. Further, it also has privacy concerns. Another way is to compute the model on edge devices in collaboration with the cloud that suffers from high latency. However, the IoT devices have low computational resources that can not train machine learning model. To address this, we propose a fog enabled cloud-IoT online training architecture to simulate a machine learning model from continuously generated data by various IoT devices efficiently. The IoT nodes generate continuous data and share it with the fog node. A fog node has sufficient computational resources to train a machine learning model. It also has storage capability to store historical data generated from IoT devices. The cloud server facilitates global model creation by aggregating all participating devices leanings. The proposed architecture is shown in Fig. 2.

The Edge layer contains a large number of resources constrained IoT devices such as cameras, watches, GPS, bulbs, sensors, radars, etc. These devices continuously generate raw data by sensing the surroundings but are limited in storage and computations. The edge layer directly connects with the fog layer and share their data to the associated fog node. With enough computational power, a fog node trains a machine learning model in collaboration with the central server. Although, a fog node stores historic data of associated devices but the training is done on recently captured periodic data frames. This simulates online training of continuously changing datasets. The Fog-Cloud layers participate in federated learning for machine learning. Once local models are trained on the fog layer, it is shared with the cloud layer. The cloud layer aggregates the local models and creates a global model. The federated learning with fog-cloud architecture is continued till the convergence of the global model. The proposed architecture is used for machine learning model training on rapidly growing data on resource constrained devices. The fog layer is responsible for data collection and federated learning training with a cloud node. While federated training fog layer only shares learning parameters to the cloud for aggregations. The raw data is stored at the fog layer, which is not shared with the central server. The proposed architecture simulates a distributed machine learning using computations of resource constrained devices.

Fig. 2. Distributed learning architecture

3.2 Online Training and Data Privacy

With the variety of digital devices, data proliferation is another challenge in machine learning training. As discussed earlier, sending this data to the server would not be an efficient way to mine it. Additionally, collecting, storing all the data, and then applying machine learning training needs huge computational resources. The data's velocity increases as the number of devices employed increases. Consequently, it increases volume of the data with a variety of data that lead to big data problems. To address this to some extent, a continuous learning approach is applied while training a model. Rather than applying training on the complete dataset, online training is done on a subset of data. The subset can be periodic data generated from a device over a fixed time interval. Once the device learns the representation from current data, it shares representations with the central curator for global modelling. The curator aggregates all the representations and creates a central model for all devices. In the next round, every device generates a new set of datasets. Subsequently, in this round, the global model is further fine-tuned with the next dataset. This is important because IoT devices are low resources devices that are incapable of machine learning training on a very large dataset. The devices continue to generate the raw data and train the model. This can simulate a global model by exploiting low constrained devices computations.

However, the IoT devices such as sensors, CCTV cameras, radars do not have enough computation resources to run on-device machine learning. This work addresses this issue by deploying a fog node near IoT devices. All data generating devices are connected to the associated fog node to share raw data. With sufficient computational and storage capabilities, a fog node stores periodic data and performs online machine learning training. The rapid growth of data accumulates a large amount on a fog node. Due to the computational limitation of a fog node, online training is done on recently captured data leaving historic data aside. At the same time, the entire data is kept at the fog layer on backup storage, which can be reproduced in future if required. Then fog node participates in the federated learning process in collaboration with the central cloud server.

IoT devices contain private data, location information, sensitive data, bank details, chat and personally identifiable data. Data privacy and security is another major challenge in cloud-IoT computation. Nowadays, there are increasing concerns for personal data sharing. The users are not comfortable in sharing personal data such as photos or chat to the cloud. The raw data contains crucial patterns that can be useful for various applications such as recommender systems, security analyses, smart homes, safety predictions, etc. Additionally, machine learning task has to follow strict data protection rules such as General Data Protection Regulation (GDPR). To address the privacy concerns, we have used a decentralized machine learning approach for model training. In the proposed work, the data is stored at the fog layer and not shared with the cloud. Fog node participates in machine learning training that only shares model parameters, not raw data. This makes the system overall privacy preserving.

4 Evaluation and Results

This section describes evaluations and experimental results of the proposed framework for machine learning model. The model training is done for radar data in IIoT setup. We simulate the proposed architecture using Docker containers. Then, training of a global model for safe distance detection is done for human position in HR workspace. To show the efficiency of the proposed work, We have trained the ANN model in distributed environments and achieved expected results.

4.1 Docker Based Fog Federation Framework

We have used docker containers to simulate the distributed machine learning. The docker engine facilitates multiple containers to run various programs independently. A container provides a run time environment for program execution. Additionally, docker creates a network of multiple containers that can communicate to others. With sufficient computational resources, we employed multiple docker containers as a fog node. Every container runs a machine learning model independently with their local data. We have used gRPC library for requests and service calls between fog and central node. The federated learning is done between cloud and fog nodes using docker containers with gRPC calls. The IoT devices generate continuous data and share it with the fog node. Fog node stores historical data at backup devices and trains the model on recent data. To simulate the continuous data generation and online training, we used a fixed set of data samples for local training. Every container has its personal data, and machine learning is done on fixed periodic sequential training samples.

We have simulated the proposed architecture to train ANN based machine learning model for human operator position detection in a human-robot workspace. The dataset is recorded from multiple Frequency-Modulated Continuous Wave (FMCW) radars. The fog nodes compile one minute of data from every device and complete one round of federating learning. We trained the model on 60 frames assuming every radar is generating 1 frame/sec. The next round of training is done on the next sequence of datasets. For this experiment, we have taken 5 fog nodes for decentralized machine learning training. The fog node trains fully connected ANN using tensorlow framework. The ANN model contains a single hidden layer with 64 neurons. The input and output layers have 512 and 8 neurons, respectively, based on data dimension and output labels.

We have trained a shallow neural network with one hidden layer. The model is a fully connected dense network with 64 units in the hidden layer. Input layer has 512 input neurons which is fully connected to the only hidden layer (dimension $512 \times 64 + 64$ bias) followed by a ReLu layer. The output layer has 8 neurons which is densely connected to the hidden layer (dimension $64 \times 8 + 8$ bias). The final output label is predicted based on sigmoid activation function at the output layer. The local training on every device is done for 5 epochs to simulate low computational resources. The network is trained by backpropogation algorithm using categorical crossentropy loss function. Further, the 'adam' optimizer is used with learning 0.001 as an optimizer to optimize the training error. The loss value of the global model is calculated on test data. Also, we have traced the accuracy performance of the global model on both personal and unknown test datasets. The fog node participates in federated learning in collaboration with a central node with its local data, where device local data is recently generated one minute data. The simulation is done to show computational intelligence of proposed work on continuously changing data.

4.2 FMCW Radar Dataset for Federated Learning

The proposed architecture is validated on a real world IIoT use case. The data is generated from FMCW radars in a human-robot workspace. FMCW radars are effective IIoT device in industrial setup for environment sensing, distance measurement, etc. These radars are placed in a shared workspace of human robot to capture human position in the environment. Detection of human position in an industrial setup is crucial for worker's safety. These radars contain data to measure the distance and position of the human operator near it. The data distribution of devices is non independent and identically Distributed(non-IID), i.e. every device has its locally generated dataset. However, each participating devices have all classes samples. We used mentioned dataset to train a machine learning model to classify human safe distance. The dataset is published by Stefano Savazzi, which can be downloaded from IEEE Dataport [18].

The radars output signals are preprocessed and converted into 512-point. The detailed methodology and collection of data are given in this paper [19].

Table 1. FMCW radars dataset

Distance (m)	Class	Critical/Safe
<0.5	1	Critical
0.5 <= d <1.0	2	Critical
1.0 <= d <1.5	3	Critical
1.5 <= d <2.0	4	Safe
2.0 <= d <2.5	5	Safe
2.5 <= d <3.0	6	Safe
3.0 <= d <3.5	7	Safe
>= 3.5	0	Safe

The input sample contains 512 points with 8 labels. The labels are characterized by different distances of human operator and radar. The dataset contains a total of 32,000 samples of FFT range measurements 521 points. The dataset is already divided into training and testing samples of 16,000 $\times$ 512 shape. Additionally, the data sample is also randomly distributed over various devices for federated learning simulation in the database. We have implemented the given data distribution over 5 devices for federated learning to learn ANN model for safe/unsafe position detection. The training is done for C = 8 classification of the potential situation in human robot workspace. The label is an integer from 0 to 7, where Class 0 represents human distance >3.5 m which is also marked as safe. Class 1 is represented as critical since the distance is <0.5 m. Other labels are marked based on different distance measures between humans and radars. Table 1 contains labels for various classes.

4.3 Results and Analysis

We trained an ANN model for human position classification in the shared HR workspace. The online training is done with 60 frames at a time with 5 fog nodes. From 16,000 training samples, every fog node receives 3200 independent samples. The local model training is done with the current 60 samples for 5 epochs only. Then central server performs aggregation of all learnt model parameters. This completes one round of federated learning. In the next round, we use next 60 samples for training by skipping previous data points. We executed such 53 rounds that exhaust entire local dataset training. At every round, we assess the model performance in terms of loss and accuracy. The global model is evaluated on the test dataset. The training loss and accuracy of the model on test data are shown in Fig. 3.

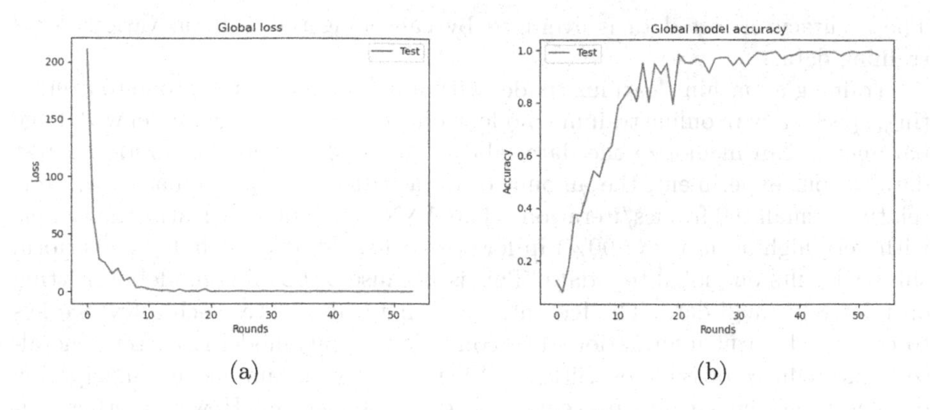

Fig. 3. Global model performance on test data Fig(a) is loss value and Fig (b) at every training round

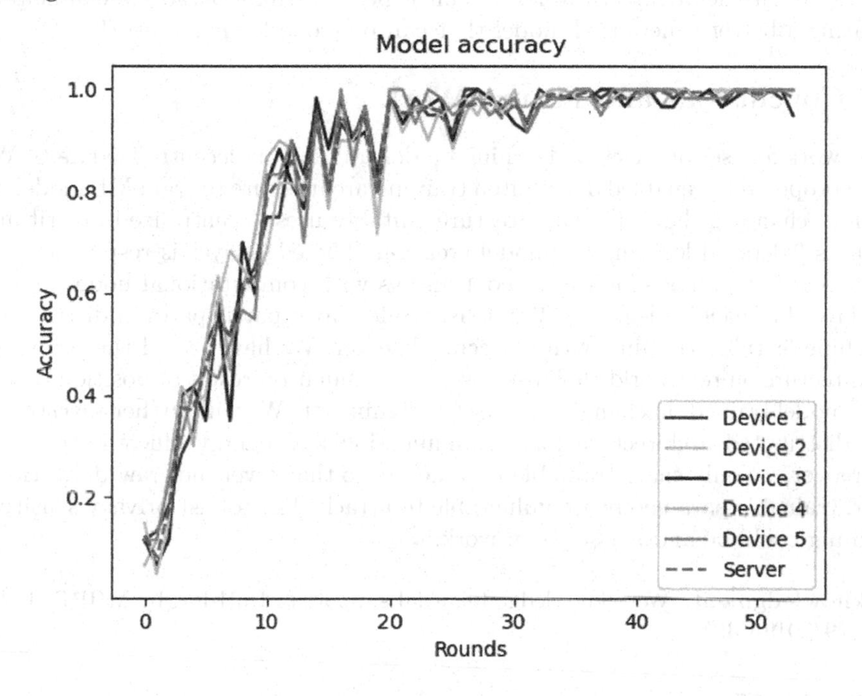

Fig. 4. Model performance on all devices

As training increases over the number of rounds, model improves its accuracy significantly. The model optimizes loss value and stabilizes the training after 30 rounds. The proposed online training performed exceptionally well as the test accuracy reached 99%. The local model is trained on multiple fog nodes parallelly. The global model is combined learning of all local models. Figure 4 shows accuracy of the global model on various local data resides in fog devices.

The accuracy of test data is averaged by cancelling the drift on various local training data.

Training a machine learning model with a neural network is prone to overfitting, specifically in online training mode. Comparatively, a large model with huge parameters can memorize the data labels that tend to perform poorly on test data. In our experiment, the amount of data passed at a particular instance is relatively small (60 frames/iteration). The ANN model quickly learnt the sample with very high accuracy (>90%) in fewer epochs. However, it failed to perform similar results on global test data. This is because of possible model overfitting on relatively small data. The federated learning consolidates such a few models to combined learnt information. The combined output model is better generalized and reduces possible overfitting. With the varying number of participating devices, federated learning prevents overfitting intrinsically. However, other generalization techniques such as dropout, normalization, regularization, etc. may be applied at the architectural level. This type of learning paradigm can help in creating a better generalized model that can be scope for future work.

5 Conclusions and Future Work

This work focuses on machine learning model training on decentralized data. We have proposed fog enabled distributed training architecture to train ML model on rapidly changing data. The architecture suitably uses decentralized algorithms such as federated learning for model creation. The edge layer is responsible for data generation. The cloud layer coordinates with computational nodes on the fog layer for machine learning. Whereas, the fog layer participates in distributed machine learning training with the central server. We have tested the proposed architecture on real world IIoT use case. The simulation result of position detection model trained on changing dataset is significant. We will further investigate the distributed architecture for communication and energy efficient training. Moreover, we only share trainable parameters to the server, not raw data. However, trainable parameters are vulnerable to attack. The robust privacy sensitive training could be another scope of work.

Acknowledgment. We acknowledge financial support to UoH-IoE by MHRD, India (F11/9/2019-U3(A)).

References

1. Al-Khafajiy, M., Baker, T., Waraich, A., Alfandi, O., Hussien, A.: Enabling high performance fog computing through fog-2-fog coordination model. In: 2019 IEEE/ACS 16th International Conference on Computer Systems and Applications (AICCSA) (2019). https://doi.org/10.1109/AICCSA47632.2019.9035353
2. Bonomi, F., Milito, R., Zhu, J., Addepalli, S.: Fog computing and its role in the Internet of Things. In: Proceedings of the First Edition of the MCC Workshop on Mobile Cloud Computing, pp. 13–16. MCC '12, Association for Computing Machinery, New York, NY, USA (2012). https://doi.org/10.1145/2342509.2342513, https://doi.org/10.1145/2342509.2342513

3. Buyya, R., Srirama, S.N.: Fog and Edge Computing: Principles and Paradigms. John Wiley & Sons (2019)
4. Chang, C., Srirama, S.N., Buyya, R.: Internet of Things (IoT) and New Computing Paradigms (2018)
5. Consortium, O.: OpenFog Reference Architecture for Fog Computing, Technical Report (February 2017)
6. Gubbi, J., Buyya, R., Marusic, S., Palaniswami, M.: Internet of Things (IoT): a vision, architectural elements, and future directions. Future Gener. Comput. Syst. **29**(7), 1645–1660 (2013). https://doi.org/10.1016/j.future.2013.01.010, https://www.sciencedirect.com/science/article/pii/S0167739X13000241
7. Hazra, A., Adhikari, M., Amgoth, T., Srirama, S.N.: Joint computation offloading and scheduling optimization of iot applications in fog networks. IEEE Trans. Netw. Sci. Eng. **7**(4), 3266–3278 (2020)
8. Holst, A.: IoT connected devices worldwide 2019–2030, August 2021. https://www.statista.com/statistics/1183457/iot-connected-devices-worldwide/
9. Kamath, G., Agnihotri, P., Valero, M., Sarker, K., Song, W.Z.: Pushing analytics to the edge. In: 2016 IEEE Global Communications Conference (GLOBECOM), pp. 1–6 (2016). https://doi.org/10.1109/GLOCOM.2016.7842181
10. Konečný, J., McMahan, H.B., Ramage, D., Richtárik, P.: Federated Optimization: Distributed Machine Learning for On-Device Intelligence. CoRR abs/1610.02527 (2016). http://arxiv.org/abs/1610.02527
11. Konečný, J., McMahan, H.B., Yu, F.X., Richtarik, P., Suresh, A.T., Bacon, D.: Federated learning: strategies for improving communication efficiency. In: NIPS Workshop on Private Multi-Party Machine Learning (2016). https://arxiv.org/abs/1610.05492
12. Li, Y., Li, H., Xu, G., Xiang, T., Huang, X., Lu, R.: Toward secure and privacy-preserving distributed deep learning in fog-cloud computing. IEEE Internet Things J. **7**(12), 11460–11472 (2020). https://doi.org/10.1109/JIOT.2020.3012480
13. Lu, Y., Huang, X., Dai, Y., Maharjan, S., Zhang, Y.: Differentially private asynchronous federated learning for mobile edge computing in urban informatics. IEEE Trans. Ind. Inform. **16**(3), 2134–2143 (2020). https://doi.org/10.1109/TII.2019.2942179
14. Luo, S., Chen, X., Wu, Q., Zhou, Z., Yu, S.: HFEL: joint edge association and resource allocation for cost-efficient hierarchical federated edge learning. IEEE Trans. Wireless Commun. **19**(10), 6535–6548 (2020). https://doi.org/10.1109/TWC.2020.3003744
15. McMahan, B., Moore, E., Ramage, D., Hampson, S., Arcas, B.A.Y.: Communication-efficient learning of deep networks from decentralized data. In: Singh, A., Zhu, J. (eds.) Proceedings of the 20th International Conference on Artificial Intelligence and Statistics. Proceedings of Machine Learning Research, vol. 54, pp. 1273–1282. PMLR (20–22 Apr 2017). https://proceedings.mlr.press/v54/mcmahan17a.html
16. Munusamy, A., et al.: Edge-centric secure service provisioning in IoT-Enabled maritime transportation systems. IEEE Transactions on Intelligent Transportation Systems, pp. 1–10 (2021). https://doi.org/10.1109/TITS.2021.3102957
17. Saha, R., Misra, S., Deb, P.K.: FogFL: fog-assisted federated learning for resource-constrained IoT devices. IEEE Internet Things J. **8**(10), 8456–8463 (2021). https://doi.org/10.1109/JIOT.2020.3046509
18. Savazzi, S.: Federated learning: example dataset (fmcw 122ghz radars) (2019). https://doi.org/10.21227/8yqc-1j15

19. Savazzi, S., Nicoli, M., Rampa, V.: Federated learning with cooperating devices: a consensus approach for massive IoT networks. IEEE Internet Things J. **7**(5), 4641–4654 (2020). https://doi.org/10.1109/JIOT.2020.2964162
20. Srirama, S.N., Dick, F.M.S., Adhikari, M.: Akka framework based on the actor model for executing distributed fog computing applications. Future Gener. Comput. Syst. **117**, 439–452 (2021)

Modular ST-MRF Environment for Moving Target Detection and Tracking Under Adverse Local Conditions

T. Kusuma(✉) and K. Ashwini(✉)

Global Academy of Technology, RR Nagar, Bangalore, Karnataka, India
dr.ashwini.k@gat.ac.in

Abstract. This paper primarily encompasses phased-implementation plan where at first it executes ST-MRF model for local conditioning, which is followed by machine learning driven MRF model to perform target detection and tracking. In other words, the proposed model at first performs local conditioning using ST-MRF, where it intends to reduce or eliminate detrimental local conditions such as haziness, fog, smoke, etc. Subsequently, over the conditioned compressed as well as moving target tracking even under adverse operating conditions. Domain video input, it executes ST-MRF (with machine learning and expectation maximization) model to perform moving target detection and tracking. To achieve it, this paper encompasses pre-processing followed by intensity value prior estimation, medium transmission modelling, dark channel estimation and maximum likelihood assisted ST-MRF model for compressed video local conditioning. Here, local conditioning (using ST-MRF) represents a task to alleviate fog, low-light condition, dehazing etc. Once performing local conditioning, the proposed model executes Edge-prior Estimation, followed by K-Means clustering for edge-preserving initial ROI labelling. Subsequently, it applied expectation maximization assisted MRF model to segment moving target and track it over consecutive frames. Thus, the proposed model exhibits both local conditioning.

Keywords: ST-MRF · ROI · Dark channel · k-means clustering · Target tracking

1 Introduction

In last few years high pace emergence of vision technologies, software computing and low-cost hardware development has opened up a broad horizon for computer vision based applications serving civil surveillance, industrial monitoring and control, defence establishment, critical infrastructure surveillance, and major security systems. Amongst major vision-based applications, visual surveillance systems have emerged as the dominating requirements which have geared up a number of research efforts across academia-corporate horizon. Computer vision technologies have been playing vital role towards surveillance systems and even the demands have been increasing exponentially.

The need of efficient surveillance system makes computer vision technology inevitable to serve major public surveillance, security, military purposes, and intelligent

S. N. Srirama et al. (Eds.): BDA 2021, LNCS 13147, pp. 93–105, 2021.
https://doi.org/10.1007/978-3-030-93620-4_8

transport systems. In practice, intelligent surveillance systems require fast and efficient approach for the moving objects detection and continuous tracking. Though, there are a number of researches or efforts made towards target detection in video; however, the efforts made towards compressed domain methods are very less, signifying a broadened horizon for the academia-industries to explore more effective spatio-temporal method for moving object detection and tracking in compressed domain. Considering it as motivation, in this paper the key emphasis has been made on developing a state-of-art new and robust moving target tracking model in compressed domain. A state-of-art new and robust "improved modular ST-MRF model assisted local conditioned environment for moving target detection and tracking" is developed. As the name indicates, the proposed model is designed as a modular solution which embodies the capacity of both ST-MRF based local conditioning model as well as ST-MRF assisted moving target detection and tracking solution. However, the strategic amalgamation of these two abilities as standalone solution makes the proposed system more robust to track moving object even under adverse background or local conditions. To achieve it, the proposed model encompasses pre-processing followed by intensity value prior estimation, medium transmission modelling, dark channel estimation and maximum likelihood assisted ST-MRF model for compressed video local conditioning. Here, local conditioning (using ST-MRF) represents a task to alleviate fog, low-light condition, dehazing etc. Once performing local conditioning, the proposed model executes Edge-prior Estimation, followed by K-Means clustering for edge-preserving initial ROI labelling. Subsequently, it applied expectation maximization assisted MRF model to segment moving target and track it over consecutive frames. Thus, the proposed model exhibits both local conditioning as well as moving target tracking even under adverse operating conditions.

The overall implementation plan of the proposed system is given in Fig. 1.

1.1 Data Collection and Pre-processing

The proposed model intends to perform moving target tracking in compressed domain video sequences. In this reference, though a number of existing benchmark data are available; however, most of the existing benchmark data are smaller in size and even don't consider highly dynamic background with adverse local conditions, such as smoke, haziness etc. On the contrary, exploring in depth one can easily found that the majority of the real-time applications including ATC, CCTV-enabled supervision systems, AVR systems etc. employ compressed domain bit-sequences or compressed domain video sequences. Considering this fact, in this phase, a primary data with aforesaid operating environment is prepared. The compressed bit-stream is reconstructed for 120-frames per second. Noticeably, the levels of significance of the frame rates as well as the operating conditions seem in sync with ATC operations, battle-field surveillance, surveillance and monitoring in forest etc. Thus, the proposed model can be of great significance towards aforesaid operating environment.

Once collecting the input compressed data (i.e., the video sequence) it has been processed for the proposed modular ST-MRF based local conditioning. This approach involved four key functions. These are:

1. Intensity Value Prior Estimation

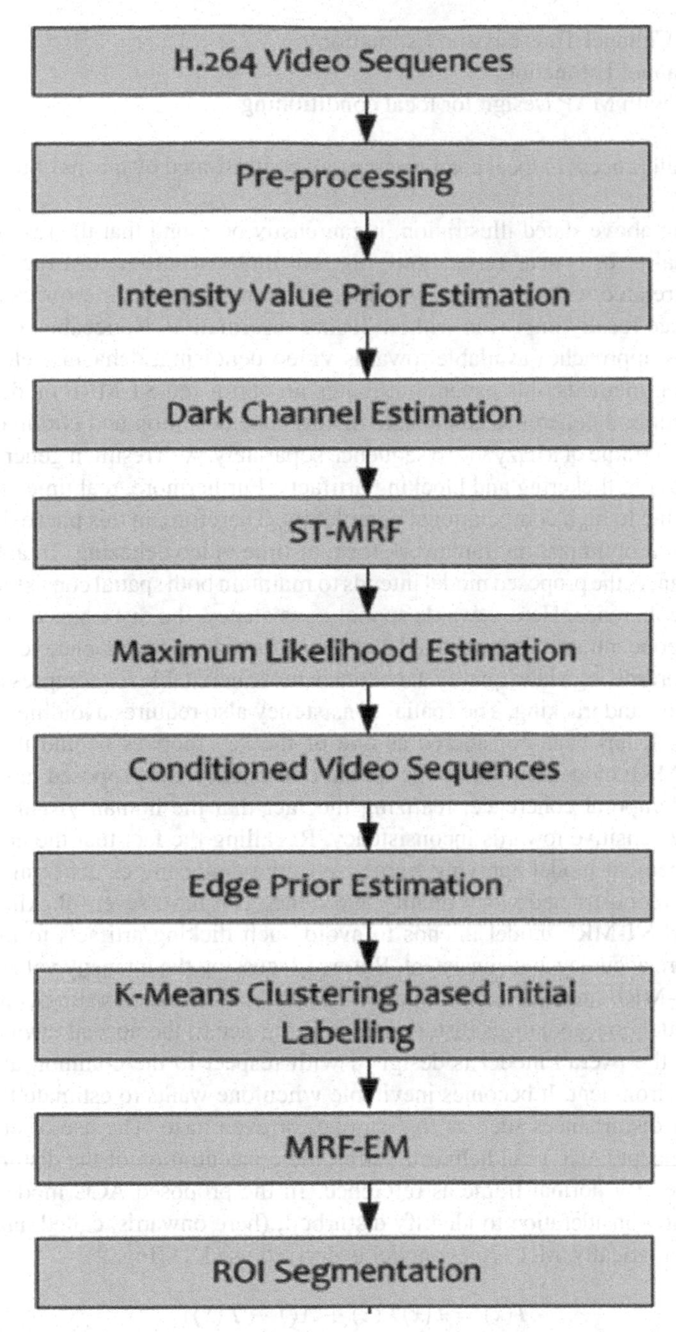

Fig. 1. Proposed local conditioned St-MRF environment for moving target detection and tracking.

2. Medium Channel Transmission Estimation
3. Dark Channel Estimation
4. ST-MRF with MAP Design for local conditioning.

The overall process of local conditioning can be illustrated by means of the following example.

Observing above stated illustration, it can easily be found that the residual energy or the residual error is near zero, signifying that the concentration of the disturbance is highly correlated with the intensity value. Thus, reducing such error can make the video sequence locally improved without losing information. Noticeably, though there are numerous approaches available towards video denoising, dehazing, etc. however, unlike existing methods, this paper contributes an optimized ST-MRF model in which the spatio-temporal coherence can enable disturbance detection and elimination frame by- frame each frame of a hazy video sequence separately. As a result, it generates recovered video having flickering and blocking artifacts. Furthermore, real time applications can prohibit due to high computational complexity. Therefore, in this phase developed a spatio-temporal optimization framework for real-time video dehazing. To achieve optimal performance, the proposed model intends to maintain both spatial consistency as well as temporal coherence. Here, towards spatial consistency, the disturbance (say, smoke, fog or haze) concentration is local fixed and the retrieved video sequence remains intact without compromising video quality. It becomes more inevitable for compressed domain target detection and tracking. The spatial consistency also requires avoiding inter-frame discontinuity. It has been considered as one of the key motives behind the proposed modular ST-MRF model. Similar to the spatial consistency, the proposed model intends to maintain temporal coherence, realizing the fact that the human visualization system is highly sensitive towards inconsistency. Recalling the fact that the conventional video enhancement model applying frame-by-frame denoising or dehazing often turn into reduced reconstructed video quality and hence can have severe flicking artifacts. The proposed ST-MRF model intends to avoid such flicking artifacts to make target detection more accuracy and consistent. Before discussing the intensity value prior estimation or ST-MRF implementation to cope up with real-world operating condition, the proposed model conceptualizes disturbance in reference to the normal environment. In other words, the overall model is designed with respect to the common atmospheric scattering environment. It becomes inevitable when one wants to estimate the concentration of the disturbances such as fog, smoke, or even haze. The use of atmospheric scattering concept (ACC) can help estimating the concentration of the disturbance and hence can identify normal frame as reference. In the proposed ACC model, standard was taken into consideration to identify disturbed, (here onwards, called hazed frame) frame. Mathematically, McCartney model is derived as (1.1) (Fig. 2).

$$I(x) = J(x)T(x) + A(1 - T(x)) \tag{1.1}$$

In the above derive ACC model (1.1) the parameter I(x) states the observed disturbed or hazed frame, while J(x) signifies the real improved (local scene) scene to be recovered. The other parameter T(x) represents the medium of transmission, which is air in the proposed operating condition. Here x defines pixels and A represents the

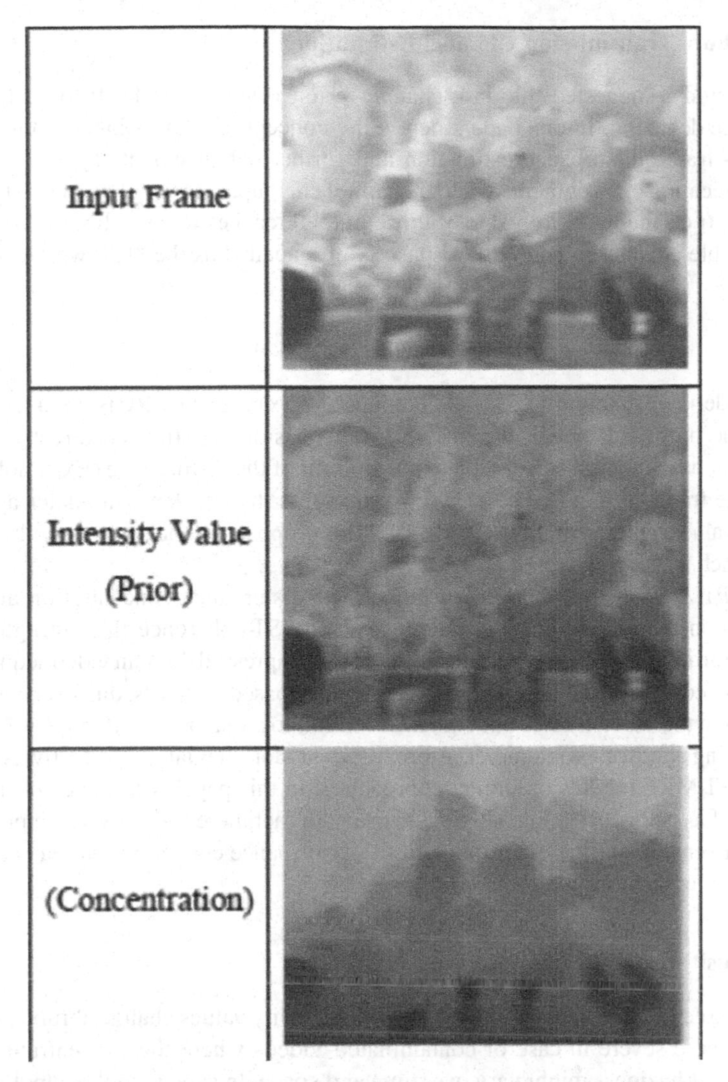

Fig. 2. Illustration of the sequential local conditioning using ST-MRF

global atmospheric light. After identifying the value of A, the improved frame J(x) can be obtained that can later help in the estimation of T(x). Noticeably, in the complete bit-stream frame the parameter A remains constant and hence it becomes easier to approximate. The parameter T(x) represents the signal which reaches the camera without scatter. Hence, the disturbance concentration map becomes easy to test mate, even without reduced accuracy.

1.2 Medium Transmission Channel Estimation

A recent study, suggested the dark channel prior, often called DCP to perform outdoor image dehazing. In this paper, merely the concept of DCP is inherited to perform frame by- frame disturbance estimation and further enhancement. Typically, in real-world application environment, a colour channel encompasses very low intensity (value) pixels that (i.e., the intensity value of those pixels) reaches to zero. In this manner, the least possible channel in RGB color space is often called as the DCP which is defined as (1.2).

$$D(x) = min_{c \in \{r,g,b\}} I^c(x) \tag{1.2}$$

In above derived function (1.2), the component Ic(x) signifies RGB color channel of I(x). In the proposed model, the medium transmission is estimated directly with the DCP encompassing high correlation to the amount of the disturbance (Ex, smoke, haze, etc.) in the frame as $\tilde{T}(x) = 1 - \omega D(x)/A$, here ! is the constant parameter applied to map DC value to the medium transmission. The proposed model considered! as a fixed value, which is assigned as 0.7.

ST-MRF Design In the proposed model, the disturbance concentration and allied concentration map is obtained in reference to the ST-coherence that strengthen it to handle the artifacts like flickering and blocking in compressed domain video sequence. To be noted, in compressed domain video, unlike pixel-based methods, due to encoding and compression the video frames undergo information loss and hence often gives flickering and blocking kind of problems. The proposed model alleviates it effectively. Unlike classical ST-MRF models or similar approaches, in this paper it has been designed on the basis of a factor called intensity value prior. In fact, here intensity value prior helps ST-MRF to enable optimal fine-tuning of the disturbance concentration map and hence achieves better accurate suppression.

1.3 Intensity Value Prior

In compressed domain videos, typically pixels intensity values change abruptly and turn out to be more severe in case of contaminated videos where the concentration of the smoke, fog or haziness might vary over time and space. In sync with the hypothesis that the $A(1 - T(x))$ represents the atmosphere scattering model, as discussed above, it can be stated that on the observed value of the concentration changes the effect of gray or white air light can be highly associated to the disturbance magnitude. In this manner, there can be the increase in the intensity value as well as the disturbance concentration due to the air light. Now, to ensure optimal performance, the proposed ST-MRF model is designed and tuned in sync with spatial consistency and temporal coherence. Spatial Consistency Considering real-time application environment, especially in the compressed domain videos, the process of estimation of pixel level concentration might not succeed. In other words, the concentration of the disturbance or allied estimation can be inaccurate because of the outlier pixel values in each succeeding frame. These kinds of the problems can be well reduced by hypothesizing the disturbance concentration as local filter which can be locally constant. However, such filters such as median, maximum or minimum often impose the blocking artifacts in the disturbance concentration map, and therefore in the proposed model a state-of-art new ST-MRF concept with intensity value prior approach is developed. Here, in ST-MRF the inclusion of intensity value prior helps retaining the locally constant and inner-frame continuity. In this method, the intensity value V(x) is linearly processed and transformed to the disturbance concentration, given as D(x) in spatial neighbourhood. Noticeably, in the proposed model, the transformation fields $W = \{w(x)\}_{x \in \forall}$ and $B = \{b(x)\}_{x \in \forall}$ are correlated to the contextual information. The spatial likelihood function defined is given as (1.3).

$$Ps = (w, b) \propto \prod_{y \in \Omega(x)} exp\left(-\frac{w(x)V(y) + b(x) - D(y)\|_2^2}{\sigma_s^2} \right) \tag{1.3}$$

In above derived function (1.3), the component (x) signifies the local patch centered at x with the size of r x r and δ_s represents the spatial parameter.

Temporal Coherence. In case of compressed domain video processing and target tracking task, the vital information existing in between the consecutive streaming frames can assist towards the flicking artifact's avoidance. Like in the previous work, where a GME was applied to alleviate the motion caused disturbance, in the proposed model the motion of camera and ROI varies the disturbance's concentration. Typically, the observed radiance gets closer to the original scene radiance as an object or ROI approaches to the camera. In fact, the observed radiance turns out to be same as the atmospheric light when the ROI moves away from the camera. Therefore, based on the changing intensity value, the haze concentration of a scene point can be modified adaptively. Thus, the use of intensity value prior as proposed can enable transformation of the neighbour (disturbance) concentration to the current frame which is very similar to the estimation of "block-matching" of the optical flow method. In the proposed model, with hypothesized spatial consistency, the temporal MRF has been employed to retain temporal coherence. At time-instance t, its likelihood function is derived as (1.4).

$$p_x(w_t, b_t) \propto \prod_{\tau \in [-f, +f]} exp\left(-\frac{w_t(x)V_t(x) + b_t(x) - D_{t+\tau}(x)\|_2^2}{\delta_\tau^2} \right) \tag{1.4}$$

In (1.4), the parameter f states the number of adjoining frames, while the temporal parameter is given by δ_s. In addition to the temporal coherence, the proposed model intends to maintain optimal spatial consistence by means of a likelihood function, derived in (1.5).

$$p_x(w_t, b_t) \propto \prod\nolimits_{\tau \in [-f, +f]} . exp\left(-\frac{w_t(x)V_t(x) + b_t(x) - D_{t+\tau}(x)\|_2^2}{\delta_\tau^2} \right) \tag{1.5}$$

In above derived uniform likelihood function, the parameter σs is removed, as it can be hypothesized to be fixed in current work. Thus, the intensity map $V_{t(x)}$ of the current video-frame would be altered to the disturbed (say, hazed, smoked, etc.) concentration map, given by $D_{t,(t-1)}(x)$ in the frames. Thus, the absolute error between the original scene and the disturbed concentration map, also called the absolute error map (AEM) in between $\tilde{D}_t(x)$ and $\tilde{D}_{t-1}(x)$ can be estimated. To perform local conditioning, the proposed model intends to make the above stated AEM parameter near zero, for which it applies Maximum Likelihood (ML)-based optimization.

Maximum Likelihood Estimation It has been noticed that, the log-likelihood function is highly suitable for the estimation of maximum likelihood estimation. In addition, it is easier to solve the parameters and obtain the derivative of log-likelihood function as compared to original likelihood function. Consider the temporal weights $\lambda_\tau = \frac{1}{\sigma_\tau^2}\left(s.t\_\sum_\tau 2\lambda_\tau = 1\right)$ to states easily and then log-likelihood function can be represented as:

$$L(w_t, b_t) = \sum\nolimits_{\tau\in[-f,+f]} n \sum\nolimits_{y\in\Omega} x - \lambda_\tau \|w_t(x)V_t(y) + b_t(x) - D_{t+\tau}(y)\|_2^2 \tag{1.6}$$

The estimation of maximum log-likelihood can be derived $(w_t, b_t) = arg\,max\,L(w_t, b_t)$ as to determine the optimal random fields W and B. The linear system is solved from $L(w_t, b_t)/w_t$ to increase the probability. Then final haze concentration map is produced by $\tilde{D}_t(x) = w_t(x)V_t(x) + b_t(x)$.

$$\begin{cases} w_t(x) = \frac{\sum_\tau n\lambda_\tau (u_\Omega[V_t(x)\, D_{(t+\tau)}(x)] - u_\Omega[V_t(x)\, u_\Omega D_{(t+\tau)}(x)])}{u_\Omega[V_t^2(x)] - u_\Omega^2[V_t(x)]} \\ b_t(x) = \sum_\tau n\lambda_\tau\, u_\Omega[]u_\Omega[D_{(t+\tau)}(x)] - w_t(x)u_\Omega[V_t(x)] \end{cases} \tag{1.7}$$

In Eq. (1.7), u[.] represents the mean filter and is given as $u[F(x)] = (1/|\Omega|)\sum_{y\in\Omega(x)} F(y)$ and $|\Omega|$ signifies local neighbourhood cardinality. Thus, applying above-stated mechanism the proposed model performs local conditioning of the compressed domain input streaming video. Subsequently, the locally conditioned video (here, it means an input video stream with dehazed, smoke-free and light-balanced feature) is passed to another ST-MRF environment, especially designed to detect and track the moving target in the compressed domain video. Being a modular design, so far developed and discussed model focused on local-conditioning, while the subsequent method intends to improve detection and tracking accuracy.

2 Machine Learning Assisted ST-MRF Environment for Moving Target Tracking

MRF has been one of the most used vision technologies for depth inference, surface reconstruction and image segmentation etc. In fact, majority of its accomplishment attributes to the well-known algorithms like Iterated Conditional Modes [3], and its consideration of both "data faithfulness" and "model smoothness" [4]. Initially, HMRF-EM was developed for the brain MR images segmentation [5]. Functionally, for a provided video frame, say $y = (y_1, \ldots\ldots, y_n)$ where each value yi signifies the pixel intensity, it turns out to be inevitable to deduce configuration of the labels $x = (x_1, \ldots\ldots, x_n)$ where $x_i \in L$ and L signifies the set of all possible labels. To be noted, the parameter L = $\{0, 1\}$ is often applied in binary segmentation task, which is the key motive behind any ROI segmentation problem. Thus, on the basis of MAP criterion, the labeling X_ can be defined as per (2.1).

$$X^* = arg_x\, max\{P(y|x, \Theta)P(x)\} \tag{2.1}$$

In above derived function (2.1), the parameter P(x) signifies the prior probability which is a Gibbs distribution. The other component $P(y|x, \Theta)$ states the joint likelihood (probability), which can be derived as (2.2).

$$P(y|x, \Theta) = \prod_i P(i|x, \theta) \tag{2.2}$$

In (2.2), $P(y_i|x_i, \theta_{xi})$ states a Gaussian distribution with parameters $\theta_{xi}(\mu_{xi}, \delta_{xi})$ Moreover, the parameter $\Theta = \{\theta_1 | l \in L\}$ states the parameter set produced by the expectation maximization method. A snippet of the expectation maximization method used in the proposed model.

2.1 Expectation Maximization Algorithm

In reference to the above discussions, the parameter set value $\Theta =$ is obtained by executing EM algorithm. The proposed model, applies the following approach to perform EM so as to estimate the value of the parameter set $\psi\boldsymbol{\Theta}\triangleright$

Step 1. Start: Let an initial parameter set $\Theta^{(0)}$.

Step 2. E-step: consider $\Theta^{(t)}$ at the ψt^{th} ψiteration and conditional expectation has been calculated as:

$$\begin{aligned} Q(\Theta)|\Theta^{(t)}) &= E\big[\ln P(x, y|\Theta)|y, \Theta^{(t)}\big] \\ &= \textstyle\sum_{x \in \chi} P(x|y, \Theta^{(t)} \ln P(x, y|\Theta)) \end{aligned} \tag{2.3}$$

In (2.3), the parameter _ states the set of all possible configurations of labels.

Step 3. M-step: $Q(\Theta|\Theta^{(t)})$ is maximized to get the next estimate:

$$\Theta^{(t+1)} = arg_x\, maxQ\left(\Theta|\Theta^{(t)}\right) \tag{2.4}$$

Subsequently, consider $\Theta^{(t+1)} \to \Theta^{(t)}$ and repeat it. Moreover, consider $G(z; \theta_1)$ be the Gaussian distribution function with parameters $\theta_1 = (\mu_1, \delta_1)$. $G(z; \theta_1)$ which can be derived as per

$$G(z; \theta_1) = \frac{1}{\sqrt{2\pi\delta_l^2}} exp\left(-\frac{(z-\mu_1)^2}{2\delta_l^2}\right) \tag{2.5}$$

Thus, the value of P(x) can be derived using (2.6).

$$P(x) = \frac{1}{Z} exp(-U(x)) \tag{2.6}$$

In (2.6), U(x) is stated as the prior energy function. In the proposed model, the function $P(y|x, \Theta)$ is obtained as per (2.7).

$$\begin{aligned} P(y|x, \Theta) &= \prod_i P(y_i|x_i, \theta_{xi}) \\ &= \prod_i G(y_i; \theta_{xi}) \\ &= \tfrac{1}{Z}, exp(-U(y|x)) \end{aligned} \tag{2.7}$$

Now, with above derived functions, the proposed model executes ST-MRF with EM to segment the ROI and track over consecutive frames. The functional steps involved in this process are given as follows:

Step 1: Start with initial parameter set $\Theta^{(0)}$.
Step 2: Estimate the likelihood distribution function $P^{(t)}(y_i|x_i, \theta_{xi})$.
Step 3: Employ θ^t to estimate the labels by performing MAP.
Step 4: Estimate the posterior distribution for all $l \in$ and all pixels y_i using

$$P^{(t)}(l|y_i) = \frac{G(y_i; \theta_l)P\left(l|x_{N_t}^{(t)}\right)}{P^{(t)}(y_i)} \tag{2.8}$$

MAP Estimation For EM-based methods it is vital to solve for X^* in which the total posterior energy is reduced iteratively. The proposed model deducts as X^*.

$$X^* = arg\ max\{U(y|x, \Theta) + U(x)\} \tag{2.9}$$

Thus, the likelihood energy with given y and Θ is obtained as per

$$\begin{aligned} U(y|x, \Theta) &= \textstyle\sum_i U(y_i|x_i, \Theta) \\ &= \textstyle\sum_i \cdot \left[\frac{(y_i-\mu_{xi})^2}{2\delta_{xi}^2} + \ln\,\delta_{xi}\right] \end{aligned} \tag{2.10}$$

Subsequently, the proposed model derives the prior energy function U(x) as (2.11).

$$U(x) = \sum\nolimits_{c \in C} V_c(x) \tag{2.11}$$

In above derived prior energy function (2.11), the parameter V c(x) states the clique potential, while C represents the set of all possible cliques. Here, the proposed model hypothesizes that one pixel has at most 4 neighbours in the frame (say, image-domain), signifying that the pixels in its 4-neighborhood. Moreover, the clique potential is obtained in reference to the pairs of neighboring pixels, using (2.12).

$$V_c\left(x_i, x_j\right) = \frac{1}{2\left(1 - I_{x_i, x_j}\right)} \tag{2.12}$$

Where

$$I_{x_i, x_j} = \begin{cases} 0\ if\ \ x_i \neq x_j \\ 1\ if\ \ x_i \neq x_j \end{cases} \tag{2.13}$$

Now, to solve target detection and tracking problem (2.11), the proposed model ex cutes an iterative algorithm, which is given as follows:

Step 1: Estimate $x^{(0)}$ from the previous loop of the EM algorithm.
Step 2: $x^{(k)}$ is provided for all $1 < i <$, then calculate

$$x_i^{(k+1)} = arg\ \min_{l \in L}\left\{U(x_i|l) + \sum\nolimits_{j \in N_i} .v_c\left(l, x_j^k\right)\right\} \tag{2.14}$$

Step 3: Repeat Step-2 till $U(y|x, \Theta) + U(x)$ converges or a maximum k is obtained.

Video segmentation or the ROI segmentation Video segmentation or the ROI segmentation in video has been an active area of research since 1980s. Current work on video segmentation can be broadly classified into two categories: frame-by-frame based approaches and volume-based approaches. Frame-by-frame approaches take as input one or two successive frames of a video, while volume-based approaches consider many successive frames of a video at once. Frame-by-frame approaches have the advantages of low memory requirement while volume-based approaches in recent years have demonstrated good coherency in object labeling over time. A video segmentation algorithm can be broadly categorized into two classes depending on the motion estimates provided:

Segmentation without correspondences: Approaches belonging to this category provide grouping of pixels as objects or object-parts across the video. Unlike classical approaches, where authors have obtained MV to be investigated in subsequent frames, this research employed a robust Machine learning assisted edge preserving segmentation model for initial ROI segmentation.

2.2 Clustering Assisted Edge-Preserving ROI Segmentation

In the proposed model the initial ROI segmentation is performed using K-Means clustering algorithm, which was executed on the pixels gray-level intensities to further implement proposed ST-MRF (EM-based ST-MRF) model for target detection and tracking. In this approach the MAP algorithm employs initial labels $X^{(0)}$ provided with initial segmentation and EM algorithm uses initial parameters $\Theta^{(0)}$. Subsequently, it executes the EM algorithm, which is followed by the refinement of the resulting label configuration x and the segmented result. Once identifying the ROI in one frame, the proposed model is executed in such manner that it keeps retaining the ROI in subsequent compressed domain videos (Fig. 3).

Realizing the fact that for compressed domain target tracking, maintaining edge information is must. There are many methods such as Berkeley's contour detection, Sarkar-Boyer edge detection, or canny edge detection. Now, that binary edge map be z, where $z_i = 1$ and for the ith pixel on an edge, while 0 for non-on edge ($z_i = 0$ if not), the ROI as detected and tracked, it is remodelled as (2.14).

$$x_i^{(k+1)} = arg \min_{l \in L} \left\{ U(y_i|) + \sum\nolimits_{j \in N_{i,z,j=0}} {}' V_c\left(l, x_j^k\right) \right\} \tag{2.15}$$

Thus, implementing above method, the proposed ST-MRF environment tracks moving target over compressed domain video streams.

3 Conclusion

To design a novel and robust ST-MRF based moving object tracking system in compressed domain, this research mainly focuses on optimizing multiple encompassing functions. Recalling the fact that to achieve accurate target tracking, segmentation enhancement is must, this approach aims to design an efficient ST-MRF environment which

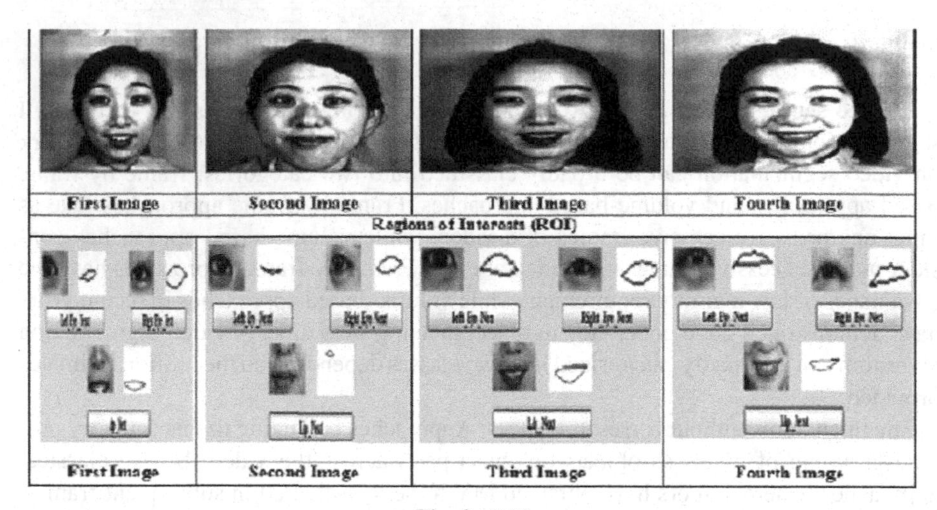

Fig. 3. ROI

could guarantee target detection and tracking even under varying background condition. Additionally, recalling the specific use of target tracking in compressed domain over batter-field, ATC etc. where the probability of haziness, low-lighting etc. can be there, the proposed work intends to address haziness problem. To achieve it, the proposed model intends to use ST-MRF concept to serve a dual purpose, first local conditioning improvement (i.e., reduce haziness or low-lighting) and second, to improve the accuracy of target detection and continual tracking over varying background. Thus, implementing above method, the proposed ST-MRF environment tracks moving target over compressed domain video streams.

References

1. Farin, D., de With, P.H., Effelsberg, W.A.: Video-object segmentation using multisprite background subtraction. In: The IEEE Conference on Multimedia and Expo (ICME), vol. 1, pp. 343–346 (2014)
2. Bluff, L., Rutz, C.: A quick guide to video-tracking birds. Biol. Lett. **4**, 319–322 (2008)
3. Chiang, J.Y., Chen, Y.C.: Underwater image enhancement by wavelength compensation and dehazing. IEEE Trans. Image Process. **21**(4), 1756–1769 (2012)
4. Kusuma, T., Ashwini, K.: Study on segmentation and global motion estimation in object tracking based on compressed domain. Int. J. Future Revolut. Comput. Sci. Commun. Eng. **4**(3), 578–584 (2018)
5. Zhu, Q., Mai, J., Shao, L.: A fast single image haze removal algorithm using color attenuation prior. IEEE Trans. Image Process. **24**(11), 3522–3533 (2015)
6. Sharma, S., Khanna, P.: ROI segmentation using local binary image. In: 2013 IEEE International Conference on Control System, Computing and Engineering, pp. 136–141 (2013)
7. Kusuma, T., Ashwini, K.: Real time object tracking in H.264/AVC using polar vector median and block coding modes. Int. J. Comput. Inf. Eng. **12**, 981–985 (2018)
8. Bibby, C., Reid, I., Zhu, Q., Mai, J., Shao, L.: A fast single image haze removal algorithm using color attenuation prior. IEEE Trans. Image Process. **24**(11), 3522–3533 (2015)

9. Bouguet, J.Y.: Pyramidal implementation of the lucas kanade feature tracker: description of the algorithm. Intel Corp. **5**, 4 (2001)
10. Ancuti, C.O., Ancuti, C., Hermans, C., Bekaert, P.: A fast semi-inverse approach to detect and remove the haze from a single image. In: Kimmel, R., Klette, R., Sugimoto, A. (eds.) ACCV 2010. LNCS, vol. 6493, pp. 501–514. Springer, Heidelberg (2011). https://doi.org/10.1007/978-3-642-19309-5_39
11. Kusuma, T., Ashwini, K.: Performance analysis of motion vector entropy, smoothed residual norm, Markov random field in the H.264/AVC compressed domain for object tracking. J. Adv. Res. Dyn. Control Syst. **10**(13), 366–372 (2018)
12. Hegde, C., Ashwini, K.: Analysis the practicality of drawing inference in automation of commonsense reasoning. In: Advances in Artificial Intelligence and Data Engineering, pp. 101–108 (2021)
13. Kim, T.K., Paik, J.K., Kang, B.S.: Contrast enhancement system using spatially adaptive histogram equalization with temporal filtering. IEEE Trans. Consum. Electron. **44**(1), 82–87 (1998)
14. Li, Z., Tan, P., Tan, R.T., Zou, D., Zhou, S.Z., Cheong, L.F.: Simultaneous video defogging and stereo reconstruction. In: IEEE Conference on Computer Vision and Pattern Recognition (CVPR), pp. 4988–4997 (2015)
15. Zhu, Q., Mai, J., Shao, L.: A fast single image haze removal algorithm using color attenuation prior. IEEE Trans. Image Process. **24**(11), 3522–3533 (2015)

Challenges of Machine Learning for Data Streams in the Banking Industry

Mariam Barry[1,2], Albert Bifet[2,3](✉), Raja Chiky[4], Jacob Montiel[2,3], and Vinh-Thuy Tran[1]

[1] BNP Paribas, IT Group, ITG Production Data, Montreuil, France
{mariam.barry,vinh-thuy.tran}@bnpparibas.com
[2] LTCI, Télécom Paris, Institut Polytechnique de Paris, Palaiseau, France
[3] AI Institute, University of Waikato, Hamilton, New Zealand
{abifet,jacob.montiel}@waikato.ac.nz
[4] ISEP, Institut Supérieur d'Electronique de Paris, Paris, France
raja.chiky@isep.fr

Abstract. Banking Information Systems continuously generate large quantities of data as inter-connected streams (transactions, events logs, time series, metrics, graphs, process, etc.). Such data streams need to be processed online to deal with critical business applications such as real-time fraud detection, network security attack prevention or predictive maintenance on information system infrastructure. Many algorithms have been proposed for data stream learning, however, most of them do not deal with the important challenges and constraints imposed by real-world applications. In particular, when we need to train models incrementally from heterogeneous data mining and deployment them within complex big data architecture. Based on banking applications and lessons learned in production environments of BNP Paribas - a major international banking group and leader in the Eurozone - we identified the most important current challenges for mining IT data streams. Our goal is to highlight the key challenges faced by data scientists and data engineers within complex industry settings for building or deploying models for real word streaming applications. We provide future research directions on Stream Learning that will accelerate the adoption of online learning models for solving real-word problems. Therefore bridging the gap between research and industry communities. Finally, we provide some recommendations to tackle some of these challenges.

Keywords: Challenges · Production · IT · Streaming · Banking

1 Introduction

A substantial volume of papers are published every year in top ranked research conferences in fundamental data science and applied data science. The first motivation of this paper is to highlight the existing challenges for mining IT data and provide a direction for **bridging the gap between academic research and industry**. The second one is about **model serving** which we define as

S. N. Srirama et al. (Eds.): BDA 2021, LNCS 13147, pp. 106–118, 2021.
https://doi.org/10.1007/978-3-030-93620-4_9

the ensemble of technical and data engineering steps needed for integrating and deploying a model into production systems, in order to deliver a service. Noah Fiedel, a Software Engineer working on *Tensorflow serving*, defines serving as *"how you apply a Machine Learning model, after you have trained it"*[1]

What if we are never done with model training? *How to process millions of transactions and continuously train and update an Anti-Money Laundering Model*[2]*? How to detect Banking Wire Transfer (BWT)*[3] *in real time?* These open questions are raised by industry data scientist and data engineers when they want to apply some of the existing research papers for real streaming applications. Other position papers exist in the literature: Zliobaite et al. [13] discuss open challenges for data streams mining, Fan and Bifet [5] focus on challenges related to big data mining. In this paper, we focus on the data modeling (related to data science) and the data engineering challenges faced by industries when deploying an online learning model for real streaming applications. To the best of our knowledge, this is the first paper covering those issues on the scope of IT data and the banking sector.

1.1 Background

Online Learning [10] refers to continuously learning from new data and making predictions. The model is dynamically updated as new data becomes available. Unlike classical batch Machine Learning (ML) models - which are trained once then deployed in production - online learning algorithms should ideally deal with the streaming setting: one pass on the data, constant or sublinear complexity, efficient resources (CPU and memory) usage, and robust to concept drift (change in data streams).

Big Data Streams refer to large volumes of data (beyond available resources for processing and storing) that are generated continuously. Big Data stream learning is the ability of extracting relevant information and patterns from big data arriving as streams. Online learning has become a promising approach for learning from continuous or evolving data streams, in particular for big data settings where streams need to be processed online and discarded in a timely manner.

Our Contributions. The main contributions of this position paper is threefolds: first we categorize IT data types and present some challenging use cases in the banking industry in Sect. 2. In Sect. 3 we discuss some relevant papers and their limits while applying their methods in IT data. We present the most important challenges of data scientists for modeling IT data streams in Sect. 4. Finally, we highlight the main challenges of data engineers for data science or AI based model deployment in industry in Sect. 5.

[1] Noah Fiedel talk at TensorFlow Dev Summit 2017 - https://www.youtube.com/watch?v=q_IkJcPyNl0 at 2,24".

[2] https://www.academia.edu/33091102/Anti_Money_Laundering_Model_in_Banking_System.

[3] BWT refers to *Transferring funds between banks to eliminate source of dirty and criminal money.*

2 Banking Information Systems

BNP Paribas is the leading bank in the Euro Zone and a key international banking group. The Group serves nearly 33 million clients worldwide in its retail banking networks and BNP Paribas Personal Finance has more than 27 million active customers within 71 countries[4]. The digitalization of Banking services is vital for the growth and sustainability of the business. This requires major changes and adaptation in the Information System that lead to important challenges: new learning methods from IT big data streams in production while maintaining a high standard for risk and security management. As an example of Big Data mining for security monitoring, a large group like BNP Paribas process more than several hundred of IT data Gigabits per day with expected growth to several Terabits processed per day in the upcoming 2 years. This paper is based on the lessons learned in IT production within BNP Paribas Information System which relies on more than 100,000 servers with about 20,000 applications.

2.1 Online Learning Use Cases in the Banking Sector

We list below some real word use cases in the banking sector which represent considerable challenges in the industry regarding data modelling and model deployment in streaming architectures.

1. Real-time anti-Money laundering detection in complex evolving network.
2. Graph streams anomaly detection within Thousand of network devices under Big Data volumetry (Millions of logs per minute).
3. Predictive maintenance of the information system to prevent critical incident in productions applications.

We aim to present some banking applications and 3 major use cases in the banking sector and highlight their challenges.

Use Case 1 - Real Time Fraud Detection. Its main goal is to prevent fraud or money laundering by detecting suspicious patterns or transactions in a timely manner. To achieve this, we deal with two challenges: first, an efficient online learning method that is adaptive to integrate different clients profiles as they are evolving over the time. The second challenge concerns the scalability of the production pipeline that should load-balance the processing to deal with critical transactions events (from more than 13 million of digital customers for BNP Paribas).

Use Case 2 - Predictive Maintenance of Network IoT Devices. The main goal of this use case is to prevent anomalies in a complex network ecosystem generating more than 10 million of events per hour with around 20 GB of *CISCO syslog* - system logs - per hour. The network ecosystem includes network devices such us Firewall, Switch, Rooter, VPN, etc. Online learning techniques are suited

[4] https://group.bnpparibas/uploads/file/bnpparibas_2019_integrated_report_en.pdf.

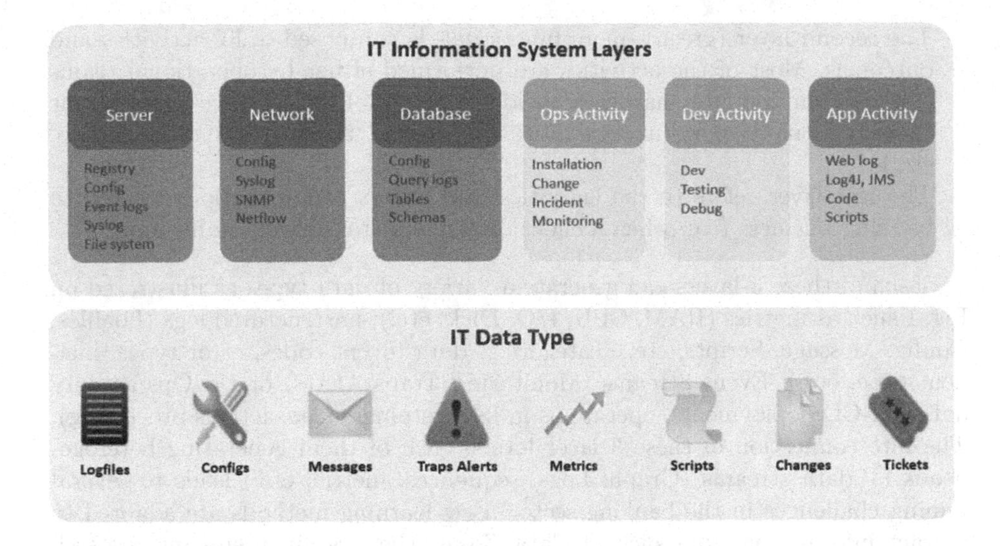

Fig. 1. Categorization of variety of IT Data from Banking Information System

to deal with such Big Data streams due to the evolving nature of network devices events. The main challenge for this use case is to provide a predictive model that can process multiple and heterogeneous data streams: raw system logs events, servers metric and traffic data of each network device. Furthermore, a flexible anomaly scoring function is also essential to extract patterns of different types of network devices.

Use Case 3 - Credit Risk Scoring for Clients. The goal of this use case is to provide a score, indication of risk level that the client will repay a given loan and propose the most suitable debt plan. To achieve this, statistical, machine learning or online learning based model can developed by mining various inputs such as client historic, revenues, transactions data to propose the most adapted loans plan. The main challenges of this use case model is to deal with GDPR, customer data privacy compliance and provide explanability for model predictions regarding regulatory obligations.

2.2 Categorization of Information System Data Sources

The goal of this subsection is to provide an overview and understanding of Information System Data sources through a categorization of different types of IT data. We propose a classification of IS data into 3 levels (colors) as shown in Fig. 1. Each block represents a specific source of IT data that is generated from different layers of a Banking Information System architecture.

- The first layer of the information system (in blue) is made up of servers, networks devices, middle-wares and databases. It constitutes the hardware layer on which the applications are based.

- The second layer (green)- more functional- is composed of IT activities and outcomes. Most of the activities are performed either by operational teams (installation, update, hardware-incident fix, etc.) or by developers teams (development of new functionalities for clients, applications maintenance, etc.).
- The last layer refers to the activities and events of applications related to banking business (e.g. when a client operates a transaction on his app).

Each of these 3 layers can generate a variety of data types as illustrated on Fig. 1 such as metrics (RAM, CPU, I/O, Disk, etc.), unstructured logs (Logfiles, Config, Message, Scripts, etc.), categorical data (event codes, error types, incident type, etc.), Event Streams (Monitoring Traps Alerts) or DevOps activity data (CI-CD deployments, operations in IS, customers transactions processing). The interconnection of these 3 layer levels, each of them generating heterogeneous IT data streams (Graph, Logs, sequences, metric, etc.) leads to several mining challenges in the banking sector. Few learning methods are adapted to extract information from such IT data. Thus, there is an interesting research opportunity for this topic. Most existing methods in the literature focus on homogeneous data structure to extract relevant patterns [16], but they are not adapted to real scenarios in the banking sector with complex architectures.

2.3 Banking Sector Applications and Use Cases

Banks provide services and business operations to their clients through applications deployed in production. Those applications are hosted and managed within the information system which data needs to be processed and mined using predictive methods. BNP Paribas IT Group managed more than 10, 000 applications and 20, 000 servers, some of them continuously generated high speed and in dependant data streams. There are the following types of applications:

- Business applications to allow customers to access their banking operations or do bank transfer.
- Network applications to transit data and information among devices (switch, rooter and firewall).
- Security applications to monitor cyber-attack in real-time or data leakage.
- Risk management application to predict fraud or detect anti-money laundering in real-time

Many of online machine learning models - which update the model incrementally by processing each new instance - can not be directly applied run in the data sources of some banking applications. For network anomaly detection, the generated instances might come from independent devices/sensors, thus streams should be pre-processed and re-ordered by sensors before pushing into the online models. For example, a fast online anomaly detection, Half-Space-Trees [21] update the mass value (count of observations that traversed a node during the training) of each node of the tree, but do not change the splitting conditions or the feature of node itself. In some banking applications, the number and type of feature is not static (IID) and might evolve over time. Two successive transactions might not have the same features thus ensemble methods based on fixed size of dimensional would be useless for such banking applications.

2.4 Challenging Use Cases of Online Learning in the Banking Sector

Use Case 1 - Real-Time Fraud Detection. The goal is to prevent fraud or money laundering by detecting suspicious patterns or transactions in a timely manner. To achieve this, we deal with two challenges: first, an efficient online learning method that is adaptive to integrate different clients whose behaviors are evolving or drifting over the time. The second challenge concerns the scalability of the production pipeline that should load-balance the processing to deal with critical transactions events (from more than 13 million of digital customers for BNP Paribas).

Use Case 2 - Predictive Maintenance in Telecommunication Network. The main goal of this use case is to prevent anomalies in a complex network ecosystem generating more than 10 million of events per hour with around 20 GB of *CISCO syslog* - system logs - per hour. The network ecosystem includes network devices such us Firewall, Switch, Rooter, VPN, etc. Online learning techniques are suited to deal with such Big Data streams due to the evolving nature of network devices events. The main challenge for this use case is to provide a predictive model that can process multiple and heterogeneous data streams: raw system logs events, servers metric and traffic data of each network device. Furthermore, a flexible anomaly scoring function is also essential to extract patterns of different types of network devices.

Use Case 3 - Credit Risk Scoring. The goal of this use case is to provide a score, indication of risk level that the client will repay a given loan and propose the most suitable debt plan. To achieve this, statistical, machine learning or online learning based model can developed by mining various inputs such as client historic, revenues, transactions data to propose the most adapted loans plan. The main challenges of this use case model is to deal with GDPR, customer data privacy compliance and provide explanability for model predictions regarding regulatory obligations.

3 Literature Review on IT Stream Learning

In this section, we provide some state of the art existing methods that can be adapted to IT Data stream mining and discuss why they can hardly be applied for real use cases. We will focus on the scope of 3 types of problems: data mining from IT logs, graph mining based methods, and frameworks for real time monitoring of IT infrastructure.

3.1 Learning Methods from IT Logs: Anomaly Detection and Log Mining

A lot of research has been carried recently for mining useful information from log data as these data represent an important and rich source of information. He et al. [9] have recently presented a rich survey on automated log analysis and various log mining tasks. Here, we focus on some challenges faced by real world applications when performing these three levels of log mining: log parsing, anomaly detection and predictive maintenance.

Log Parsing: Recently Zhu et al. [23] presented a comprehensive evaluation study on automated log parsing, benchmarking 13 log parsers considered state of the art log parsing methods. They also shared success stories and lessons learned in an industrial application of an automated log parser at Huawei. We tested one of the most recommended log parser DRAIN [8], in a benchmark on real big data streams of *Syslogs*. We figured out that an additional pre-processing of logs using expert knowledge is needed before feeding the log parser. Indeed, many existing methods in the literature, need an adaption - requires domain expert knowledge - to deal with real use cases.

Log Based Anomaly Detection: DEEPLOG, a recent state of the art algorithm by Du et al. [3], is a deep neural network model that uses Long Short-Term Memory (LSTM), to model a system log as a natural language sequence. Logs are converted into sequences of normal events, and then a LSTM is trained to remember what should be a normal execution flows, to capture the long-term dependencies over these sequences. In the predict phase, new entries of logs are compared to a normal sequence of logs to predict anomaly, with the possibility to integrate feedback to adjust its weights dynamically online. Their approach demonstrated relevant results compared to other state of the art methods. However, applying such methods for IS settings is not trivial. Like many neural networks or batch machine learning approaches, DEEPLOG requires a training data produced by normal system execution to learn normal sequences, which is not always available in real world applications. It is hard to provide such labeled logs data for model training as the big data nature of the logs (several Terabits of log per day) does not allow to label each of them manually.

3.2 Pattern Mining from Graph Data Streams

Massive graphs arise in real world applications generating big data streams of events that need to be processed online and discarded in timely manner. In particular in the banking sector, graph streams mining is a suited technique to extract complex patterns from both complex anomalies, money laundering or cyber-attack detection in streaming.

In recent works, some advanced methods have been proposed to extract patterns from dynamic graph streams. We refer to the Mc Gregor survey [17] on Graph stream algorithms for a detailed and complete survey. Manzoor et al. algorithm STREAMSPOT [16] efficiently detects anomaly graphs with desirable properties such as constant space and constant time to update the graph online. This approach is promising, yet the model need to be extended to work on IT data where anomalies are correlated to a variety of data types (see Fig. 1) that can be connected through connected graph data.

3.3 Streaming Frameworks for Mining IT and DevOps Events

Researchers from IBM presented an unsupervised learning method based on a drift analysis tool named CHASER [18] to detect and analyze abnormal changes. This tool provides learned change models, root cause analysis and detects patterns in real time. Their work is one of few that are well adapted to our scope for learning from IT Data streams. Yet, the tool CHASER has not been open sourced yet, which make it difficult to experiment with it. Kuruba et al. work named *"Real-time DevOps Analytics in Practice"*[5] is about designing, developing and using a DevOps analytics system using open source tools for a product team in an industrial setting. Their approach is well suited for IT production monitoring to deal with the impact DevOps practices. The future work direction could be to empower such analytics pipeline with powerful streams learning methods to provide recommendations to DevOps to accelerate their activities.

4 Data Science Challenges for IT Data Stream Learning

4.1 Multiple Data Streams Mining for Anomaly Detection

Data streams mining has gained popularity over recent years. Most of the existing methods focus on mining an unique source and type of data streams. However, few research has been done in mining Multiple Data Streams (MDS), in particular for IT operations. Wu and Gruenwald [22] define MDS as a set of flows generated by corresponding sources satisfying the following properties: continuous, one-pass, sequential and self-functional, distributed, diversified, and autonomous. The complexity of IT data streams generated by Banking Information System (IS) can be viewed as multiple data streams where each data stream source is a source of IT data presented in the previous section. Some critical anomalies can be detected using techniques of MDS learning. For example, if we consider an anomaly where a customer could not connect or make a bank transfer using his application (IS front layer), that anomaly can be caused by a succession of inter-dependant micro-anomalies located in different layers of the information system (IS middle-ware, hardware, Network). For example, those micro-anomaly events can be: an error in the application transaction logs, a failure in network rooter, a Database file system is full or server CPU is to high. Thus, one need a data science model that can process multiple events and data streams at the same time (from different data sources within the information system) in order to mine the evolving dynamics under the data sources and achieve predictive maintenance on IT assets (servers, network devices, applications.)

[5] Presented at the 6th International Workshop on Quantitative Approaches to Software Quality - 2018.

4.2 Online Learning from Heterogeneous Data Streams

Banking Information System management is essential for the growth and sustainability of the bank business. Data generated by IS have a variety of typologies as shown on Fig. 1, and there is a crucial need to process such data streams online to extract valuable patterns to deal with critical business applications. IT data streams can take several forms: raw and unstructured log data, time series of numeric metrics, sequence of process and activity flow, IT error categories, and Knowledge Graph Data base. Thus, IT data can be viewed as heterogeneous Data Streams and need appropriate learning methods that can deal with this variety of IT data types at the same time to extract complex patterns. There is a challenge of providing such learning methods that deal with bare business-oriented under streaming constraints. An example of two real applications of such learning methods are: first, suspicious path-finding in dynamic network Graph to detect the network of money launderers (who is involved and how the dark money is dispatched) and second, to perform predictive maintenance from information systems events.

5 Data Engineering in Applying Models in Production

5.1 Model Governance Challenges Regarding Banks Regulations

The European Banking Federation has recently recall in his position paper [4][6] that banks are subject to legal and regulatory obligation which oblige them to provide high standards of risk management, data quality and governance of solutions in production. The regulatory obligations include putting in place a governance for models in production and GDPR compliance regarding customers data processing. In particular, depending on the use case, Banks are expected to follow these two point: first, put in place tools and processs to guaranty that any data manipulation (data processing, features engineering, modelisation) is GDPR compliant [1]. And secondly provide a governance and risk management policy for models in productions and auditable process.

These two requirements generate considerable challenges for models governance and data management within the Banking sector. Most of these constraints are not taken into account when building data mining models by the scientific community.

5.2 Engineering Challenges for Deploying Online Learning Models

Deploying and serving a batch machine learning model is a complex procedure for data engineers. Yet, some tools have been proposed in the market to facilitate this integration such as TensorFlow serving or Kubeflow [14]- a toolkit to deploy batch Machine Learning models on Kubernetes [15]. However, regarding **serving online learning models** (from the RIVER library for example) which are

[6] https://www.ebf.eu/wp-content/uploads/2020/03/EBF-AI-paper-_final-.pdf.

more sophisticated and constrained than batch models - processing streams and updating models online - it arises a **variety of technical challenges, issues** and open questions from model scaling to model monitoring.

1. Which architecture deploy to process online and real time training of incremental models for every new streams?
2. How to scale, load-balance and parallelize model runs to handle big data streams of events with reasonable latency? Using Flink and Kafka.
3. When & How to retrain and update the online learning model to deal with evolving data streams? Using River.
4. How to replace the current model and switch to new version without stopping the production service (no downtime)? Using Jenkins.
5. How to save predictions and monitor the performance of online machine learning model when labels are not always available in production?
6. Which windows strategy use to retrain and validate streaming model?
7. Who will be accountable if the model recommendations or decisions lead to critical impact incident or business loss?

Based on IT applications and lessons learned, we identified six issues and challenges that online learning and AI-based algorithms should take into account to be easily integrated in IT production. For example, some fully unsupervised models which algorithm is not generic enough to integrate model updates and feedback from real users, are less likely to be deployed in banking production. In particular, when we need to deploy the model using a CI/CD (Continuous Integration/Continuous Deployment) tool such as Jenkins, models' code should be flexible.

We finally propose in Fig. 2 a high level technical functional architecture to deal with the first data engineering challenges (questions 1 to 5) by deployment online learning models using state of art and scalable big data streaming tools[7] from Apache Foundation [2] **Kafka** [12], **Flink** [6], **NIFI** [20]). We recommend

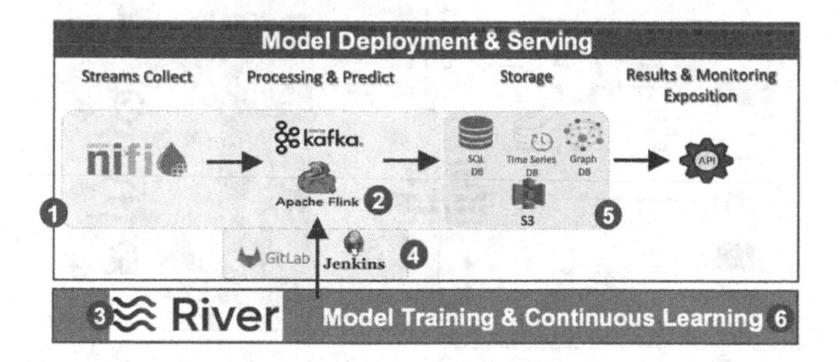

Fig. 2. Overview of proposed architecture to tackle major Data engineering challenges

[7] Due to space restrictions, for more details, we refer the reader to the official documentation of technologies of Apache Kafka, Apache Nifi, Apcahe Flink, and Jenkins.

using **Jenkins** [11] and Gitlab [7] to deploy automatically code and resources using Continuous Integration and Continuous Delivery pipelines, and **River** [19] - main online learning framework in python that includes a variety of online methods such as incremental learning for data streams - to process streaming data and continuously train and update online learning incrementally.

6 Conclusion

In this paper, we provide a categorization of the major challenges faced by data science and data engineering teams in the banking industry, based on real word applications within an international banking group, BNP Paribas.

Data Modelling Challenges

1. Extraction of relevant information from **heterogeneous events** (logs, graphs)
2. Change or drift detection in **multiples data streams** from IoT.
3. Transparency or **explainability** for online model predictions.

Data Engineering Challenges

1. **Online** learning model training and **serving** to predict using big data.
2. Building a **seamless bridge between AI models** and streaming platforms.
3. **Models monitoring & risks management** to be compliant with banking regulations on automated data processing and data privacy.

The aforementioned challenges and their interactions with corresponding actors are illustrated in Fig. 3.

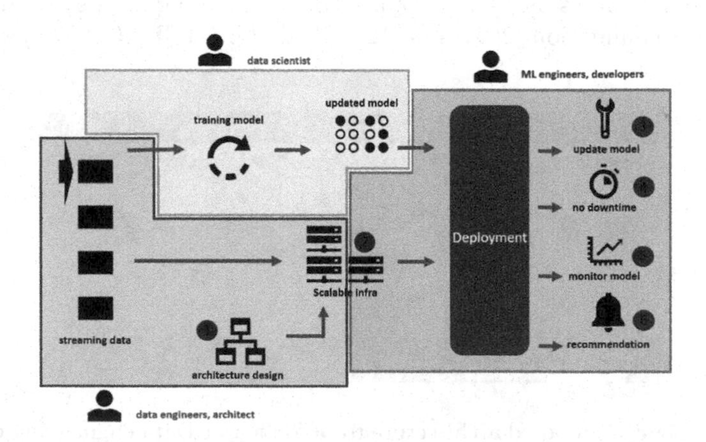

Fig. 3. Overview of 6 Challenges for IT Data Streams Learning for Data Modelling and Data Engineering - Dealing with smooth transition between each block is critical for a stable data pipeline.

These industry constraints should be taken into account when building model algorithms in order to facilitate their integration into a streaming setting for IT use cases. Thus, further research developments would need to be made over the next few years to address these challenges. Finally, to bridge the gaps between current research and real applications, we recommend more collaboration between researchers and industry R&D teams since they are complementary in many aspects to deliver valuable insights from real data.

References

1. AEPD: GDPR compliance of processings that embed Artificial Intelligence An introduction. The Spanish Data Protection Agency (2020). https://www.aepd.es/sites/default/files/2020-07/adecuacion-rgpd-ia-en.pdf. Accessed 10 Dec 2020
2. Apache: The Apache Software Foundation (2021). https://www.apache.org//. Accessed 10 May 2021
3. Du, M., Li, F., Zheng, G., Srikumar, V.: Deeplog: anomaly detection and diagnosis from system logs through deep learning. In: Proceedings of the 2017 ACM SIGSAC Conference on Computer and Communications Security, CCS 2017, pp. 1285–1298. Association for Computing Machinery, New York (2017). https://doi.org/10.1145/3133956.3134015
4. EBF: European Banking Federation, EBF position paper on AI in the banking industry. EU Transparency Register, ID number: 4722660838–23 (2019). https://www.ebf.eu/wp-content/uploads/2020/03/EBF_037419-Artificial-Intelligence-in-the-banking-sector-EBF.pdf. Accessed 15 May 2021
5. Fan, W., Bifet, A.: Mining big data: current status, and forecast to the future. ACM SIGKDD Explor. Newsl. **14**(2), 1–5 (2013). https://doi.org/10.1145/2481244.2481246
6. Flink: Apache Flink, a framework and distributed processing engine (2021). https://flink.apache.org/. Accessed 04 Mar 2021
7. Gitlab: Gitlab, Create a Jenkins Pipeline (2021). https://about.gitlab.com/handbook/customer-success/demo-systems/tutorials/integrations/create-jenkins-pipeline/. Accessed 10 May 2021
8. He, P., Zhu, J., Zheng, Z., Lyu, M.R.: Drain: an online log parsing approach with fixed depth tree. In: 2017 IEEE International Conference on Web Services (ICWS), pp. 33–40 (2017)
9. He, S., He, P., Chen, Z., Yang, T., Su, Y., Lyu, M.: A survey on automated log analysis for reliability engineering (September 2020)
10. Hoi, S., Sahoo, D., Lu, J., Zhao, P.: Online learning: a comprehensive survey (February 2018)
11. Jenkins: The leading open source automation server for deploying projects (2021). https://www.jenkins.io/. Accessed 16 Apr 2021
12. Kafka: Apache Kafka, an open-source distributed event streaming platform (2021). https://kafka.apache.org/. Accessed 5 May 2021
13. Krempl, G., et al.: Open challenges for data stream mining research. ACM SIGKDD Explor. Newsl. **16**, 1–10 (2014). https://doi.org/10.1145/2674026.2674028
14. Kubeflow: The Machine Learning Toolkit for Kubernetes (2021). https://www.kubeflow.org/. Accessed 01 Apr 2021

15. Kubernetes: Automating deployment containerized applications (2021). https://kubernetes.io. Accessed 04 Mar 2021
16. Manzoor, E., Milajerdi, S., Venkatakrishnan, V., Akoglu, L.: Fast memory-efficient anomaly detection in streaming heterogeneous graphs (February 2016)
17. Mcgregor, A.: Graph stream algorithms: a survey. SIGMOD Rec. **43**, 9–20 (2014)
18. Meng, F.J., Wegman, M.N., Xu, J.M., Zhang, X., Chen, P., Chafle, G.: It troubleshooting with drift analysis in the DevOps era. IBM J. Res. Dev. **61**(1), 6:62-6:73 (2017)
19. Montiel, J., et al.: River: machine learning for streaming data in Python (2020)
20. NIFI: Apache NIFI, an easy to use, powerful, and reliable system to process and distribute data (2021). https://nifi.apache.org/. Accessed 10 May 2021
21. Tan, S., Ting, K., Liu, F.T.: Fast anomaly detection for streaming data, pp. 1511–1516 (January 2011). https://doi.org/10.5591/978-1-57735-516-8/IJCAI11-254
22. Wu, W., Gruenwald, L.: Research issues in mining multiple data streams, pp. 56–60 (July 2010)
23. Zhu, J., et al.: Tools and benchmarks for automated log parsing (November 2018)

A Novel Aspect-Based Deep Learning Framework (ADLF) to Improve Customer Experience

Saurav Tewari[1(✉)], Pramod Pathak[2], and Paul Stynes[2]

[1] The Harker School, 500 Saratoga Ave, San Jose, CA 95129, USA
22SauravT@students.harker.org
[2] National College of Ireland, Mayor Street, Dublin 1, Ireland
{Pramod.Pathak,Paul.Stynes}@ncirl.ie

Abstract. Restaurateurs manage the customer experience of a restaurant through the overall rating of reviews on platforms such as Yelp, Google, and TripAdvisor. The challenge is to identify aspects of the restaurant to improve based on a deeper analysis of restaurant reviews. This research proposes a Novel Aspect-Based Deep Learning Framework (ADLF) to improve the customer experience of restaurants based on the value of Key Performance Indicators (KPIs) derived from the sentiment of restaurant reviews. The proposed framework combines an information retrieval algorithm, *Okapi BM25* and a deep learning model, *word2vec-cnn*. The model is trained on the Yelp dataset that consists of 600,000 reviews. Key Performance Indicator's (KPIs) are identified to help a restaurateur improve customer experience based on the sentiment of restaurant reviews. Five predetermined aspects namely flavor, cost, ambience, hygiene, and service are used to create the KPIs. Results demonstrate that diners express positive sentiment about "service" and negative sentiment about "cost". The proposed framework achieved an accuracy of 94% and AUROC of 0.98. This novel framework, ADLF, shows promise for providing restaurateurs with a way to mine the unstructured textual opinion of their customers into KPIs that allows them to improve the customer experience of a restaurant.

Keywords: Sentiment analysis · Deep learning · Customer experience · Restaurant reviews · Yelp reviews · Information retrieval · Key performance indicators · Aspect-based sentiment analysis

1 Introduction

Online review analysis allows businesses to discover customer satisfaction based on restaurant reviews. The reviews provide insights into the business that allow the restaurateur to develop strategies to retain a competitive edge [1–5]. However, online reviews come in the form of unstructured text which the restaurateur needs to read in order to understand the opinions expressed. Customers usually assign a rating between 1 to 5 to assess their overall satisfaction which may not capture all the opinions expressed in

S. N. Srirama et al. (Eds.): BDA 2021, LNCS 13147, pp. 119–130, 2021.
https://doi.org/10.1007/978-3-030-93620-4_10

the underlying review text. Identifying the aspects that customers appreciate and specific areas that could be improved is quite challenging as it requires deeper analysis of reviews. Techniques like natural language processing, information retrieval, data mining and deep learning can be applied to discover the opinions expressed.

The aim of this research is to investigate to what extent an aspect based deep learning framework can help a restaurateur improve the customer experience of a restaurant based on the values of the Key Performance Indicators (KPIs) derived from the sentiment of restaurant reviews. The major contribution of this research is a novel Aspect-Based Deep Learning Framework (ADLF) to improve the customer experience. The framework combines information retrieval algorithm, *Okapi BM25* and a deep learning model *word2vec-cnn* in order to derive the Key Performance Indicators (KPIs) values to improve the customer experience of a restaurant. The Information retrieval algorithm is used to identify the reviews that are relevant to an aspect or a combination of aspects such as flavor, cost, ambience, hygiene, and service. The different combinations of these aspects include "flavor:cost", "flavor:cost:hygiene", "flavor:cost:hygiene:ambience", and so on. The *word2vec-cnn* deep learning model is used to predict the aspect-based review sentiment. The framework provides restaurateurs with KPIs that allow them to track and improve aspects of their business thereby positively impacting customer experience. This paper discusses related work on sentiment analysis in Sect. 2. The research methodology is discussed in Sect. 3. Section 4 discusses the design components of the novel ADLF framework. The implementation of this research is discussed in Sect. 5. Section 6 presents and discusses the results. Section 7 concludes the research and discusses future work.

2 Related Work

Online review analysis allows restaurateurs to manage a restaurant primarily through KPIs that are categorized as Sales and Profitability KPIs, Marketing KPIs and Customer Experience KPIs. The majority of customers will read online reviews before they buy and so Customer Experience KPIs are critical to the business. Restaurateurs manage the customer experience through the overall rating of reviews on platforms such as Yelp, Google, and TripAdvisor. It is important for restaurateurs to understand public opinion on the different aspects of their business which is embedded in the unstructured textual content of reviews. Sentiment analysis provides a way to classify opinion expressed in unstructured text. Using sentiment analysis to classify opinion on predefined aspects provides insight to restaurateurs to manage their Customer Experience KPIs. Sentiment analysis has been used for detecting mismatched reviews, Lou et al. [24] and the sentiment of aspects of the review, Tang et al. [3]. This research focuses on the aspect-based sentiment analysis.

Sentiment analysis techniques can be categorized into three categories: lexicon based, machine learning-based, and deep learning-based models. Deep learning-based models, especially those based on Convolution Neural Networks (CNNs) and Recurrent Neural Networks (RNNs) have significantly outperformed classical lexicon based or machine learning-based models such as Naive Bayes and Support Vector Machines [6–9]. The fundamental requirement for applying neural networks to Natural Language Processing tasks is to convert words into numbers. One such framework that represents words as vectors is word embeddings. Word embeddings capture both syntactic and semantic relationship between words. Mikolov et al. [15] propose Word2Vec which is a group of shallow two-layer neural networks which learn word associations from a large corpus of text. It can detect synonymous words or suggest additional words for a partial sentence. Word2Vec outperforms other models when measuring syntactic and semantic word similarities over a large corpus of 1.6 billion words. According to Kim [11] a simple CNN with one layer of convolution and hundreds of filters built on top of word2vec can perform well for sentiment analysis. Their approach achieved an accuracy of 81% on online MR (movie reviews), an accuracy of 86.8% on SST (Stanford Sentiment Treebank), and an accuracy of 84.7% on CR (customer reviews on various products) [11]. Le et al. [16] demonstrated that the shallow-and-wide structures (neural networks with one layer of convolution and multiple filters) outperform deep models (neural networks with multiple layers of convolution) on word-level inputs. They also found that word-level shallow models outperform deep char-level networks.

Seungwan et al. [10], conducted a comparative study of different deep learning-based sentiment classification models. They compared eight deep learning models, three of which are CNN based and five are RNN based. They concluded that RNN-based models outperform CNN-based models both for word-level and character-level input. They did this work to address a practical issue where best model structure depends on the characteristics of the data and the knowledge of an expert. Panthati et al. [19] used a *word2vec-cnn* model to predict the sentiment of Amazon phone reviews. They used the neural network proposed by Kim [11] and a pre-trained word embedding, word2vec, by Mikolov et al. [15]. Their approach achieved an accuracy of 90.697% and AUROC of 0.97 which is better than other similar models. They had to restrict the size of the data due to the limitation of the available computational power. They recommend training the model on a much larger dataset using a GPU. Their work was limited to static data. They recommend using the model on real-time e-commerce reviews. For service-based organizations knowing that a review is positive or negative is not enough. It is important to know which aspects of the service or experience are positive or negative. Information Retrieval (IR) is the science of searching for material (usually documents) which are unstructured (usually text) that satisfies a need from within large collections [21]. *Okapi BM25* is a ranking function used by search engines to estimate the relevance of documents to a search query. It is based on the probabilistic framework developed by Robertson et al. [22]. In this research *Okapi BM25* is used to find relevant reviews for an aspect.

In conclusion, Panthati et al. [19] indicates that a *word2vec-cnn* model achieves the best accuracy to predict the sentiment of amazon phone reviews. This demonstrates the potential for this model to analyze the sentiment of aspects of a restaurant review. This research focuses on the sentiment analysis of five predetermined aspects namely flavor, cost, ambience, hygiene, and service, which are part of the customer experience KPIs in the restaurant industry. This research proposes to build on the work of Robertson et al. [22] and Panthati et al. [19] to combine an information retrieval algorithm, *Okapi BM25*, and a deep learning model *word2vec-cnn* in order to derive values for the Key Performance Indicators (KPIs) from a restaurant review dataset consisting of 600,000 reviews using a GPU.

3 Methodology

The research methodology consists of five steps namely data gathering, data pre-processing, data transformation, data modelling, and model evaluation as shown in Fig. 1.

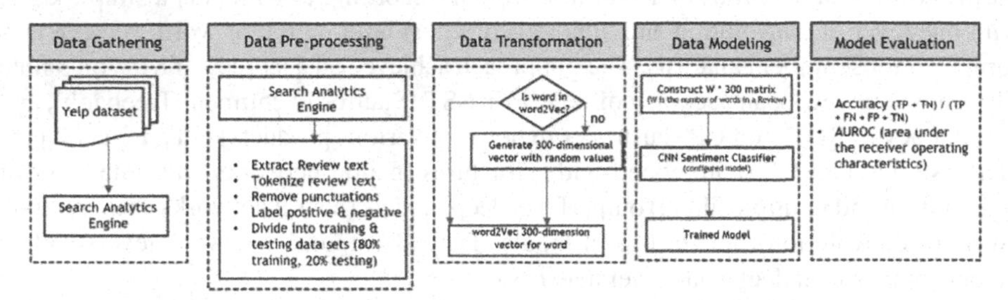

Fig. 1. *word2vec-cnn* training and testing pipeline for Restaurant Reviews

The first step, Data Gathering, involves loading 600,000 Yelp JSON restaurant reviews [20], into the Search Analytics Engine.

During the second step, Data Pre-processing, the reviews are extracted, tokenized, cleaned by removing punctuations and converted to lowercase. Reviews that have a rating of 5 are labeled positive and reviews with a rating of 1 are labelled as negative. 96,000 reviews equally split between positive and negative reviews are used for training. 20,000 reviews equally split between positive and negative reviews are used for testing.

The third step, Data Transformation, maps each word in the review text to a word embedding. Each word is mapped to one real-valued vector in a predefined vector space. Google's word2vec [15] pretrained word embeddings are used where each word in the review is mapped to a 300-dimensional vector.

The Data Modelling step involves neural network model configuration, model training and optimization. The neural network model configuration consists of a one-layer CNN with multiple filters. A filter is applied to a window of words to produce a feature map. In the next layer max pooling operation [17] is applied to extract the maximum value of the feature corresponding to this filter to capture the most important feature with the highest value for each feature map. The model uses multiple filters to obtain multiple features. These features form the penultimate layer and are passed to a fully connected SoftMax layer which provides as output the probability distribution over labels. Regularization dropout is applied on the penultimate layer with a constraint on l2-norms of the weight vectors [18].

The fifth step, Data Modelling, uses the *word2vec-cnn* neural network configuration proposed by [19]. Mobile phone reviews from Amazon were downloaded [23]. The *word2vec-cnn* neural network model was modified to use 8 $\times$ 400 kernel size, Adam updater, and maximum review size of 1000. The model was trained on 34,000 Amazon reviews equally split between positive and negative reviews and tested on 10,000 reviews equally split between positive and negative reviews. Hyperparameters used for training are ReLU activation function, 8 $\times$ 400 kernel, 1 $\times$ 300 stride, dropout rate 0.5, and mini-batch size of 64. Training is done over shuffled mini batches with the Adam updater.

The Model Evaluation step involves assessing the performance of the model based on accuracy and the area under the receiver operating characteristics curve (AUROC).

4 Design

The Aspect-Based Deep Learning Framework, ADLF, architecture consists of a KPI Service, Scoring Service, Analytics Search Engine, Sentiment Classifier and a KPI dashboard as shown in Fig. 2.

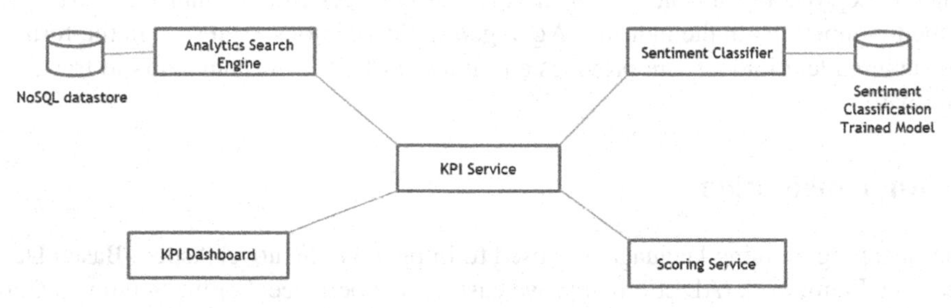

Fig. 2. KPI Engine

The design is based on the model-view-controller design pattern. The model is stored in the NoSQL datastore. The KPI dashboard is made up of multiple views and the KPI Service is the controller. The KPI service is responsible for extracting, transforming, and loading the restaurant reviews into the Analytics Search Engine. The KPI Service requests the Analytics Search Engine to perform a search for aspects of the customer experience namely, flavor, cost, ambience, hygiene, and service. The search query is automatically expanded to include synonyms for the specified aspects. The Analytics Search Engine applies the *Okapi BM25* algorithm to score and return relevant reviews that match the aspects. *Okapi BM25* is a ranking function used by search engines to estimate the relevance of documents to a search query. It is based on the probabilistic framework developed by Robertson et al. [22]. The KPI Service tags each relevant review with the relevant aspects and stores the computed tags along with the review in the Analytics Search Engine.

Once the relevant reviews have been tagged with the aspects, the KPI Service sends a request to the Sentiment Classifier which uses the trained *word2vec-cnn* model to predict the positive and negative sentiment probabilities for each review. This is achieved using the trained *word2vec-cnn* model. The Sentiment Classifier tags a positive sentiment probability of 0.8 or greater as a positive review and a positive sentiment probability of less than 0.2 is tagged as a negative review. The review data in the Analytics Search Engine is updated to include the predicted positive and negative sentiment probabilities.

The Scoring Service is responsible for calculating the "positivity" and "negativity" score for each customer experience aspect. The "positivity" score is calculated as the ratio of the number of positive tagged reviews to the total number of reviews in the relevant review set for the aspect being considered. Similarly, the "negativity" score is computed as the ratio of number of negative tagged reviews to the total number of reviews in the relevant review set for the aspect being investigated. Subjectivity such as extremely positive and positive comments are handled within the 20% band. Thus, extremely positive and positive are considered alike. Likewise, for extremely negative and negative comments.

The "positivity" and "negativity" scores for flavor, cost, ambience, hygiene, and service represent the KPI values for customer experience. These scores are calculated for a specific restaurant and are also calculated across restaurants. The restaurateur is presented with a dashboard where they can visualize their KPI scores compared to the industry benchmark. Comparing the individual restaurant KPI scores with the industry benchmark provides valuable insight on where the business needs to improve with respect to their competitors in the industry. Aggregating the opinions expressed in the form of unstructured text into a score provides an important KPI for restaurateurs to track and improve.

5 Implementation

The Java programming language was used to implement the novel Aspect-Based Deep Learning Framework (ADLF) to improve customer experience. For the Sentiment Classifier, the neural network model configuration, model training and evaluation was done using the Deep Learning for Java Framework (DL4J). The Nvidia CuDNN (Cuda Deep

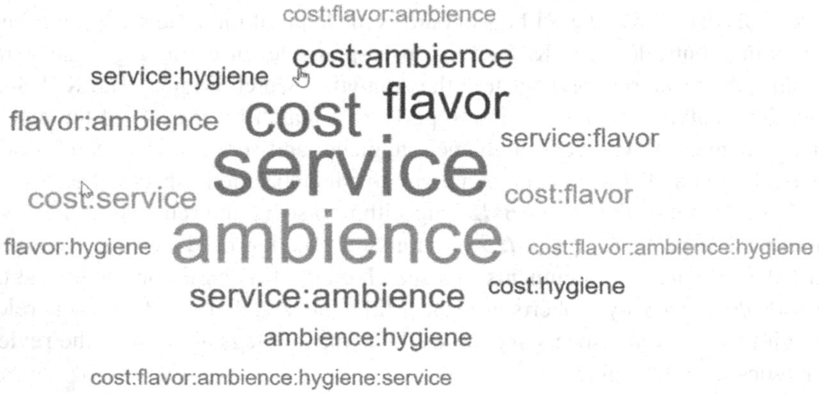

Aspect	Count
service	148,450
ambience	125,148
cost	98,951
flavor	68,321
service:ambience	31,973
cost:ambience	27,546
cost:service	24,426
flavor:ambience	19,187
cost:flavor	14,562
service:flavor	13,099

Fig. 3. Aspect Tag Cloud by number of reviews

Neural Network) library was used on NVIDIA GeForce GTX 1070 GPU which resulted in significant performance gain during model training and evaluation.

The analytics search engine is implemented using Elasticsearch which also acts as the NoSQL datastore. Kibana is used to create a dashboard to visualize the data. The end user interacts with the system through this dashboard. The KPI Service, Sentiment Classifier and Scoring Service are implemented as microservices using Spring Boot which is part of the Spring Framework. Microservices based architecture is a well-established scalable and fault tolerant architecture.

The number of opinions expressed by diners for different aspects of the restaurants is shown in Fig. 3. The highest number of diners at approximately 150 thousand expressed an opinion on the service aspect of the restaurants. This was followed by ambience, cost, flavor, and combinations of aspects such as service and ambience.

The positivity/negativity score of the sentiment that relates to the aspect of the review is shown in Fig. 4. Blue indicates a positive sentiment and red indicates a negative sentiment.

The aspects illustrated in Fig. 3 form the KPIs. The values for these KPIs calculated across all the restaurants form the benchmark as shown in Fig. 4. Diners expressed the highest positive sentiment for the service they received. Diners expressed the highest negative sentiment for the cost of food/beverages.

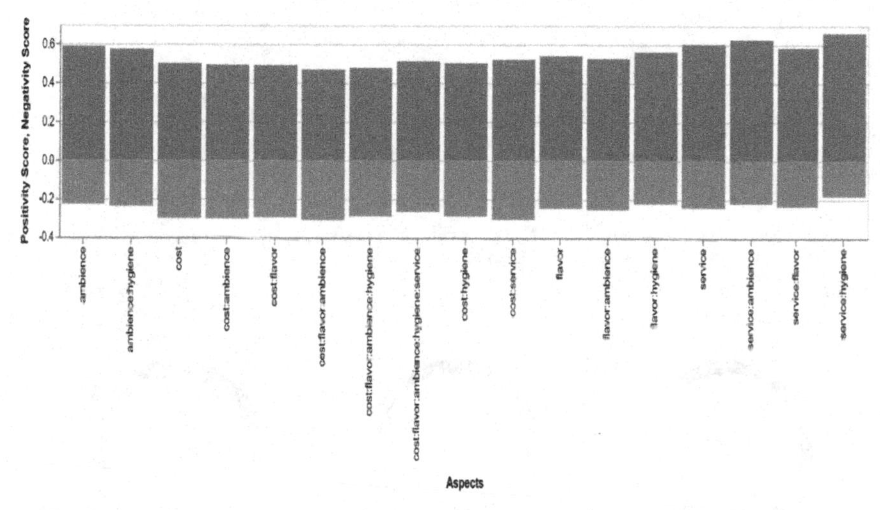

Fig. 4. Customer experience aspect positivity negativity scores (Color figure online)

6 Results and Discussion

The aim of the experiment is to evaluate the proposed novel Aspect-Based Deep Learning Framework (ADLF) to improve Customer Experience which combines an information retrieval algorithm, *Okapi BM25*, and a deep learning model *word2vec-cnn* using accuracy and area under the receiver operating characteristics curve (AUROC). This research focuses on available online reviews.

cost:flavor:ambience:hygiene
flavor:ambience
cost:hygiene
flavor:hygiene
flavor
cost:service
cost:flavor
cost
service
service:flavor
ambience
service:ambience
ambience:hygiene
cost:ambience
cost:flavor:ambience
service:hygiene

Aspect	Count
cost	1,094
ambience	778
flavor	664
service	619
cost:ambience	242
cost:flavor	208
flavor:ambience	200
cost:service	143
service:ambience	123
service:flavor	83

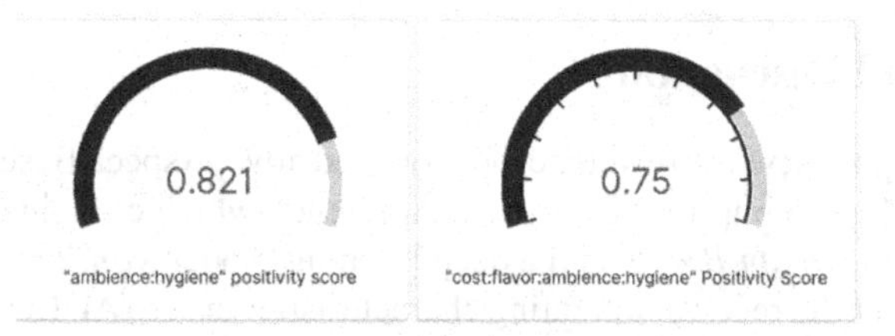

Fig. 5. KPI dashboard for a specific restaurant

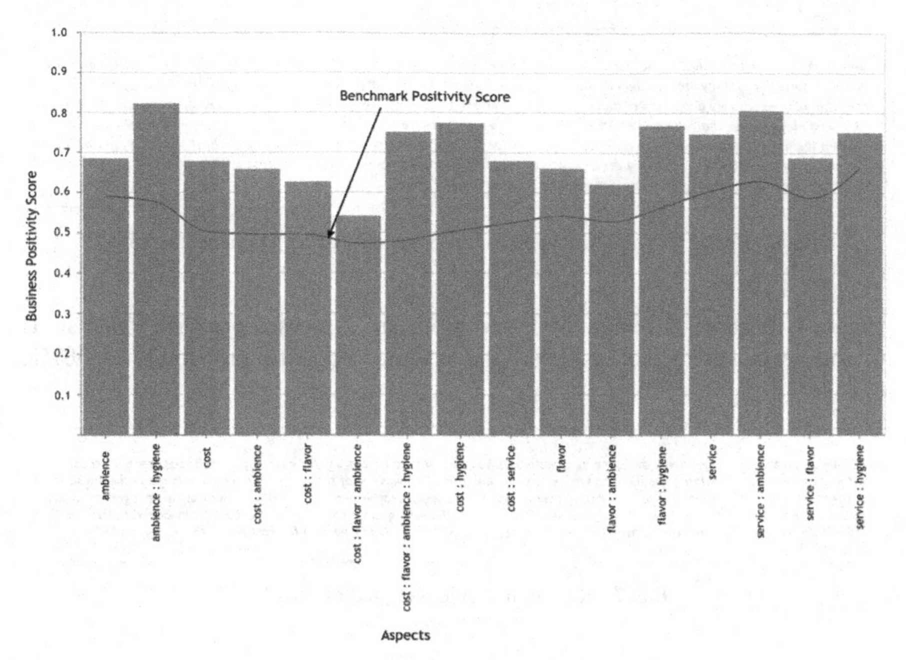

Fig. 5. continued

In this novel research the *word2vec-cnn* neural network model proposed by [19] was modified to use 8 × 400 kernel size, Adam updater, and maximum review size of 1000. Hyperparameters used for training are ReLU activation function, 8 × 400 kernel, 1 × 300 stride, dropout rate 0.5, and mini-batch size of 64. Results demonstrate that the model achieved an accuracy of 94% and area under the receiver operating characteristics curve (AUROC) of 0.98.

The KPIs for a specific restaurant were also calculated by running the scoring algorithm on only the reviews for this restaurant. In the positivity and negativity score calculation reviews of other restaurants were filtered out (Fig. 5).

The Restaurant KPIs on the dashboard illustrates how a specific restaurant is doing in comparison to other restaurants in the industry. The red line corresponds to the benchmark values for each of the KPIs for all restaurants. The specific restaurant in this analysis is performing better than the industry benchmark on all the KPIs. On the "ambience:hygiene", "cost:hygiene" and "cost:flavor:ambience:hygiene" KPIs this restaurant is performing significantly better than the industry benchmark.

Figure 6 illustrates a review where opinions are expressed on hygiene, cost, flavor, ambience, and service. The framework tagged this review as "hygiene:cost:flavor:ambience:service" and assigned a positive probability of 99.99%.

"Although I was a bit trepidacious at first after seeing the web site with the rather chic decor (I tend to favor family-style high-value places), I gotta say that my meal at the Rice Barn is the best Thai food I can remember. My previous favorites, the King and I and the Amarin, are both very good (and perhaps a slightly better value), but for sheer quality the Rice Barn is tops. The ambiance is clean and elegant. The Thai Spring rolls were hot, crispy, and very, very fresh tasting. The Pad Thai was rich, with layered flavors. Chicken slightly grilled. A good size portion too. My wife's Jungle Tofu was spicy, with wonderfully fresh basil flavors. Even the Thai iced tea was excellent - the tea flavor was not lost in the condensed milk. The only downside is that quality comes with prices about 40% higher than other Thai restaurants. Sometimes it's worth paying extra."

Fig. 6. Restaurant review tagged as "hygiene:cost:flavor:ambience:service"

Figure 7 illustrates a review where opinions are expressed primarily on cost. The framework tagged this review as "cost" and assigned a positive probability of 99.7%.

"I'm always on the lookout for new/new to me local businesses & restaurants to try out. I got the craving for delicious breakfast tacos without the delicious breakfast tacos lines. So I looked on Yelp and this place was one of the first ones to pop up. Its inside the gas station in the shopping center. They have a little restaurant set up and the manager (maybe owner) said hello and welcomed me. I got the migas taco and chorizo taco. Oh... my... gosh. They were delicious. There is something about knowing your food is legit made right after you order it. Hearing all the ingredients sizzling on the grill was pretty cool. Oh, and did I mention the cost is super reasonable? I highly recommend!"

Fig. 7. Restaurant review tagged as "cost"

7 Conclusion and Future Work

The aim of this research is to investigate to what extent an aspect-based deep learning framework can help a restaurateur improve the customer experience of a restaurant based on the values of the Key Performance Indicators (KPIs) derived from the sentiment of restaurant reviews.

The proposed novel framework, ADLF, combines an information retrieval algorithm, *Okapi BM25* and a deep learning model *word2vec-cnn*. Results demonstrate that the ADLF framework achieved an accuracy of 94% and area under the receiver operating characteristics curve (AUROC) of 0.98. The novel framework, ADLF, was able to identify the values for the KPIs and provide a comparison to the benchmark of KPIs from restaurants in the industry. The ADLF framework shows promise for restaurateurs that would like to improve the customer experience aspects most important to diners of their restaurant.

ADLF can be extended by including feature extraction to identify KPIs from the corpus of customer reviews. In addition, future work could include a recommender system that recommends activities a restaurateur would need to complete in order to improve their KPIs.

References

1. Dellarocas, C., Zhang, X.M., Awad, N.F.: Exploring the value of online product reviews in forecasting sales: the case of motion pictures. J. Interact. Marketing **21**(4), 23–45 (2007)

2. Zhu, F., Zhang, X.M.: Impact of online consumer reviews on sales: the moderating role of product and consumer characteristics. J. Market. **74**(2), 133–148 (2010)
3. Tang, F., Fu, L., Yao, B., Xu, W.: Aspect based fine-grained sentiment analysis for online reviews. Inf. Sci., **488**, 190–204 (2019). http://www.sciencedirect.com/science/article/pii/S0020025519301872
4. López, M., Valdivia, A., Martínez-Cámara, E., Luzón, M.V., Herrera, F.: E2SAM: Evolutionary ensemble of sentiment analysis methods for domain adaptation. Inf. Sci. **480**, 273–286 (2019). http://www.sciencedirect.com/science/article/pii/S0020025518309873
5. Hyun, D., Park, C., Yang, M.-C., Song, I., Lee, J.-T., Hwanjo, Y.: Target-aware convolutional neural network for target-level sentiment analysis. Inf. Sci. **491**, 166–178 (2019). https://doi.org/10.1016/j.ins.2019.03.076
6. Heikal, M., Torki, M., El-Makky, N.: Sentiment analysis of Arabic Tweets using deep learning. Procedia Comput. Sci. **142**, 114–122 (2018)
7. Ghulam, H., Zeng, F., Li, W., Xiao, Y.: Deep learning-based sentiment analysis for roman urdu text. Procedia Comput. Sci. **147**, 131–135 (2019)
8. Alharbi, A.S.M., De Doncker, E.: Twitter sentiment analysis with a deep neural network: an enhanced approach using user behavioral information. Cognit. Syst. Res. **54**, 50–61 (2019)
9. Chakraborty, K., Bhattacharyya, S., Bag, R., Hassanien, A.E.: Comparative sentiment analysis on a set of movie reviews using deep learning approach. In: Hassanien, A.E., Tolba, M.F., Elhoseny, M., Mostafa, M. (eds.) AMLTA 2018. AISC, vol. 723, pp. 311–318. Springer, Cham (2018). https://doi.org/10.1007/978-3-319-74690-6_31
10. Seo, S., Kim, C., Kim, H., Mo, K., Kang, P.: Comparative study of deep learning-based sentiment classification. IEEE Access **8**, 6861–6875 (2020)
11. Yoon K.: Convolutional neural networks for sentence classification. arXiv preprint arXiv:1408.5882 (2014)
12. Ouyang, X., Zhou, P., Li, C.H., Liu, L.: Sentiment analysis using convolutional neural network. In: Proc. IEEE 2015 IEEE International Conference on Computer and Information Technology; Ubiquitous Computing and Communications; Dependable, Autonomic and Secure Computing; Pervasive Intelligence and Computing, pp. 2359–2364 (2015)
13. Kalchbrenner, N., Grefenstette, E., Blunsom, P.: A convolutional neural network for modelling sentences. arXiv:1404.2188 (2014). https://arxiv.org/abs/1404.2188
14. Young, T., Hazarika, D., Poria, S., Cambria, E.: 'Recent trends in deep learning based natural language processing.' IEEE Comput. Intell. Mag. **13**(3), 55–75 (2018)
15. Mikolov, T., Chen, K., Corrado, G., Dean, J., Sutskever, L., Zweig, G.: word2vec. Google Scholar (2014)
16. Le, H.T., Cerisara, C., Denis, A.: Do convolutional networks need to be deep for text classification? In: Proceedings of Workshops 22nd AAAI Conference of Artificial Intelligence. pp. 1–8 (2018)
17. Collobert, R., Weston, J., Bottou, L., Karlen, M., Kavukcuglu, K., Kuksa, P.: Natural language processing (almost) from scratch. J. Mach. Learn. Res. **12**, 2493–2537 (2011)
18. Hinton, G., Srivastava, N., Krizhevsky, A., Sutskever, I., Salakhutdinov, R.: Improving neural networks by preventing co-adaptation of feature detectors. CoRR, abs/1207.0580 (2012)
19. Panthati, J., Bhaskar, J., Ranga, T.K., Challa, M.R.: Sentiment analysis of product reviews using deep learning. In: Proceedings of the International Conference on Advances in Computing, Communications and Informatics (ICACCI) (2018)
20. Yelp restaurant review dataset [Online]. https://www.yelp.com/dataset/documentation/main
21. Manning, C.D., Raghavan, P., Schutze, H.: Introduction to Information Retrieval," Cambridge, United Kingdom: Cambridge University Press, ISBN 978–0–521–86571–5 (2008)
22. Robertson, S., Walker, S.: Okapi/keenbow at trec8. In: The Eighth Text Retrieval Conference (TREC8), page 151162. Gaithersburg, MD, NIST (2000)

23. Amazon phone review dataset [Online]. https://www.kaggle.com/PromptCloudHQ/amazon reviews-unlocked-mobile-phones/data
24. Luo, Y., Xiaowei, X.: Comparative study of deep learning models for analyzing online restaurant reviews in the era of the COVID-19 pandemic. Int. J. Hospital. Manag. **94**, 102849 (2021). https://doi.org/10.1016/j.ijhm.2020.102849

IoTs, Sensors, and Networks

Routing Protocol Security for Low-Power and Lossy Networks in the Internet of Things

Akshet Patel[1], D. Shanmugapriya[2], Gautam Srivastava[3](✉), and Jerry Chun-Wei Lin[4]

[1] Department of Mechatronics Engineering, Manipal University Jaipur, Jaipur, India
[2] Department of Computer Science, Vivekanandha College of Engineering for Women, Elaiyampalayam, India
[3] Department of Math and Computer Science, Brandon University, Brandon, Canada
srivastavag@brandonu.ca
[4] Department of Computer Science, Electrical Engineering and Mathematical Sciences, Western Norway University of Applied Sciences, Bergen, Norway

Abstract. The Internet of Things (IoT) is a two-decade-old technology that allows the interconnection of various cyber-physical objects located anywhere in the world over the internet. A cyber-physical object is an inanimate object with the ability to connect itself to the internet and communicate with other similar objects. This communication is enabled by communication protocols that bridge the gap between hardware devices and the user by ensuring secure communication with optimum efficiency. As the number of IoT devices is increasing exponentially every year with the current estimation of more than 35 billion IoT devices according to a study conducted by findstack.com, there is a need for a lightweight and highly secure communication protocol for IoT devices. This creates a need for handling all the dynamic data exchanges in real-time. Hence, Big Data and IoT technologies are often proposed together for processing and analyzing data streams. Incorporating Big Data analytics in IoT applications enables real-time information transfer, data monitoring and processing easily. To facilitate efficient communication in IoT networks, the Routing Protocol for Low-Power and Lossy Networks (RPL) comes into the picture which is a low power consuming routing protocol that operates on IEEE 802.15. This paper aims to encrypt the RPL Protocol using cryptographic algorithms and compare the results with a non-encrypted RPL Protocol using Cooja Simulator so that the data is securely communicated for Big Data analysis. The (Secure Hash Algorithm) SHA3 algorithm had been used to encrypt the sink nodes of the RPL protocol and the parameters such as average power consumption, network graph, latency is analyzed and compared. The results of this experiment show a considerable drop in latency and power consumption after deploying cryptographic encryption.

Keywords: Big Data · Big Data analytics · RPL protocol · Contiki OS · Cooja simulator · SHA · Encryption algorithms · Cryptography

1 Introduction

The fourth industrial revolution will immensely impact the daily lives of everyone as almost every inanimate object will be able to communicate with each other and other

S. N. Srirama et al. (Eds.): BDA 2021, LNCS 13147, pp. 133–145, 2021.
https://doi.org/10.1007/978-3-030-93620-4_11

cyber-physical objects over the internet. According to [1], a large number of applications including technical and non-technical, social, industrial, agricultural, and smart cities and healthcare are encompassed by IoT which makes it a multidisciplinary domain. Wireless Sensor Networks (WSN) are a crucial part of the Internet of Things because it is a special kind of Low-Power Wireless Personal Area Network (LoWPAN) consisting of many nodes which are equipped with various types of sensors and data acquisition devices. The processing power and the energy requirements of these LoWPAN. Due to the lack of infrastructure for interoperability of the Internet and the Wireless Sensor Network, in October 2004, the Internet Engineering Task Force (IETF) formed a working group known as the 6LoWPAN which is short for IPv6 over Low Power Wireless Personal Area Networks. The main objective of the 6LoWPAN Working group was to develop a protocol enabling the use of IPv6 over IEEE 802.15.4 networks. After a few years, a new working group was formed, and their main objective was to develop an efficient routing solution for Low power and lossy networks (LLN).

1.1 The Role of Big Data in IoT

The development of the streaming data collected from different sensors from all the IoT devices, however, gives rise to the "Big Data" phenomena, which is defined by the four V's: Volume, Velocity, Variety, and Value. Several Internet of Things (IoT) and Big Data technologies have recently been created to enable the gathering, processing, and analysis of massive amounts of streaming data. The goal is to extract current and future circumstances in real-time to predict changes and adapt services to better meet the requirements of users. For the simple deployment of context-aware apps, many IoT platforms have been suggested. Current IoT apps, such as Thingspeak23, may be used for data monitoring and analysis (e.g., traffic monitoring) without any assurances on transmission and processing [15]. As a result, feeding data from an embedded system into a data analytics cloud as well as analyzing that data may take tens of seconds or less.

1.2 The RPL Protocol

The Routing Protocol for Low Power and Lossy Networks also known as the RPL was developed in March 2012 by Routing over Low-power and Lossy networks (RoLL) working group for LLNs as a proactive protocol based on tree architecture which builds an acyclic graph between the sink node (border router) and the leaf nodes. The low power and lossy networks are classified by their limited power of processing, power consumption and energy requirements. It is generally used for wireless networks with low bandwidth and low power and in connecting Wireless Sensor Networks (WSN) because of its low resource utilization feature.

The RPL protocol is vulnerable to many attacks such as [2]:

1. **The Rank Attack:** This attack, when triggered injects a malicious node which will change the processing of all the other nodes ultimately changing the flow and increasing the network traffic which increases the latency and power consumption.
2. **The Neighbor Attack:** This attack, when triggered replicates any of the DODAG information objects and rebroadcasts them. This attack changes the route of the messages and hence increases the power consumption and decreases reliability.

To secure the RPL protocol from such and many more attacks and as there is an exponential increase in the network traffic, securing the network has become the most important aim for researchers all around the globe. This has motivated us to develop and encrypt the RPL protocol using the Secure Hash Algorithm (SHA3) and simulate the network on Cooja Simulator.

1.3 The Cooja Simulator

COOJA is a simulator that blends low-level sensor node hardware simulation with high-level behaviour simulation in a single simulation. COOJA is adaptable and extensible because all tiers of the system, including sensor node platforms, operating system software, radio transceivers, and radio transmission models, can be altered or relocated. The simulator is designed in Java, making it easy for users to extend it, while the Java Native Interface allows sensor node software to be written in C. Furthermore, the sensor node software can be run as compiled native code for the simulator's platform or in a sensor node emulator, which simulates an actual sensor node at the hardware level.

There are several tools to control the network simulation like:

1. **Network-** The location and status of each node can be visualized.
2. **Simulation Control-** This control panel is used to control the speed of the simulation and to start, pause, reload, or execute steps of the simulation of the network.
3. **Mote Output-** The output of every mote can be visualized in a serial interface.
4. **Timeline-** The log of every message signal and event change can be visualized in the timeline.

2 Related Works

There have been various qualitative surveys of different communication conventions which could be suitable to IoT. Though, only some papers are related to the quantitative comparison of IoT protocols have been published up to now. There are a few works in which the RPL, 6lowpan and CoAP protocols are evaluated independently, or evaluation was compared with other protocols [3].

In [4], Pongle *et al.* presented the most common attacks on RPL and 6LoWPAN like the Selective forwarding attack, sinkhole attack, Sybil attack, hello flooding attack, etc. and its consequences and existing solutions to mitigate those vulnerabilities. The existing solutions presented in the paper were deploying the intrusion detection system (IDS), Artificial intelligence-based IDS Bayesian model and data mining-based IDS.

In [5], Zaminkar *et al.* proposed the Detection of sinkholes in RPL protocol communication in a four-step process. The steps are creating the reliable RPL and then detecting the sinkhole. The third steps involve quarantining the detected malicious nodes and the fourth phase transmits the encrypted data.

The authors of the paper [6] presented a survey on different lightweight encryption and authentication schemes for low power and lossy networks and explained the three modes of security of the RPL protocol which are the Authenticated mode, the pre-installed mode, and the unsecured mode.

In [7], the authors discussed the gaps in the deployment of the RPL protocol and surveyed to find out the reasons for security attacks and proposed countermeasures to mitigate those attacks. Various encryption algorithms like the PRESENT, CLEFIA and ChaCha20 were discussed.

In [8], Gawde *et al.* proposed authentication and developed encryption for low power and lossy networks (LLNs) especially the RPL protocol. They successfully incorporated the SQUASH algorithm to authenticate the RPL protocol and the PRESENT algorithm having an encryption key of 80 bits.

The authors offer an equal access approach in [17]. It's been set aside for an Internet of Things system that allows many people and organizations to access each other's data. All members were linked to the same blockchain network, which allowed for data access control. Every network member has a "purse" that contains access information for other members of the blockchain network as well as all of the keys towards the information that is allowed to be seen. In such systems, if X wants a resource that is being handled by Y, the latter must broadcast the request to the blockchain network along with all of the relevant keys. Following that, the blockchain network checks X's authorization to use Y's resources. If X is authorized to receive the data, Y transmits it to that individual.

For WSN networks, the authors of the paper [18] offer FlexCrypt, an automated, flexible, and lightweight encryption method. To efficiently rotate cluster headship across network nodes, the proposed approach employs a dynamic clustering technique that enables mobility in WSN.

Furthermore, the suggested approach employs a lightweight encryption mechanism that is adaptable to cryptographic parameters in sensor nodes with limited resources. They also offer a lightweight and energy-efficient key management and authentication system for establishing a secure connection for data and symmetric key exchange in WSNs.

RPL is a distance vector-based routing protocol that is optimized for low-power, resource-constrained networks. The whole routing protocol is made up of ICMPv6 control messages exchanged between smart devices. Functions restrict the best path selection of the RPL routing protocol; multiple functions may be developed for different circumstances to make the structure more flexible, the network more efficient, and network communication more stable. The RPL routing protocol can handle rapidly changing network topologies in a fixed state at a given location because the whole network structure is a destination-oriented directed acyclic graph (DODAG). The network topology can change at any time, and the RPL can react fast to changes in network status and choose the best path depending on functions. Due to the enormous number of threats to the internet of things, the RPL protocol has been extensively researched.

3 Problem Statement

With the number of attacks and assaults on network protocols increasing every passing day, there is a need to encrypt them. In this paper, we have encrypted the RPL Protocol using SHA3 encryption which stands for Secure Hash Algorithm. The SHA algorithm is based on the MD4 encrypting algorithm. SHA256 has a hash length of 256 bits, and it is known as the SHA-256. As it is evident from the contributions of [9] and [10] that latency and power consumption are severely compromised under the influence of an attack, we have deployed the SHA encryptions on 8 sink nodes communicating via the RPL protocol and the latency and the power consumption is analyzed and tabulated with graphical representations of the same.

4 Methodology

4.1 Implementing the SHA Encryption

SHA is an acronym for Secure Hashing Algorithm. SHA is a hashing algorithm that is based on MD5. It is used to hash data and certificates. Using bitwise operations, modular additions, and compression functions, a hashing algorithm compresses the input data into a smaller form that cannot be comprehended.

SHA-1, SHA-2, SHA-256, SHA-512, SHA-224, and SHA-384 are some examples of SHA names, although only two types of algorithms exist, i.e. SHA-1 and SHA-2. The higher values, such as SHA-256, are just SHA-2 variants that indicate the bit lengths of the SHA-2. The first secure hashing method, SHA-1, produced a 160-bit hash digest after hashing.

SHA-2 can generate hash digests with a range of bit lengths ranging from 256 to 512 bits, enabling it to give fully distinct values to each hash.

Version Number and Rank Authentication (VeRA) is an RPL protocol that uses one-way hash chains to secure version numbers and rankings against spoofing attacks. VeRA is still subject to rank spoofing topological assaults. An attacker can use the first vulnerability to build a counterfeit rank hash chain and claim whatever rank in the topology. In a second strike, the opponent replays his parent's rank, moving one rank step up the hierarchy. An encryption chain is added to prevent a fabricated rank hash chain from being created by an attacker (Fig. 1).

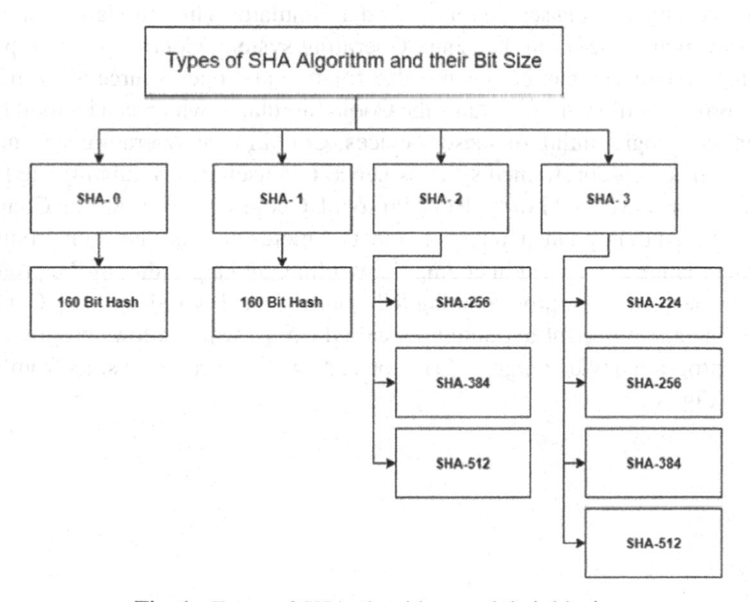

Fig. 1. Types of SHA algorithms and their bit size

The current study applies different types of Dynamic Hash-Based Security to an IoT situation. The main variations used in this study are DynamicHashAlgorithm-1, DynamicHashAlgorithm-2-256, and DynamicHashAlgorithm-3-256, as well as their effects on various parameters and resource optimization variables in the Internet of Things environment [20].

4.2 Methodology Followed

The methodology for the entire experiment began with launching the Cooja simulator which runs on Contiki OS which is a Linux-based operating system. The Cooja simulation is run using the ant run command and Sky Motes are created and the Source and Sink of data are defined. The SHA algorithm is implemented in the UDP-sink.c file and the simulation is started. The same code is run without any encryption and the results are logged and analyzed (Fig. 2).

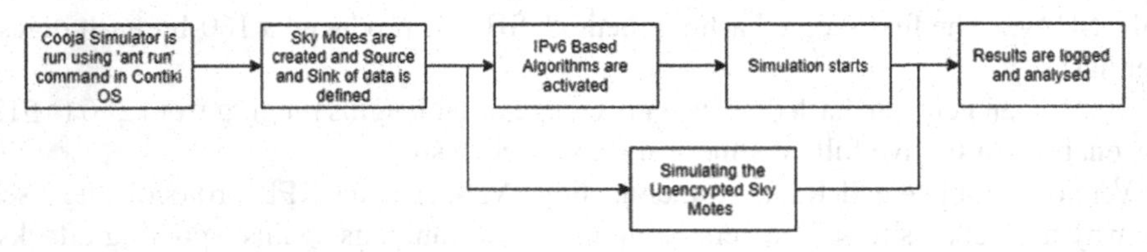

Fig. 2. Methodology followed

4.3 Running the Cooja Simulator

Figure 3 depicts the loading screen for the Cooja Simulator. This simulator works on the Contiki OS which is based on the Linux Operating system. Contiki is a core platform or operating system that can be downloaded for free and open source at (http://www.contiki-os.org). Contiki comes features the Cooja Simulator, which can be used for both simulation and programming of sensor devices. Contiki is an operating system that is designed for memory-constrained systems hence it is useful in simulating the Routing Protocol for Low power and Lossy (RPL) Protocol. Cooja simulator runs on Contiki OS, and it can be used to implement a large number of motes in a simulated environment. C programming language is used in coding the working and algorithm of the protocol. It has a lot of possibilities for programming IoT nodes for real-world use [19]. Contiki has an outstanding and powerful IoT simulator called Cooja, which allows programmers to import and program a wide range of IoT motes, as well as acquire results from various algorithms (Fig. 4).

```
akshetpatel@akshetpatel-HP-Pavilion-Notebook: ~/contiki/tools/cooja
akshetpatel@akshetpatel-HP-Pavilion-Notebook:~/contiki/tools/cooja$ ant run
Buildfile: /home/akshetpatel/contiki/tools/cooja/build.xml

init:

compile:

copy configs:

jar_cooja:
      [jar] Building jar: /home/akshetpatel/contiki/tools/cooja/dist/cooja.jar

jar:

init:

compile:

jar:

init:

mspsim:
```

Fig. 3. Running the Cooja Simulator using the ant run command

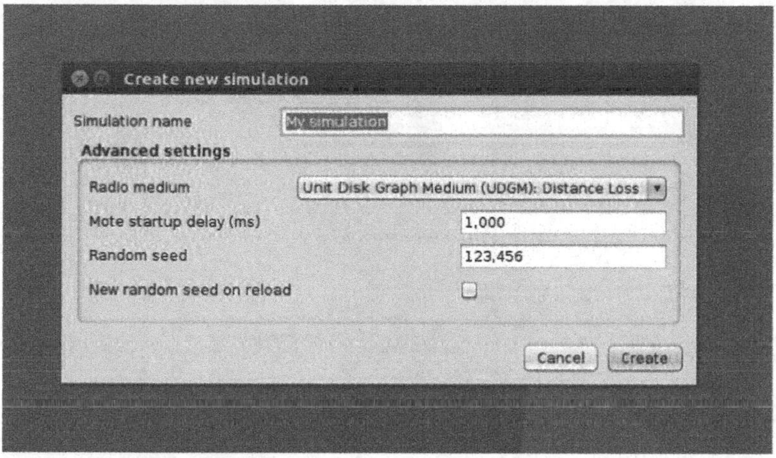

Create new simulation
Simulation name: My simulation
Advanced settings
Radio medium: Unit Disk Graph Medium (UDGM): Distance Loss
Mote startup delay (ms): 1,000
Random seed: 123,456
New random seed on reload
Cancel Create

Fig. 4. Creating a New Simulation on Cooja Simulator

4.4 Simulating the Unencrypted RPL Protocol

In terms of turnaround time, power consumption, and energy in IoT networks, the proposed implementation with a secured hash-based key method results in a higher degree of security and performance [11]. The work that has been completed is quite effective in terms of a variety of factors, including low overheads and complexity, as well as a greater degree of performance for numerous and increasing numbers of nodes. In RPL, SHA-based security improves efficiency and integrity while also increasing overall security [12] (Fig. 5).

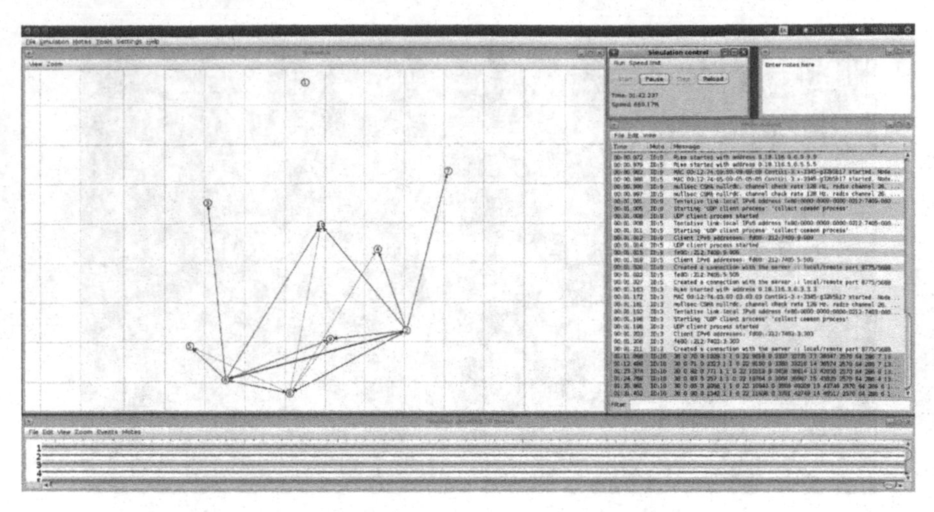

Fig. 5. Simulating the RPL Protocol without any encryption

File Edit View Search Tools Documents Help

Open

```c
/*
 * Define SHA3_USE_KECCAK to run "pure" Keccak, as opposed to SHA3.
 * The tests that this macro enables use the input and output from [Keccak]
 * (see the reference below). The used test vectors aren't correct for SHA3,
 * however, they are helpful to verify the implementation.
 * SHA3_USE_KECCAK only changes one line of code in Finalize.
 */

#if defined(_MSC_VER)
#define SHA3_CONST(x) x
#else
#define SHA3_CONST(x) x##L
#endif

/* The following state definition should normally be in a separate
 * header file
 */

/* 'Words' here refers to uint64_t */
#define SHA3_KECCAK_SPONGE_WORDS \
        (((1600)/8/bits to byte/)/sizeof(uint64_t))
typedef struct sha3_context_ {
    uint64_t saved;             /* the portion of the input message that we
                                 * didn't consume yet */
    union {                     /* Keccak's state */
        uint64_t s[SHA3_KECCAK_SPONGE_WORDS];
        uint8_t sb[SHA3_KECCAK_SPONGE_WORDS * 8];
    };

#define UIP_IP_BUF   ((struct uip_ip_hdr *)&uip_buf[UIP_LLH_LEN])

#define UDP_CLIENT_PORT 8775
#define UDP_SERVER_PORT 5688

static struct uip_udp_conn *server_conn;

PROCESS(udp_server_process, "UDP server process");
AUTOSTART_PROCESSES(&udp_server_process,&collect_common_process);
/*---------------------------------------------------------------------------*/
void
collect_common_set_sink(void)
{
}
/*---------------------------------------------------------------------------*/
void
collect_common_net_print(void)
{
  printf("I am sink!\n");
}
/*---------------------------------------------------------------------------*/
void
collect_common_send(void)
{
  /* Server never sends */
```

Fig. 6. Editing the source code of the sink node by inserting SHA in C language.

Figure 6 demonstrates a text editor view with C code for programming the IoT sink nodes using SHA. The SHA code is entered, and the code for the new results is compiled.

4.5 Simulating the Unencrypted RPL Protocol

The secured hash method is implemented in the simulated environment as shown in Fig. 7. With the key authentication process and the opportunity to copy the network logs, the wireless sky motes can be viewed in the output section [13].

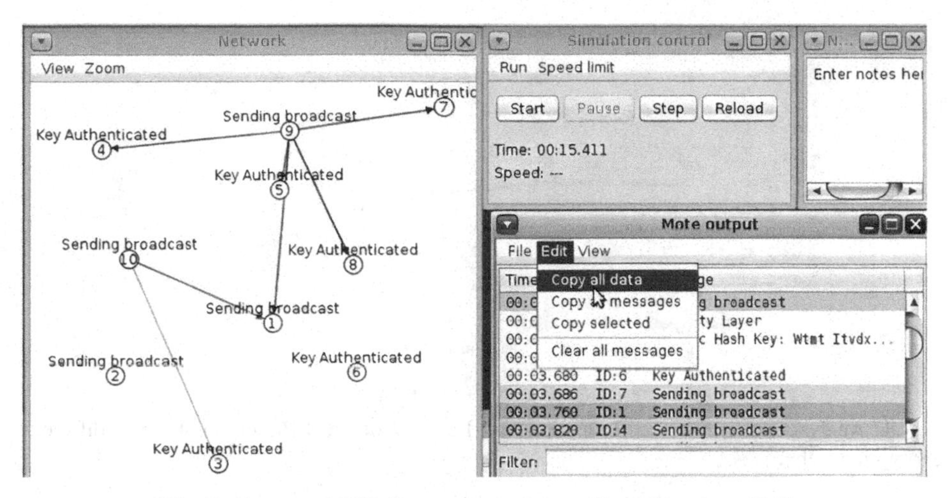

Fig. 7. Encrypted RPL Protocol running on Cooja Simulator [14]

The experiment is conducted by deploying 8 sink motes and 2 source motes. 4 scenarios are used for testing, i.e., Unencrypted RPL motes and encrypted using SHA1, SHA2 and SHA3 algorithms.

5 Results and Discussions

Routing information replay attacks may also be carried out by an RPL node. It keeps track of valid control messages from other nodes and sends them to other nodes in the network later. Because the architecture and routing patterns of dynamic networks are frequently modified, this assault is particularly destructive. Nodes update their routing tables with outdated data because of replay assaults, resulting in a misleading topology. To maintain the freshness of routing information, the RPL protocol employs sequence counters such as the Version Number for DIO messages and the Path Sequence in the Transit Information option of DAO messages.

After the successful deployment of the SHA algorithms, the motes were set up in the Cooja Simulator. There is a significant drop in the latency observed when the RPL protocol is encrypted using the SHA3 algorithm. There is a significant drop of more than 2.5 times in latency. A drop of almost 10% is observed in the power consumption in the encrypted protocol compared to the unencrypted. The results after the simulation are tabulated in Table 1 (Fig. 8).

Table 1. Results of the simulation

Sr. No.	Algorithm	Latency observed (ms)	Consumed power (μW)
1	Unencrypted	480	820
2	SHA1	490	830
3	SHA2	410	950
4	SHA3	190	750

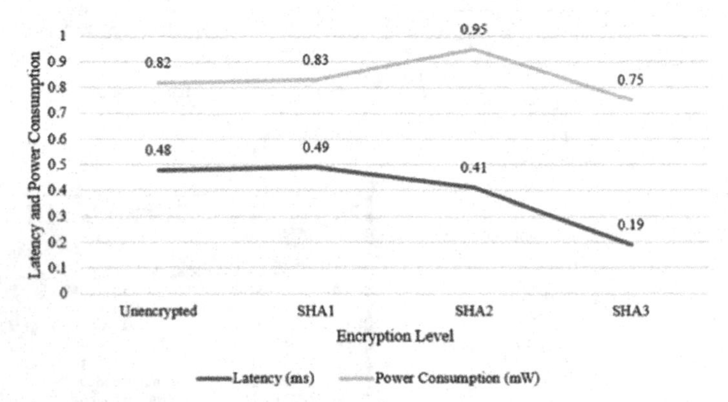

Fig. 8. Analyzing the Power consumption and Latency of the RPL Protocol using different encryption techniques.

The overall time it takes for a packet to travel from the client node to the root node is referred to as network latency (end-to-end delay). As it can be observed from the graph plotted, the latency marginally increases with the level of encryption and reduces significantly when the SHA3 algorithm is deployed (Fig. 9).

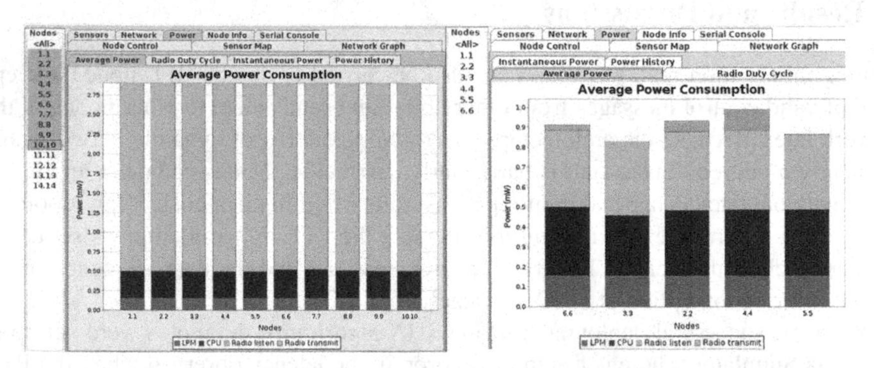

Fig. 9. Comparing the average power consumed by unencrypted RPL protocol with SHA3 encrypted RPL Protocol [14]

Energy Consumption in lossy networks and estimating power is critical since it determines the lifespan of the WSN. It has been observed that power is consumed more than usual while retransmitting and listening idly. The average power consumed by the nodes without encryption is less than that of the average power consumed by the nodes which are encrypted. This proves that the SHA algorithm is energy efficient.

6 Future Work

Future work linked with this study could include the deployment of soft computing technologies as well as the development of new algorithms that can provide a higher level of security than previous work. Soft computing technologies can give a greater level of accuracy and security for overall performance and in a generalized network environment, as well as other associated factors. Machine learning, evolutionary computation (EC), probability, and many other implementation components are related to soft computing [16].

7 Conclusion

This study provides a comprehensive analysis of RPL-based routing protocols, as well as technical perspectives and recommendations for various RPL implementations and optimization techniques. It also looks at where RPL stands now in terms of its usability and efficiency in IoT applications. There is a lot of work going on these days in the IoT segment, which is the advanced wireless network environment, but there are a lot of vulnerability elements that need to be handled, and that is the main purpose of this effort. With the examination of many factors, this work implements several hash techniques and security protocols. According to the data, DynamicHashAlgorithm-3 has the largest block size with the fewest rounds and is effective in collisions. This SHA variant produces results that can be used in a variety of IoT protocols and key generation methods. The following essential factors connected to the efficiency of DynamicHashAlgorithm-3 are offered based on the results and intrinsic aspects of SHA:

- Lower Overhead and Complexity.
- Greater Security of Larger Block Size and Output Size.
- Greater Integrity.
- Lower Latency.

As IoT technology advance, a significant number of applications in a variety of sectors have emerged. This article is a timely study that examines present and future IoT big data storage solutions in cloud computing while also surveying the state-of-the-art literary works from the perspective of data processing. The IoT storage system allows critical information about objects to be tracked as they pass via cloud platforms. It adds considerable value to IoT applications by giving a precise knowledge of current IoT processing data, resulting in increased availability and resource flexibility. Big Data storage systems that enable IoT devices may be used to boost data processing performance and provide IoT applications with a significant competitive edge. Semantic connections between IoT data have been demonstrated to lead to increased global intelligence and interoperation abilities. Enterprises will be able to gain this capability thanks to IoT data storage solutions.

References

1. Gluhak, A., Krco, S., Nati, M., Pfisterer, D., Mitton, N.: Razafindralambo T. Challenges for IoT Experimentation, November, pp. 58–67 (2011)
2. Le, A., Loo, J., Luo, Y., Lasebae, A.: The impacts of internal threats towards routing protocol for low power and lossy network performance. In: Proceedings of the International Symposium on Computing and Communication, pp. 789–794 (2013)
3. Mayzaud, A., Sehgal, A., Badonnel, R., Chrisment, I., Schönwälder, J.: Using the RPL protocol for supporting passive monitoring in the Internet of Things. In: Proceedings of the NOMS 2016 - 2016 IEEE/IFIP Network Operations and Management Symposium, no. 4, pp. 366–374 (2016)
4. Pongle, P.: Computer Engineering Department, Sinhgad College of Engineering, Pune IGC, Computer Engineering Department, Sinhgad College of Engineering, Pune I. A Survey: Attacks on RPL and 6LoWPAN in IoT (2015)
5. Zaminkar, M.: A Method Based on Encryption and Node Rating for Securing the RPL Protocol Communications in the IoT Ecosystem, October, pp. 1–18 (2020)
6. Razali, M.F., Rusli, M.E., Jamil, N., Ismail, R.: The Authentication Techniques for Enhancing the RPL Security Mode: A Survey, no. 119, pp. 735–743 (2017)
7. Gawade, A.U.: Lightweight secure RPL: a need in IoT. Int. Conf. Inf. Technol. **2017**, 214–219 (2017)
8. Gawde, P.A.: Specifications A. Lightweight Authentication and Encryption mechanism in Routing Protocol for Low Power and Lossy Networks (RPL), ICICCS, pp. 226–229 (2018)
9. Avila, K., Jabba, D., Gomez, J.: Security aspects for RPL-based protocols: a systematic review in IoT. Appl Sci. **10**(18), 6472 (2020)
10. Kamble, A., Malemath, V.S., Patil, D.: Security attacks and secure routing protocols in RPL-based Internet of Things: Survey. In: 2017 International Conference on Emerging Trends and Innovations ICT, ICEI 2017, pp. 33–39 (2017)
11. Dhanda, S.S., Singh, B., Jindal, P.: Lightweight cryptography: a solution to secure IoT. Wirel. Pers. Commun. **112**(3), 1947–1980 (2020). https://doi.org/10.1007/s11277-020-07134-3
12. Kharrufa, H., Al-Kashoash, H.A.A., Kemp, A.H.: RPL-based routing protocols in IoT applications: a review. IEEE Sens J. **19**(15), 5952–5967 (2019)
13. Tang, W., Wei, Z., Zhang, Z., Zhang, B.: Analysis and optimization strategy of multipath RPL based on the COOJA simulator. IJCSI Int. J. Comput. Sci. Issue. **11**(5), 27–30 (2014)
14. Aggarwal, R.K., Rathee, G.: Security Analysis of RPL Protocol, p. 141405 (2018). http://www.ir.juit.ac.in:8080/jspui/bitstream/123456789/17454/1/SP13345_Rajat_Kumar_Aggarwal_IT_2018.pdf
15. Herden, O.: Architectural patterns for integrating data lakes into data warehouse architectures. In: Bellatreche, L., Goyal, V., Fujita, H., Anirban Mondal, P., Reddy, K. (eds.) Big Data Analytics: 8th International Conference, BDA 2020, Sonepat, India, December 15–18, 2020, Proceedings, pp. 12–27. Springer, Cham (2020). https://doi.org/10.1007/978-3-030-66665-1_2
16. Singh, S., Singh, J.: Analysis of GPS trajectories mapping on shape files using spatial computing approaches. In: Bellatreche, L., Goyal, V., Fujita, H., Anirban Mondal, P., Reddy, K. (eds.) Big Data Analytics: 8th International Conference, BDA 2020, Sonepat, India, December 15–18, 2020, Proceedings, pp. 91–100. Springer, Cham (2020). https://doi.org/10.1007/978-3-030-66665-1_7

17. Ouaddah, A., Abou Elkalam, A., Ait, O.A.: FairAccess: a new Blockchain-based access control framework for the Internet of Things. Secur. Commun. Netw. **9**(18), 5943–5964 (2016)
18. Khashan, O.A., Ahmad, R., Khafajah, N.M.: An automated lightweight encryption scheme for secure and energy-efficient communication in wireless sensor networks. Ad Hoc Netw. **115**, 102448 (2021). https://doi.org/10.1016/j.adhoc.2021.102448
19. Randhawa, P., Shanthagiri, V., Kumar, A.: Violent activity recognition by E-textile sensors based on machine learning methods. J. Intell. Fuzzy Syst. **39**(6), 8115–8123 (2020)
20. Alkhamisi, A.O., Buhari, S.M., Tsaramirsis, G., Basheri, M.: An integrated incentive and trust-based optimal path identification in ad hoc on-demand multipath distance vector routing for MANET. Int. J. Grid. Util. Comput. **11**(2), 169–184 (2020)

MQTT Protocol Use Cases in the Internet of Things

D. Shanmugapriya[1], Akshet Patel[2], Gautam Srivastava[3](✉), and Jerry Chun-Wei Lin[4]

[1] Department of Computer Science, Vivekanandha College of Engineering for Women, Elaiyampalayam, India

[2] Department of Mechatronics Engineering, Manipal University Jaipur, Jaipur, India

[3] Department of Mathematics and Computer Science, Brandon University, Brandon, Canada
srivastavag@brandonu.ca

[4] Department of Computer Science, Electrical Engineering and Mathematical Sciences, Western Norway University of Applied Sciences, Bergen, Norway

Abstract. In the imminent generations, the Internet of Things (IoT) will have a vital role in 'Networking'. There are a lot of subscribing/publish protocols used in IoT like MQTT, AMQP, XMPP, HTTP etc., but every protocol used here will not have the same features required for the Internet of Things. This paper helps you to find which is the best and efficient protocol through the use cases and the Big Data analysis used in IoT. To choose the best protocol, it should be used in the real-world application and analyzed through some statistical measurements, as like in this study two use cases are taken to experiment with the protocol. Lightweight MQTT (Message queuing telemetry transport) protocol is vastly used in IoT, So MQTT protocol is used to subscribe/publish in these use cases, using the use cases measurements (speed of the transmission of messages and throughput of the message transmission) are measured and the security systems are either explained. Mainly for the use cases like home automation and vehicular network, there are a huge number of resources like sensors are used, these sensors collect massive data, so for storing these massive data big data is used. This paper majorly focuses on the use cases of MQTT protocol in IoT, the advantages of using MQTT protocol and big data analytics used in IoT.

Keywords: MQTT protocol · Big data · IoT · Node-red · AWS · Home automation · Vehicular network

1 Introduction

The Internet of Things (IoT) and big data creates an innovative evolution of the technological world with a new era of mature intelligence computing [12], so there should be an in-depth knowledge and clarity in the systems and protocols used in IoT and also about the big data which is used to collect the huge amount of data from the IoT devices and perform big data analysis. Whenever someone needs in-depth knowledge in IoT, there should be research done and should even try using that technology's application

S. N. Srirama et al. (Eds.): BDA 2021, LNCS 13147, pp. 146–162, 2021.
https://doi.org/10.1007/978-3-030-93620-4_12

and analyze them, so in this study, there are two use cases used to evaluate the efficiency of the MQTT protocol. Another thing to be considered in this technological era is, everything has become easy, anybody who knows a programming language can create a communication protocol but there comes an issue "which is best? which is secure? And why should I choose this?". As for IoT, there are a lot of protocols being used and this paper will explain with the help of two use cases to determine "which is the best protocol for IoT ? ". There are three parts explained in this paper. 1. **Use Case 1**: Home Automation using Node-red, 2. Use case 2: Vehicular network and 3. Justification to prove MQTT is more efficient than other protocols.

This study demonstrates two use cases

1. Home automation
2. Vehicular networks

Using these use cases and measurements the efficiency and the security systems of MQTT are explained. The reason for MQTT being used vastly in IoT is explained by comparing with IoT protocols and listing the features of MQTT.

2 Use Case 1: Home Automation Using Node-Red

Internet of things with Artificial intelligence brings intelligence computing, where the manpower is reduced, and automation begins. At the beginning of the technological era, the industries started reducing the manpower and now they are making the manpower to be zero and everything should be automated. For example, in neural science, there is a reflex reaction, where there is an automatic reaction that takes place without the knowledge of the human, as in the industries they are implementing a reflex reaction for the machines. Automation has been started in the manufacturing industries like cars, food, etc., and now the people's expectation tends to automate the whole system of a home. Here in this study, Node-red software is used to create the home automation platform, where the user can control the home using the dashboard created in the Node-red or even through the mobile application like IFTTT or MIT app inventor. In some research they have made a website to operate the home devices, the user interface of the website is designed in a very simple way, and everybody can access it [12]. There are three steps to set up a secured home automation simulation and to calculate the measurements of the MQTT protocol.

2.1 Setup of Virtual Server in AWS and Interconnecting Node-Red, MQTT Box, Mosquitto Broker and AWS

AWS Setup. Every IoT network should be secured, so the first step here is to create instances, security groups and key pairs in AWS. In the AWS management console, an instance should be launched at the EC2 means creating a virtual server to activate the interconnected devices. Here an Ubuntu server is installed which will be accessed through Putty software. At next a security group is created in instances, where only certain people who do have access can use this virtual server. Then a private and a public key should be generated, it authenticates itself and sends the public key to the certificate authority for the verification and at last, gives access to use the virtual server. Add inbound rules for 1883 and 1880 port ranges as shown in Fig. 1. After these steps are done user would get a DNS IP address and IPV4 address.

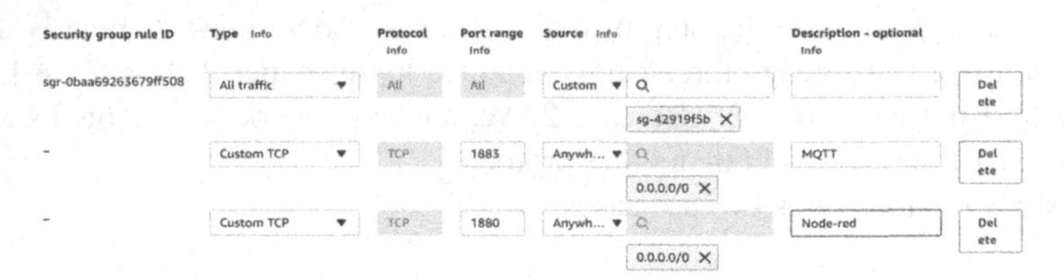

Fig. 1. Addition of Inbound rules in AWS.

Putty Software and MQTT Box. In putty software, a Mosquitto broker is installed to handle the MQTT broker. There is some sort of coding used to install Mosquitto broker and for fixing a username and password for the MQTT box. In the beginning, whenever the user needs to use the Ubuntu server in putty, the user should have the private key for the authentication and the IP address which is created in the AWS security groups.

```
Coding to install the Mosquitto
$sudo apt install mosquitto mosquitto-clients
Coding to set a password for MQTT box
$sudo mosquitto_passwd -c /etc/mosquitto/passwd xxxx
$sudo nano /etc/mosquitto/conf.d/default.conf
$allow_anonymous false
$password_file /etc/mosquitto/passwd
$sudo systemctl restart mosquito
```

After running the coding, the user can create a client in the MQTT box, to create a client the user should enter the username and password with the IP address and port number as a host as given in Fig. 2.

Fig. 2. Creation of new client and setting up username and password.

2.2 The Home Automation System in Node-Red

IBM Node-RED is a flow-based programming tool, for wiring together hardware devices, APIs and online services and is supported by the JS Foundation [13]. In this case, Node-red is used for simulating a home automation setup. Node-red is a platform where the user can create an IoT simulation using various nodes. In this home automation, there are four flows created [6–10]. The user can use his/her Internet protocol (IP) address and port number to get into Node-red software (IP: port number) as the website address. To login to the dashboard of Node-red, the user must give IP, port number and /UI (IP: port number/UI). Here there are various nodes to create a flow of IOTsimulation, each node has specific functions. The user can create various flows required, after constructing a flow the user can deploy and get his/her result. MQTT is the protocol used in this programming tool, there are two nodes in Node-red used to publish and subscribe in MQTT, those are MQTT IN and MQTT OUT. MQTT IN is used to publish the message and MQTT OUT is used to subscribe to the topic of the message. In this way, MQTT plays a vital role in Node-red. The first flow is all about the room temperature of each room at home. There are three nodes connected here (MQTT In, function and a gauge) as given in Fig. 3. This flow collects the data through the MQTT box by publishing a message as shown in Fig. 4 and shows the temperatures of four rooms on the node-red dashboard as shown in Fig. 5.

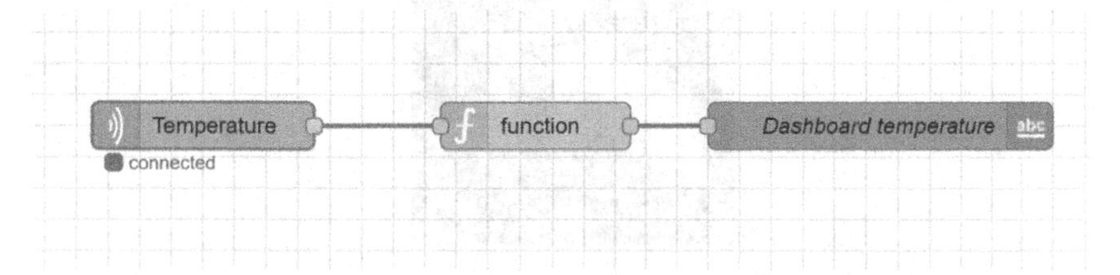

Fig. 3. The flow of temperatures of each room at Node-red.

The second flow consists of the activation of lights of every room at home. There are two types of nodes used at this flow (MQTT out and switches). The user can on or off the lights through the Node-red dashboard as shown in Fig. 6 and can check them in the MQTT box by subscribing to the topic as shown in.

The third flow consists of the activation of fans of every room at home. There are two types of nodes used at this flow (MQTT out and switches). The user can on or off the fans through the Node-red dashboard as shown in Fig. 7 and can check them in the MQTT box by subscribing to the topic.

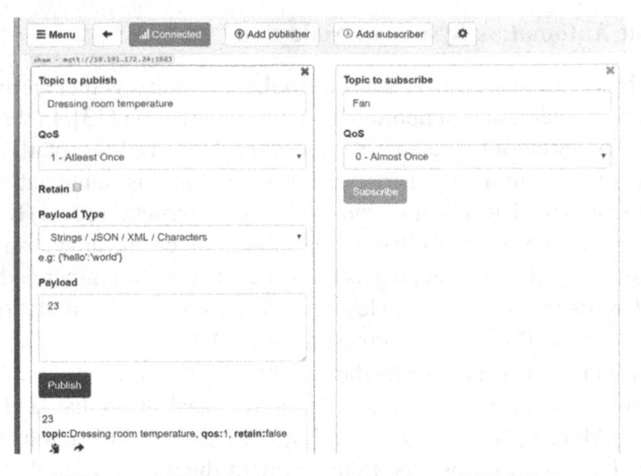

Fig. 4. Publishing the temperature in an MQTT box.

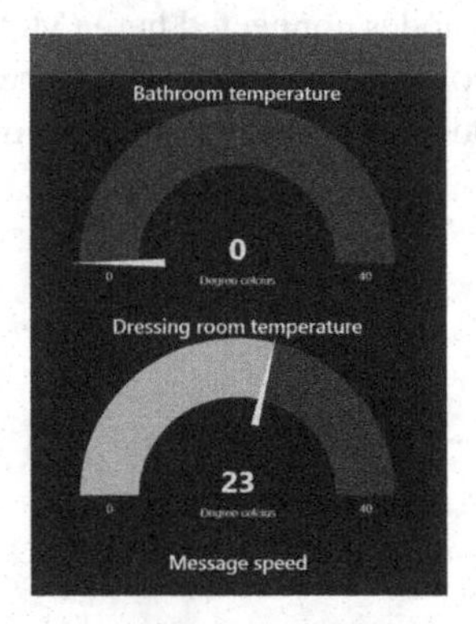

Fig. 5. Dashboard of temperature

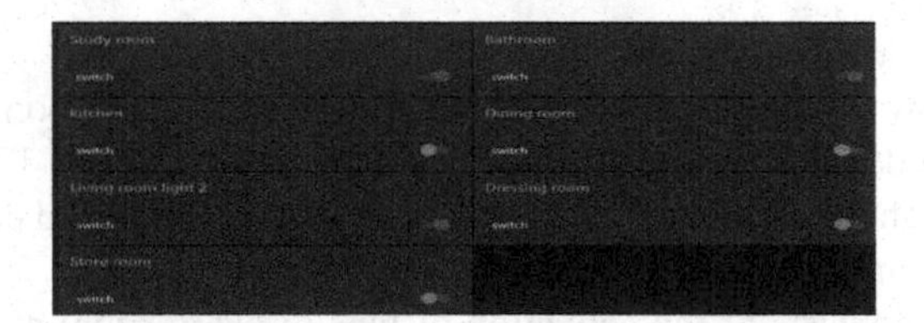

Fig. 6. Dashboard of lights of each room.

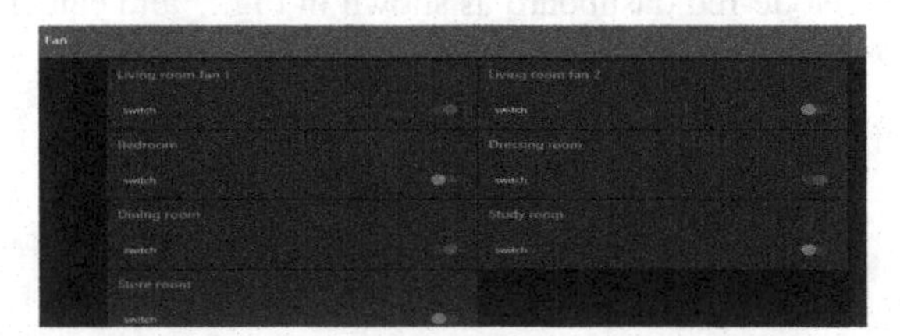

Fig. 7. The dashboard of fans belongs to each room.

The fourth flow is important and so useful because it consists of devices seven devices (Television, home theatre, air conditioner, washing machine, car shed gate opener, water tank level indicator and pet feed device).

This flow is made up of seven nodes (MQTT in, MQTT out, inject, slider, button, switch, and gauge). In this flow either the same way is brought up as the lights and fans are operated by publishing and subscribing in the MQTT box and the dashboard as shown in Figs. 8, 9 and 10.

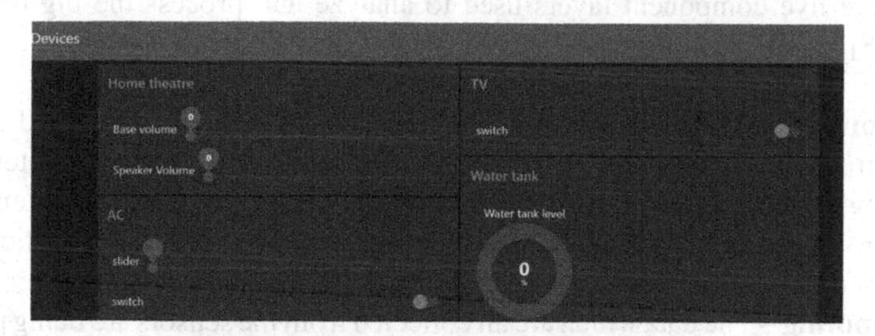

Fig. 8. Dashboard of home theatre, TV, AC and water tank.

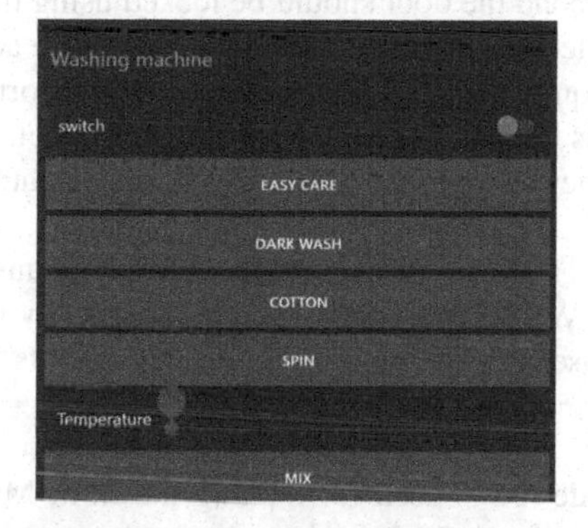

Fig. 9. Dashboard of the washing machine.

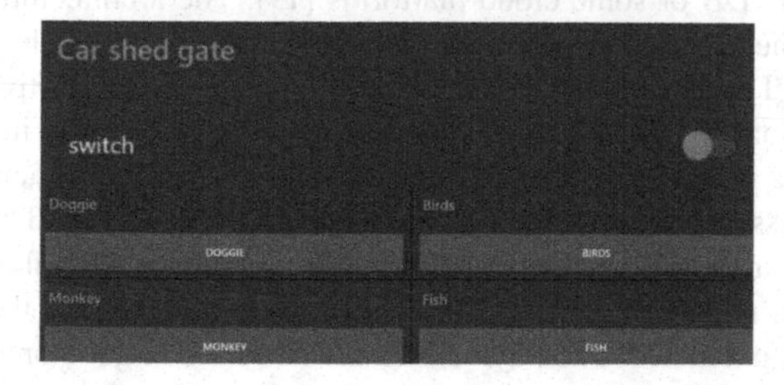

Fig. 10. Dashboard of car shed and pet foods.

2.3 Big Data in Home Automation

In-home automation system the data are collected from the sensors and performs the tasks by analyzing spontaneously. Here mostly non-instructive sensors are used like the motion sensors, fire alarms, pressure sensors, temperature sensors, etc...., by setting up the proper sensors and actuators the home devices have been given an intelligence through big data stream of activities, so by this way the home devices can communicate each other easily.

There are five component layers used to analyze and process the big data in IoT systems [18].

(i) **Sensors** – Sensors are the hardware component that is used to sense and collect the data from the environment or the user according to the given parameter, it's like getting the input from the user. In this home automation system at most temperature sensors, level indicators using ultrasonic sensors, fire alarms and motion sensors are used.
(ii) **Computing** – The data which are all collected from the sensors are being processed in this layer. Here machine learning, cloud and big data analysis are used. This process is responsible for spontaneous reactions, when we request to close the door, at the next second the door should be locked using the motion sensor.
(iii) **Collectors** – It collects the data from the user about the action to be taken and it will be sent as a request for the action needed to be performed to the actuators.
(iv) **Actuators** – This is the layer of triggers the device to act. The data from the user is collected, then the data is being analyzed and the actuators trigger the device to perform the task.
(v) **Device** (Client) – This is a so-called edge device because it performs the task instead of the user. Whatever works to be done physically without the guidance or knowledge of the user is being done by the device with its intelligence by big data analysis.

In this use case, Node-red is used for making the flow, MQTT broker is used to subscribe/publish the messages, Grafana is used to make analytics and monitoring for the database collected in the big data and at last, the data are being stored in some database like InfluxDB or some cloud platforms [19]. The architecture of this system either works in the same way, like there will be a message published or subscribed through the MQTT broker, then the flow from the Node-red gets activated, the Node-red dashboard visualizes the activities, then the data is being stored in the database at InfluxDB, the data is being transferred from the database to the Grafana, Grafana does the analysis process and the decision makings and at last the task will be done.

Whenever the user creates a flow and leaves from Node-red the flows will be there as it is till the user makes changes in it, even the dashboard remains the same with the values subscribed/published. The payloads in the MQTT box get stored automatically. These data from the flows and payloads are collected and stored in the data warehouse. It can handle the generated Big Data from residential units as well as scale up to more residential areas that can be added in future [16]. These data would also be needed in some way in some emergencies and used to analyze the data. In this case, we use big data analytics, an off-the-shelf business intelligence software tool was utilized to make smart decisions from the received big data [16]. This tool classifies the data according to its units or format, for example in this home automation system calculates the temperatures of each room, the water level of the water tank, washing machine temperatures and mode of washing, so there are varieties of data collected from different systems, this tool would sort the data according to its parameters. In this way, IoT is interconnected with big data here.

2.4 Measurement of Message Throughput and Message Speed Through Nodes

The main reason for taking measurements or statistical values is to evaluate the process and to analyze its efficiency. For an accurate value and perfect evaluation, the measurements should be taken for a long-time span. In this case study, there are two measurements taken for a total of five hours.

Message Transmission Speed. Message transmission speed defines the number of messages transferred for a particular time.

$$\text{Message transmission speed} = \text{No. of messages/Time (s)}$$

According to this formula, there is a reading taken using some nodes in node-red. A flow is designed with four nodes. The timestamp node is set to continuously send messages for a particular interval (0.0001 s). Function node is attached to activate a separate operation. There is a new node installed node-red-contrib-message-speed which is specially designed to calculate the speed of the message transmission. In the end, a chart is attached which has the X-axis values as the number of messages and the y-axis values as timing per second for a whole of five hours. These nodes are configured accordingly as shown in Fig. 11.

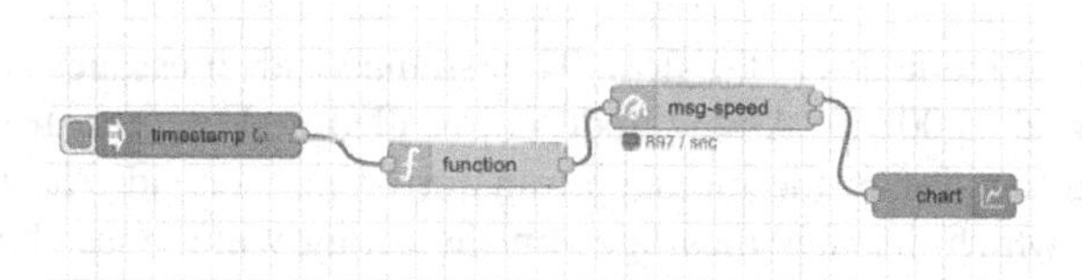

Fig. 11. Flow for calculation of message speed.

Here are the statistical values of message transmission speed. These three graphs show the reading of a sum of five hours as shown in Fig. 12. The peak value reached is 950 messages/s, 950 messages/s is the maximum value obtained in the reading taken in five hours and the average value is around 900 messages/s to 920 messages/s.

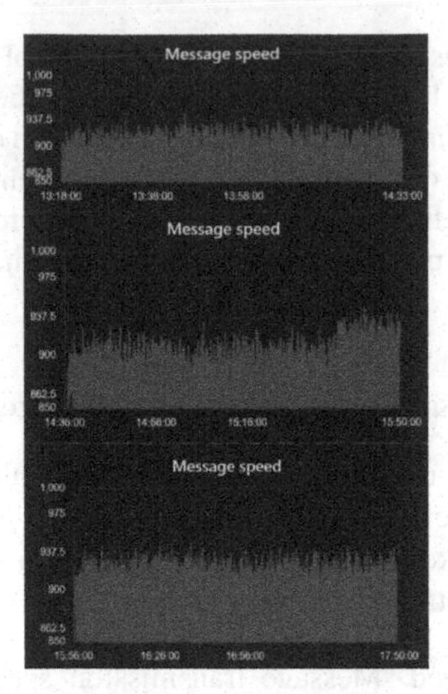

Fig. 12. 1.18 pm to 5.50 pm

Figure 12 charts of five-hour analysis of the message speed.

2.5 Throughput of the Message Transmission

Throughput of the message transmission defines the size of the message transferred at a particular time. This reading is taken for the minimum size that is for just activating the system like on or off the lights. The formula for throughput is

$$\text{Throughput of message transmission} = \frac{\text{Size of the message (bytes)}}{\text{Time (s)}}$$

Here there are three nodes in the flow. Timestamp sends messages continuously for a particular interval (0.001 s). A new node is installed node-red-contrib-message-size which calculated the size of the message transferred per second. At an end, a debug node is attached which displays messages on the debug window as shown in Fig. 13. The maximum throughput obtained is 73 bytes/s.

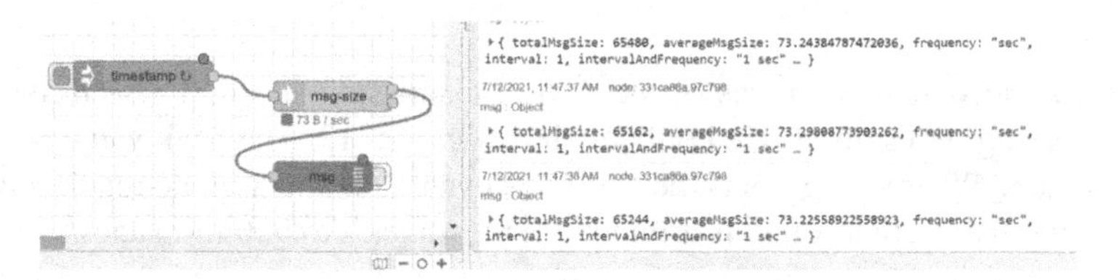

Fig. 13. The flow of calculating the message size.

3 Use Case 2: Vehicular Network

A vehicular network is a sensational technology because it is a big deal to build a vehicular network and it requires security and privacy. This technology is related to our life, if there is a single fault in this network, it may even cause death, so the network should be with 100% security. The decentralization process is used for spatial intelligence, a multi-tier multi-access edge computing framework that generates different levels of clusters for more efficient integration of different types of access technologies including licensed/unlicensed long-range low-throughput communications and unlicensed short-range large-throughput communications [15]. People think that vehicular network and automatic car without a driver is impossible, this makes the researchers do more research in a vehicular network. The next generation is built in a way that everything is automated like a smart city, to build a smart city vehicular network and vehicle monitoring is most important. More connected vehicles signify more resources to be utilized and better network performance when the cars are treated as the potential communication resources and networked for gathering IOTdata [14]. Here there are a huge amount of data collected from the vehicles and those data are collected and stored using big data and cloud technology. Hive MQ (MQTT broker) has already experimented with the system of the connected car with the features of the MQTT 5 version. There is no need for proprietary libraries in Hive MQ and can deploy the cloud platforms in AWS or Microsoft azure. In this use case, there is an inbuilt template for the simulations of devices on the AWS simulator. In that case, vehicular monitoring is created over here. There are two points explained here.

3.1 Connecting 100 Vehicles and Analysis of Statistics in the Dashboard in AWS Simulator

This simulation monitors the connected vehicles and their parameters like fuel level, speed, RPM etc..... In this AWS device simulator, it is very easy to create a simulation. First, in the AWS platform, an AWS cloud formation template should be downloaded to own an AWS device simulator account. This platform mainly uses MQTT for message transmission. In the device simulator, the automotive platform is used to create a vehicular network. The user can add vehicles over there and can monitor them all at a time. Here a sum of 100 vehicles is added as given in Fig. 14.

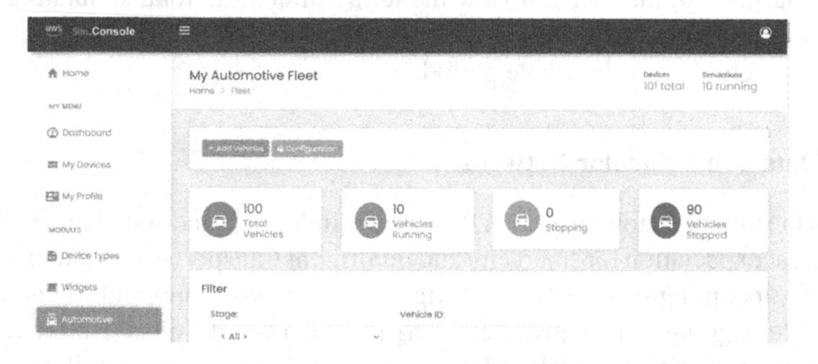

Fig. 14. Creating the vehicular network in AWS device simulator.

Each vehicle has its ID, and the user can modify the motion of the vehicle like start or stop and can delete the vehicle from the network. The timing and the date of the motion of the vehicle are noted as shown in Fig. 15.

Fig. 15. The activity of each vehicle.

The user can view the details of each vehicle. The parameters the user can measure and view on the dashboard are throttle, speed, RPM, oil, transmission torque, fuel level, odometer readings and transmission gear as shown in Fig. 16.

Fig. 16. Parameters measured in each vehicle.

On the dashboard, the user can view the total simulation, total simulation minutes, whether the simulation runs or not, device breakdown and the statistical view of the motion of the vehicles for the whole period.

3.2 Big Data in a Vehicular Network

Big data is the vital technology used in large-scale industries and social media. In future, there is a great expectation for smart city and intelligent transportation system, especially for these two upcoming systems IOT and Big data is the most important technology used. Every vehicle is going to be networked using the internet of things and the data or the parameters are collected and analyzed using big data. These data are collected from the sensors and devices and stored in the data warehouse. Big data collects the vehicle's information using GPS (Global positioning system).

There are few analyses taken from the GPS data set. At first, the analysis is based on the complexity of the network, where there is an analysis taken for the special mobility of the vehicles. For example, it is similar to checking the traffic of the city on the daily basis at each traffic area and having to analyze the complexity of the network. Next, the mobility model design analysis is done, under the inspiration of the e-commerce recommended system, we design some latest vehicular mobility models, which approximates actual data and is easy to explain [17]. The third analysis is about the time-variant, the vehicular network or the traffic will not be the same at every time, it's a dynamic network, so this analysis is taken to design the network based on time and the special characteristics. These are the three-analysis taken through the GPS data set collected in the big data.

Intelligence in the vehicle is about dealing with plenty of data from varieties of resources and there are four major sources of data acquired from the vehicle and the environment of the smart city. Those are,

(i) Vehicle sensing data – As mentioned in the above use case built in the AWS device simulator, the sensors fixed in the vehicles acquire the parameters like throttle, speed, RPM, oil, transmission torque, fuel level, odometer readings and transmission gear are calculated and represented in the dashboard and stored in the cloud platform or at the database.
(ii) GPS data – The GPS is the system used to track the vehicle's location and its movement. A separate dataset is collected from the GPS either. GPS data can be used for diversified goals, such as navigation, traffic management, communication routing optimization, vehicular content caching and sharing, etc. [20]. It shows the perfect altitude and the geographical location of the vehicle through satellite communication.
(iii) Self-driving-related data –The camera and LiDAR sensors collect data from the environment for the self-driving car. Here the distance of the vehicles around the specific vehicle, the traffic signals, 3D view of the environment around the vehicle and the climatic conditions are even sensed by the sensors for using the glass wipers in the vehicle.
(iv) Vehicular mobile service data - In-vehicle infotainment is becoming more crucial for improving the experience of both drivers and passengers. Mobile applications such as video/audio streaming, online gaming, social networks, and user-generated content (UGC) require or generate a huge amount of data [20].

These are the sources from where various types of data are collected from and next there is a process of storing and processing the data. The data is being collected from the vehicle and the environment around the vehicle, then the data is being transferred to the software platforms which are the very raw data. Next, the raw data is being processed, which means it filters or sorts the data according to the parameters and the specifications using big data analytics software, then the cleaning process will be done, the data are being sorted out from the whole processed packets of data. At last feature, extraction is done like collecting the exact human-understandable language at the user interface and stored as a dataset the extraction of vehicular user interests needs to establish a functional relationship between previously visited data and future actions, and thus further predict the selection similarities of infotainments for vehicular users and achieve social relation establishment [21].

4 Justifications to Prove MQTT is More Efficient than Other Protocols

MQTT is a standardized publish or subscribe protocol, which is light weighted and MQTT was planned to send data accurately under the long network delay and low bandwidth network condition [8]. Nowadays MQTT is vastly used in IoT networks for message transmission. It is easy to transfer the data and make connectivity to the devices in a fast, reliable and efficient manner through MQTT. It uses TCP as a transport protocol and TLS/SSL for security. Thus, communication between client and broker is connection-oriented [11]. Automation industries demand standard digitalization and connectivity. Consumers are also expecting a secured machine-to-machine communication system. These all could only be possible through the best and efficient protocol. Through focused research on MQTT protocol, justification is given here to prove MQTT is the best protocol for IoT and is already vastly used. So in this paper by using the MQTT protocol in applications, by checking its performance and by comparing the features of other protocols with MQTT.

4.1 Use Cases Basis

The first important thing noticed in the MQTT broker and AWS is security. A virtual server was created and accessed through the private key which is implemented in the security groups. To use the Mosquitto broker the user needs to have the private key file and the DNS IP address. Only with the username and password, the user could connect to the MQTT box. Even when the user needs to connect the MQTT broker with any other devices or tools the user needs to know the IP address, port number, username, and password. Nowadays security and privacy are the major concern for people. An Asian company Orvibo was easily accessible through HTTP, over 3 billion records have been ceased like the account details, passwords, etc. [1], so here while setting up the whole simulation the security is given more importance. Therefore, the MQTT protocol is highly secured.

In the home automation system, there are a lot of messages transferred at a time and it even needs a spontaneous response for the actions requested. So, in this case, study there was a testing process done by activating every node in this system through subscribing and publishing the message, calculations are either made for checking the speed and the throughput for five hours. The maximum speed was 950 messages per second and the average was around 900 to 920 messages/s. The throughput is different according to the type of messages like it differs for the normal signals, text messages and videos. In this study, according to the normal messages the throughput was 73 bytes/s. The speed and the throughput are higher in the MQTT protocol than the other protocols. If the buttons or switches are activated in the MQTT box within a fraction of a second it will be activated in the Node-red. The efficiency was also high while handling the Node-red home automation system.

The internet of vehicles is quite a risky area where a large amount of data is collected through a lot of sensors and its security concerns will be the main issue [1], but it will only be possible through the MQTT protocol. The automotive in the AWS device simulator would be very useful for the travel services like Uber, Ola etc. which tracks every detail of each vehicle? In this system for every second, the vehicles should be tracked, and their parameters are sent to the server. This system uses MQTT for the message transmission because it is light weighted and only then it will be easy to receive a lot of messages with a short period. Collection of huge amounts of data at a time with security is only possible in MQTT protocol.

4.2 Comparative Analysis of MQTT, CoAP and HTTP

4.2.1 MQTT

Message queue telemetry transport was implemented to publish/subscribe messages between the client and the broker. Whenever they say MQTT, the first thing that sparks in mind is a lightweight protocol. The reasons why MQTT is lightweight are it has a lean header structure so; it reduces packet parsing, which is suitable for IoT, and the other reason is it is easy to implement [4]. Advantages of MQTT:

(i) Extremely lightweight overhead.
(ii) MQTT provides reliability by supporting QoS.
(iii) Increase scalability.
(iv) Consumes low energy.

Disadvantages of MQTT

(i) All nodes are connected to the broker, there are chances for the messages to collide.
(ii) It uses TCP/IP, so it requires more communication capabilities.

4.2.2 CoAP

A constrained application protocol is a transfer protocol. It is a specialized internet application protocol for RFC 7252. The communication between the server to the client is a peer. It is a server layer protocol used for wireless networks, CoAP can run on devices that support UDP and/or UDP analogues. Advantages of CoAP:

(i) Operates fast communications by sending small packets [3].
(ii) Long battery life for IoT.
(iii) It provides IPSEC for security.

Disadvantages of CoAP

(i) CoAP is unreliable because of UDP.
(ii) It acknowledges every receipt of the message, so it takes more process time.
(iii) It is said as an unstandardized protocol among other protocols.

4.2.3 HTTP

Hypertext transfer protocol is an application layer protocol. It is used to transmit hypermedia documents like HTML. This is mainly helpful for the communication between web browsers and web servers. HTTP must transfer a large number of tiny packets and the protocol overhead of HTTP causes serious issues, so HTTP is not much suitable for IoT [7]. Advantages of HTTP

(i) HTTP uses an advanced scheme of addressing.
(ii) It can download extensions if it needs additional capability.
(iii) The handshaking process is held only if the connection is enabled.

Disadvantages of HTTP

(i) There is no encryption process.
(ii) It uses a greater number of system resources for working with IoT.
(iii) Administrative overhead in the communication.

5 Features of MQTT

5.1 Security

There are a lot of factors that mainly focus on security. Authentication, authorization, TLS/SSL, OAuth 2.0, payload encryption and message data integrity [2]. In the authentication process, the username and password will be used for authentication. In the process of authorization, the broker will control the topic to be subscribed or published. Encryption of the data is handled through the TLS/SSL. X.509 certificate is provided for the client, it's like the private key for authentication, so only with that certificate, the client can access the server. The payloads will be encrypted and throughout the transmission, they cannot be decrypted, only at the other end, they could be decrypted. Digital signatures are used to publish/subscribe to the packets. These are all the processes used to secure the data in the MQTT protocol.

5.2 QoS

MQTT provides a quality of service that comes under the filtering process of the topics. In this process, there are three levels of messages. In level 0 the message is delivered at most once or not at all delivered, level 1 messages are sent at least once, and level two is to verify the message is sent exactly at least [9].

5.3 Last Will Message

Each client specifies a last-will message to the broker and that message will be stored by the broker until the client has disconnected gracefully.

6 Conclusion

This paper majorly focused on "how MQTT is used in use cases?", "why MQTT is dominant among other protocols in IoT?" and "How Big data is interconnected with IoT". With the help of a home automation system in Node-red and vehicular network in AWS. The efficiency of the MQTT protocol is tested, either the speed or throughput is calculated for message transmission. In the end, the comparison among protocols and features of the MQTT protocol proved that MQTT is the best protocol IoT. Even though the whole system is made and networked with IoT, the data are stored and analyzed using big data, so big data will always be necessary when we use IoT. This paper will be useful for the upcoming researchers who research MQTT protocol and IoT.

References

1. Malina, L., et al.: Post-quantum era privacy protection for intelligent infrastructures. IEEE Access. **24**(9), 36038–36077 (2021)
2. Malina, L., Srivastava, G., Dzurenda, P., Hajny, J., Fujdiak, R.: A secure publish/subscribe protocol for internet of things. In: Proceedings of the 14th International Conference on Availability, Reliability, and Security, 26 Aug 2019, pp. 1–10 (2019)
3. Çorak, B.H., Okay, F.Y., Güzel, M., Murt, Ş., Ozdemir, S.: Comparative analysis of IoT communication protocols. In: 2018 International Symposium on Networks, Computers, and Communications (ISNCC), 19 Jun 2018, pp. 1–6. IEEE (2019)
4. Gündoğan, C., Kietzmann, P., Lenders, M., Petersen, H., Schmidt, T.C., Wählisch, M.: NDN, CoAP, and MQTT: a comparative measurement study in the IoT. In: Proceedings of the 5th ACM Conference on Information-Centric Networking, 21 Sept 2018, pp. 159–171 (2018)
5. Asghar, M.H., Mohammadzadeh, N.: Design and simulation of energy efficiency in node based on MQTT protocol in Internet of Things. In: 2015 International Conference on Green Computing and Internet of Things (ICGCIoT), 8 Oct 2015, pp. 1413–1417. IEEE (2015)
6. Kodali, R.K., Anjum, A.: IoT based home automation using node red. In: 2018 Second International Conference on Green Computing and Internet of Things (ICGCIoT), 16 Aug 2018. pp. 386–390. IEEE (2018)
7. Yokotani, T., Sasaki, Y.: Comparison with HTTP and MQTT on required network resources for IoT. In: 2016 International Conference on Control, Electronics, Renewable Energy, and Communications (ICCEREC), 13 Sept 2016, pp. 1–6. IEEE (2016)
8. Soni, D., Makwana, A.: A survey on MQTT: a protocol of internet of things (IoT). In: International Conference on Telecommunication, Power Analysis and Computing Techniques (ICTPACT-2017), Apr 2017, vol. 20 (2017)
9. Atmoko, R.A., Riantini, R., Hasin, M.K.: IoT real time data acquisition using MQTT protocol. J. Phys. Conf. Ser. **853**(1), 012003 (2017). IOP Publishing
10. Lekić, M., Gardašević, G.: IoT sensor integration to Node-RED platform. In: 2018 17th International Symposium INFOTEH-JAHORINA (INFOTEH), 21 Mar 2018, pp. 1–5. IEEE (2018)
11. Naik, N.: Choice of effective messaging protocols for IoT systems: MQTT, CoAP, AMQP and HTTP. In: 2017 IEEE International Systems Engineering Symposium (ISSE), 11 Oct 2017, pp. 1–7. IEEE (2017)
12. Mahmud, S., Ahmed, S., Shikder, K.: A smart home automation and metering system using internet of things (IoT). In: 2019 International Conference on Robotics, Electrical and Signal Processing Techniques (ICREST), 10 Jan 2019, pp. 451–454. IEEE (2019)

13. Kodali, R.K., Rajanarayanan, S.C., Boppana, L., Sharma, S., Kumar, A.: Low-cost smart home automation system using smart phone. In: 2019 IEEE R10 Humanitarian Technology Conference (R10-HTC) (47129), 12 Nov 2019, pp. 120–125. IEEE (2019)
14. Li, H., Liu, Y., Qin, Z., Rong, H., Liu, Q.: A large-scale urban vehicular network framework for IoT in smart cities. IEEE Access **28**(7), 74437–74449 (2019)
15. Wu, C., Liu, Z., Zhang, D., Yoshinaga, T., Ji, Y.: Spatial intelligence toward trustworthy vehicular IoT. IEEE Commun. Mag. **56**(10), 22–27 (2018)
16. Al-Ali, A.R., Zualkernan, I.A., Rashid, M., Gupta, R., AliKarar, M.: A smart home energy management system using IoT and big data analytics approach. IEEE Trans. Consum. Electron. **63**(4), 426–434 (2017)
17. Sun, R., Ye, J., Tang, K., Zhang, K., Zhang, X., Ren, Y.: Big data aided vehicular network feature analysis and mobility model's design. Mob. Netw. Appl. **23**(6), 1487–1495 (2018)
18. Asaithambi, S.P., Venkatraman, S., Venkatraman, R.: Big data and personalisation for non-intrusive smart home automation. Big Data Cognit. Comput. **5**(1), 6 (2021)
19. Nițulescu, I.V., Korodi, A.: Supervisory control and data acquisition approach in node-RED: application and discussions. IoT **1**(1), 76–91 (2020)
20. Cheng, N., et al.: Big data driven vehicular networks. IEEE Netw. **32**(6), 160–167 (2018)
21. Xu, C., Zhou, Z.: Vehicular content delivery: a big data perspective. IEEE Wirel. Commun. **25**(1), 90–97 (2018)

Large-Scale Contact Tracing, Hotspot Detection, and Safe Route Recommendation

Chandresh Kumar Maurya(✉), Seemandhar Jain, and Vishal Thakre

IIT Indore, Indore, India
cse180001062@iiti.ac.in

Abstract. Recently, the COVID-19 pandemic created a worldwide emergency as it is estimated that such a large number of infections are due to humanto-human transmission of the COVID-19. As a necessity, there is a need to track users who came in contact with users having travel history, asymptomatic and not yet symptomatic, but they can be in the future. To solve this problem, the present work proposes a solution for contact tracing based on assisted GPS and cloud computing technologies. An application is developed to collect each user's assisted GPS coordinates once all the users install this application. This application periodically sends assisted GPS data (coordinates) to the cloud. To determine which devices are within the permissible limit of 5 m (tunable parameter), we perform clustering over assisted GPS coordinates and track the clusters for about t mins (tunable parameter) to allow the measure of spread. We assume that it takes around 3–5 mins to get the virus from an infected object. For clustering, the proposed M-way like tree data structure stores the assisted GPS coordinates in degree, minute, and second (DMS) format. Thus, every user is mapped to a leaf node of the tree. The crux of the solution lies at the leaf node. We split the "seconds" part of the assisted GPS location into m equal parts (a tunable parameter), which amount to d meter in latitude/longitude. Hence, two users who are within d meter range will map to the same leaf node. Thus, by mapping assisted GPS locations every t mins (usually t = 2:5 mins), we can find out how many users came in contact with a particular user for at least t mins. Our work's salient feature is that it runs in linear time O(n) for n users in the static case, i.e., when users are not moving. We also propose a variant of our solution to handle the dynamic case, that is, when users are moving. Besides, the proposed solution offers potential hotspot detection and safe-route recommendation as an additional feature, and proof-of-concept is presented through experiments on simulated data of 2/4/6/8/10M users.

Keywords: COVID-19 · Contact-tracing · M-way like tree · Clustering · Assisted GPS

1 Introduction

Contact tracing is the problem of identifying users who have come in contact with a user within a certain distance for a specific amount of time. This problem came

S. N. Srirama et al. (Eds.): BDA 2021, LNCS 13147, pp. 163–182, 2021.
https://doi.org/10.1007/978-3-030-93620-4_13

into the limelight during the COVID-19 pandemic, which is thought to spread from humans to humans. The main problem is that users are asymptomatic and can pass the virus to other users unknowingly for weeks or months. In such a case, it is challenging to identify users who came in contact with a particular user, given the users dynamic and complex movement patterns. Another prevalent problem during the pandemic situation is that users are interested in knowing the hotspot areas to avoid them while visiting. Therefore, there is a growing need for a recommendation of a safe route for travel.

Current solutions for contact tracing problems can be categorized into (a) human-based approach, (b) Bluetooth (BT)-based approach, and (c) GPS-based approach. In the first approach, police personnel are deployed to follow the user's movement trail and find out users who were present in close proximity of the particular user for a specific duration. This approach is *costly, cumbersome, and time-consuming* [11]. In the second approach, each user is asked to install an app such as India's Aarogya Setu app or Google-Apple privacy-preserving app. Such apps use BT signals for exchanging data between devices. There are certain limitations of these apps. For example, the limitations of Aarogya Setu are: (a) it does not tell if we were in contact with a user a week or month ago who recently got diagnosed with COVID-19, (b) it also does not tell which areas are hotspots and should be avoided, (c) it is based on self-assessment only, and hence reliability is a concern.

The solutions for hotspot detection are currently based on the government's rules and regulations. For example, sites like COVID tracker[1] mark geographical areas as hotspot-based on the directions received from the government. They do not provide real-time hotspot information, such as alerting the user that they are moving through hotspot/containment zones specially in villages where monitoring is difficult. Further, the existing solutions lack the feature of recommending safe routes for travel, such as Google map.

To overcome the limitations of the existing solutions, we propose an assisted GPS location-based solution that receives the assisted GPS coordinates of the user in the backend and without forcing the user to keep on Bluetooth time and again. This solution does not require any user interventions and sends the assisted GPS coordinates to the cloud periodically. Received coordinates are mapped to the proposed **M-way like tree data structure** in degree, minute, and second (DMS) format. Once the mapping is over, the proposed clustering algorithm executes to track the history of users who spent at least t minutes with other users. **Our solution is scalable and can find users who happen to make a contact event in 17 mins among a population of 10M users.**

Major differences between our solution based on assisted GPS and BT-based solution for contact tracing are: (i) our approach can provide location information where the user got **potentially infected** whereas BT-based approaches fail to do so since they do not provide location information. Why this information may be crucial is that let us suppose that the user visits a dairy milk shop every day besides tons of other places. On one day (s)he got COVID +ve.

[1] https://www.COVIDhotspots.in/?city=Delhi.

In this case, (s)he might be interested in knowing where (s)he got the virus so that they can inform their family members to be vigilant. In this scenario, the BT-based approach has no clue of the location, (ii) assisted GPS-based solution is centralized where BT-based one is decentralized, (iii) our solution is more robust in the sense that BT-based solution communicates with **many heterogeneous devices**. In contrast, ours does not, (iv) BT always needs to be turned on, which usually user forget or do not turn on for power saving purposes. As for assisted GPS is concerned, once the app is installed, it will keep using the location sharing in the background and does not ask the user to turn it on time and again. This option is not available in BT-based solutions currently, (v) the scalability of BT-based apps is also a concern since these apps can communicate to only eight other BT devices at a time. Further, note that we are using *assisted* GPS which is more reliable than GPS, where location information is calculated using satellite signals, mobile towers, and wifi beacons. Therefore, our solution will work in indoor settings as well. From now on, whenever GPS is mentioned will mean assisted GPS. To emphasize to our readers that our solution offers an alternative to BT-based solutions and does not replace them.

In short, we make the following contributions:

- Propose a solution for contact tracing that runs in the time linear to the numbers of users i.e. $O(n)$. the space complexity of the solution is also linear in the number of user identifiers, i.e., $O(n)$.
- Our solution can handle static as well as dynamic case. That means if users stay at the same location (static case) or the users are moving together (dynamic case).
- The proposed solution relies only on location sharing information, which is the property of most android applications such as Google news, maps, etc.
- Our solution can additionally provide hotspot information (areas of large gatherings) and recommend a safe route for travel.

2 Related Works

Contact tracing and regular monitoring of infectious diseases (COVID-19) are essential for public health. Utilizing emerging technologies such as remote sensing, internet-based surveillance, telecommunications, infectious disease modeling, Global Positioning System (GPS), IoT devices, and mobile phones, such contagious diseases can be monitored, prevented (by generating alarms after real-time predictions), and controlled [2].

Few research works [1,12] have introduced the stochastic model used to reduce a deterministic approach for attaining the fundamental dynamics of the epidemics and other associated measures. Mostly, previous models deal only with generic networks. To measure the precision of the trace contact models, there is a need to account for the network of contacts. For example, Huerta and Tsimring [8] present a stochastic model to estimate the effect of random screening and contact tracing as part of the epidemic control strategy in complex networks. This work indicates that a significant outbreak can significantly be reduced via tracing contacts at a low additional cost. A similar approach is also used by Farrahi et al. [5].

Ferretti et al. [6] state that isolation and contact tracing are being practiced but not preventing the COVID-19, which is why the same high number of asymptomatic infected individuals remain undetected, and cases continue to spread. Therefore, the authors propose mobile apps to trace previous mobile contacts and show mathematically that epidemics can remain even when not all population use the application. Hellewell et al. [7] also propose similar inference through the simulated model. In most scenarios, highly effective case isolation and contact tracing are sufficient to control a new outbreak of COVID-19 within three months; however, only 79% of the contacts are traced. Nevertheless, such conditions create smartphone-based tracing, which is far from a realistic solution. One of the first efforts through mobile phones to trace the contacts was the FluPhone application developed by Cambridge University [18] which uses Bluetooth-based wireless signals to estimate the physical contact. It also asks the users to report on flu-like symptoms to appraise the risk of infections. Next, the Singapore Government also developed a mobile app called Trace-Together for COVID-19[2]. This also uses Bluetooth technology, and it had already been utilized to control the disease spread. Few similar works also focused on privacy as the Pan-European Privacy-Preserving Proximity Tracing (PEPP-PT) [15]. Finally, Google and Apple have teamed up to design, develop and integrate a similar approach into the Android operating Systems and iOS. The proposed solution is more efficient and ubiquitous to users. Aarogya Setu app[3] from the Indian government works on location sharing and Bluetooth. Secondly, it tells us that we are in contact with a COVID +ve person based on Bluetooth proximity. There are certain limitations of Aarogya Setu. For example, it does not tell if we were in contact with a person a week or month ago who recently got diagnosed with COVID-19. It also does not reveal which areas are hotspots and should be avoided. It is based on self-assessment only, and hence reliability is a concern. Recently, Mahapatra et al. [10] present a digital contact tracing solution based on a dynamic graph streaming algorithm. Concretely, they use an index-based adjacency list to store graphs whose nodes are users and edges are close contacts. However, they do not mention if they use Bluetooth or GPS to find direct and indirect close contacts. The proposed solution runs in $O(q^L|I|)$ where $q, L, |I|$ denote the number of contacts, level of tracing, and the number of infected users, respectively. All Bluetooth (even GPS) based apps for contact tracing suffer from a severe drawback: they may be subjected to high "false alarm". For example, when two users are standing opposite side of a wall, contact tracing apps will signal that they are within the infection-proximity [16]. There have been reports of delayed notification (1+ days) to individuals regarding contacts with COVID +ve patients. Such delays can create depression in users later on. Further, many of these apps do not show hotspot areas nor recommend a safe route. To ameliorate/minimize *some of the above problems* and provide additional features, our

[2] https://www.tracetogether.gov.sg/.

[3] https://www.mygov.in/aarogya-Setu-app/.

solution relies on location sharing via GPS/assisted GPS such as magnetic fingerprints and the BLE beacon RSSI (Received Signal Strength Identifier) nearby, for localization [14] in an indoor setting. Unlike existing solutions, our approach provides *potential hotspot* information as well as *safe route* recommendation which is presented in Sects. 4 and 5 respectively.

3 Contact Tracing

As discussed in the introduction section, our solution for contact tracing relies on the location sharing data, i.e., GPS location of the users collected through our app. The naive solution will be to compute the Euclidean distance (or *Haversine distance* as shown in (1), where ϕ_i and λ_i for $i \in \{1, 2\}$ are lat/long of two users, if one wants to be more exact) every t mins for finding the contact event.

$$d = 2r \arcsin\left(\sqrt{\sin^2\left(\frac{\phi_2 - \phi_1}{2}\right) + \cos(\phi_1)\cos(\phi_2)\sin^2\left(\frac{\lambda_2 - \lambda_1}{2}\right)}\right) \tag{1}$$

A contact event is defined as when two users are within a distance of d meter for at least t mins. The naive solution for contact tracing requires computing pairwise distances and has the complexity of $O(n^2)$ for n users and hence not feasible for large n. Therefore, we discuss a data structure which is M-way *like* tree data structure for storing GPS locations (latitude/longitude) as shown in Fig. 1. In other words, we map latitude/longitudes to the proposed tree as discussed later in details. Further, latitude and longitudes are mapped to two separate trees for parallel computation of the contact distances. How do we arrive at the direct distance between two users from latitude and longitude distances is presented in detail in Sect. 3.2. Note that the M-way like tree is not the same as k-d tree which is a binary tree and requires comparison for insertion. Whereas M-way like tree involves look-up operation and no comparison. In this spirit, it is a kind of multi-level hashing.

A GPS coordinate that consists of latitude and longitude is usually expressed in the degree, minute, and second format (DMS), e.g., (28°50′30.12462"). For latitude and longitude, degree ranges in $[0, 1, \ldots, 89]$ and $[0, 1, \ldots, 179]$ respectively while minutes and seconds vary in $[0, 1, \ldots, 59]$ for both which are shown in the top 3 gray levels in Fig. 1. For notational brevity, we do not show degree (°), minute (') and seconds (") symbols in the fig. The last gray level is the partition of the seconds into m equal parts. As the most interesting property of M-way like tree data structure is the partition at the leaf, i.e., partition every *second* of GPS (which is equal to approx. 30 m) to m equal parts so that $m = 30/d$. Here, d is our contact distance which is a tunable parameter ($d = 5$ in our case). The benefit of the partition is that all users within a distance of d meters on latitude will fall into the same leaf (bucket) and form natural clusters. For example, consider $d = 5$ and hence $m = 6$, then the intervals for the fractional part of the seconds (which varies in [0,1]) at the leaf node will be $[0, 1/6), [1/6, 2/6), \ldots, [5/6, 1]$. Users u_1 and u_2 whose latitudes are (28°50′30.12462") and (28°50′30.05462") respectively will map to the same leaf node because their fractional part of

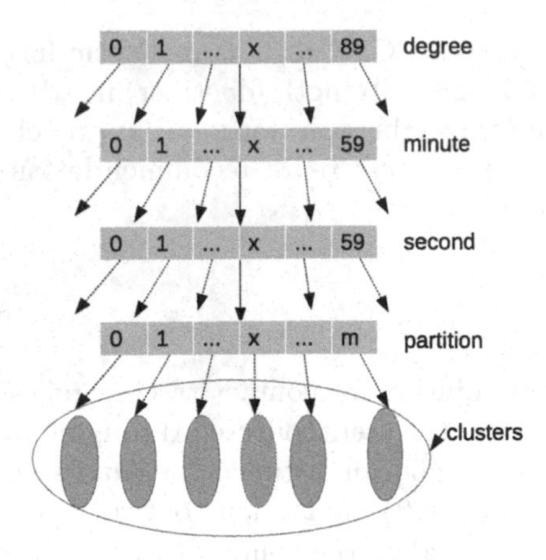

Fig. 1. M-way tree like data structure for storing coordinates

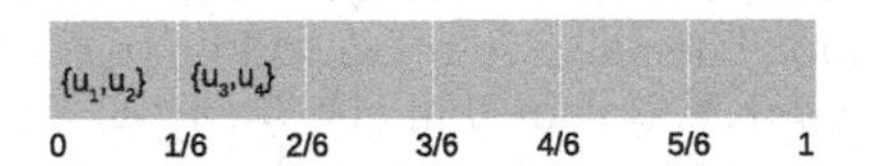

Fig. 2. Example to handle cases when users in the neighboring clusters may be within contact event distance d and not captured during mapping of lat/long to the M-way like tree.

the second (.12462 and .05462) will map to the same leaf and thus are in the same cluster. The approach presented above misses users who are within contact event distance d but fall in the neighboring clusters. For example, users u_1 and u_3 whose latitudes are (28°50′30.12462") and (28°50′30.17462") will map to different clusters but are within the contact event distance $d = 5m$. An example of such a case is shown in Fig. 2 with four users. Out of these four users, two fall in one cluster, and the other two fall in a different neighboring cluster. To capture missed cases during the mapping of lat/long to the tree, we perform pairwise distance computation. For example, we need 4 distances computed ($\{u_1, u_3\}, \{u_1, u_4\}, \{u_2, u_3\}, \{u_2, u_4\}$) in the worst case (worst case occurs when half of the users fall in one cluster and the other half fall in the neighboring cluster). Had we not mapped users to the tree, $\binom{n}{2}$ computations are required with quadratic complexity $O(n^2)$ and not preferred for large n. To see this in Fig. 2, instead of 6 pairwise distance computations in our running example, we needed only 4 pairwise distance computations in the worst case and hence saving two distance computations. Such a saving becomes paramount for large n.

We collect the GPS locations using our in-house app and send it to the cloud every $t/2$ mins for tracking users at least for t mins ($t = 5$ mins in our experiments). That means we map GPS locations to M-way like tree structure every $t/2$ mins. We have two trees for every interval of $t/2$ mins: one at time step $t/2$ and the other at t. At this step, we perform the intersection of leaves, and the ID of the users found in the intersection is stored. This method guarantees that two users in the same cluster have been in contact for at least $t/2$ mins. It also ensures that it will capture all users who have spent at least t mins together within d meter distance. One limitation is that it might miss users who have been in contact for 3 or 4 mins. Since our goal is to find all users who came in contact for at least 5 mins or more, we can ignore such corner cases.

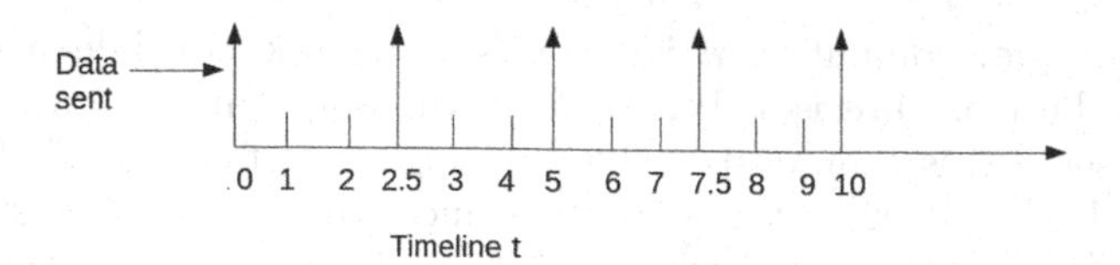

Fig. 3. Timeline to show when to send GPS coordinates to the cloud

3.1 Intuition Behind $t/2$ Mins

It is not necessary to send GPS coordinates every $t/2$ mins. However, to guarantee that two users make contact for at least t mins for the infection to occur, it is necessary to send GPS data every $t/2$ mins as shown in Fig. 3 for $t = 5$ mins. We can see in the Fig. 3 that if two users met at time $t_1 = 1$ and remain in contact until $t_2 = 6$ ($\Delta t = t_2 - t_1$) then we can see that we have sent their data two times (at $t = 2.5$ and $t = 5$ mins). As a result, we can capture the contact event. On the other hand, if we are sending data every 5 mins, it is not possible to capture the aforementioned contact event. This approach guarantees that all users whose contact event duration is at least 5 mins will be definitely captured. However, it may miss contact events of duration 3 or 4 mins.

3.2 How Lat/long Distances Map to Circular d m?

We want to compute the distance $d = AB$ as shown in the Fig. 4. That is, all users at a distance d from user A. For small distances ($d < 10$ or 15 m in our case), we can safely assume that the points A and B lie on Euclidean coordinate system. Therefore, we can draw a simple circle at A whose radius is d (otherwise, it is **great circle**).

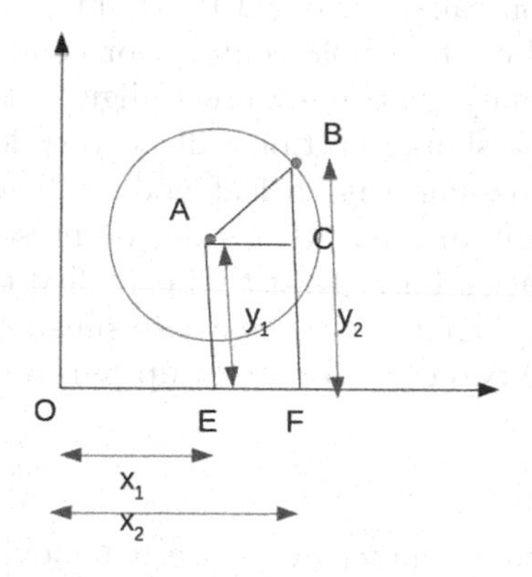

Fig. 4. Mapping lat/long to circular distances.

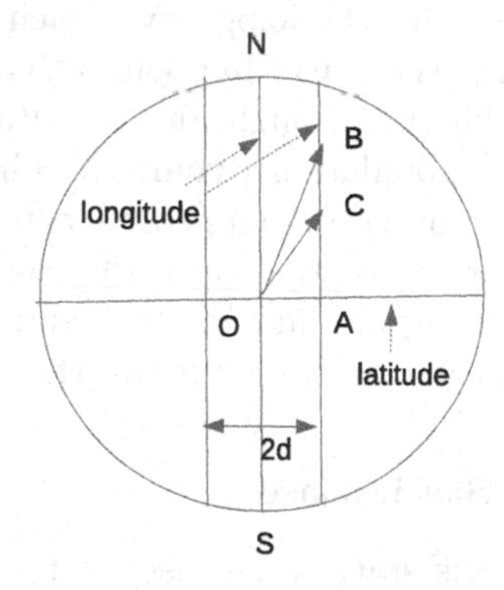

Fig. 5. Figure showing the case when many users are within contact distance d along latitude but can be for away along longitude.

Main Challenge: computing d is easy (we can take Euclidean distance or Haversine to be more precise). But, Euclidean computing distance for a large number of users (n) is computationally challenging. It takes $O(n^2)$ for pairwise distances. When n = 1M, it takes 10^{12} distances to compute and storing them requires $O(n^2)$ space. Thus, pairwise Euclidean distance computation is impractical since we want a scalable solution where n can be as large as 1.3B. Therefore, our main idea is to avoid computing d directly and instead compute AC and BC distances, which are lat/long distances between A and B.

Pythagorean theorem says that $AB^2 = AC^2 + BC^2$. Thus, if we can compute lat/long distances, we can find d. For $d = 5$ m. We can find either (lat, long) = (3 m, 4 m) or (4 m, 3 m). That means we find all users who are 3 m away from user in question on latitude and 4 m on longitude or vice-versa. Choosing (lat, long) = (3 m, 4 m) gives a more accurate solution since computing distance along longitude depends on the accuracy of the distance along latitude [17]. Further, notice that we map latitude/longitude of a user to two separate M-way like trees. That means we create a tree for mapping latitude and a tree for longitude. Such trees are created every $t/2$ mins, and users making contact events are found by the intersection of leaf nodes. More details of the procedure is described in the next section. Approximating diagonal distance via distances along latitude/longitude does not always hold. In the previous example, when $(lat, long) = (3m, 4m)$, we are finding users who are within 3 m (4 m) along latitude (longitude) and clustering them through mapping their latitude (longitude) on the M-way like tree data structure. However, the catch here is that there is a possibility that users within the contact distance along latitude but far away along longitude (or vice versa) are mapped to the same leaf node. As shown in Fig. 5, users at locations B and C will fall in the same leaf node when we map longitudes to the tree; however, locations B and C can be miles away along latitude (parallel lines passing through B and C but not shown in the figure). This can degenerate into a case when many users are mapped to the same leaf node of either tree. Such a situation can be avoided by running our proposed methodology over each city instead of the whole country for contact tracing. What will happen is that the probability that many users align to the same latitude (longitude) simultaneously for a short duration will be very low due to population dynamics. Therefore, we assume that a leaf node will not have many users mapped to it in the worst case. In the next section, we present solutions when the users stay at the same location for at least t mins (called the static case) and moving (called dynamics case). Though dynamic case subsumes the static case, we bifurcate the mapping into two cases for speed-up purposes.

3.3 Static Case

A case is static when users who participate in a contact event are not moving at least for t mins. A static case can be easily handled by mapping lat/long to the M-way like tree every $t/2$ mins. The intuition is that two users who stay within a distance of d m for at least t mins will map to the same leaf node of the M-way like tree. We then find the intersection of the two M-way like trees so obtained (intersection of two corresponding leaves to be more precise since users are static). Users in the intersection will be our output candidates (see Fig. 6 where leaves from two trees are shown facing each other).

Illustration of the static case: In the Fig. 6, lat/longs are mapped to two trees as discussed before. The user u_1, u_2 and u_3 were at a location at time t (top leaf node). After $t + \Delta t$ ($\Delta t = 2.5$ mins), users u_1, u_3 and u_3 were found at the same location resulting in the leaf node shown in the bottom. When the intersection is performed, u_1, u_2, and u_3 are found to be present at the same location. That means, user u_1, u_2 and u_3 participated in a contact-event. The entire process of the static case is described step by step in Algorithm 1.

Algorithm 1: Static Case. [] denotes indexing in an array/dictionary

```
Input: T_t - M-way tree at time t, T_{t+1} - M-way tree at time t + Δt
Output: CT - Contact Tracing Vector
1  for d in degrees do
2      for m in minutes do
3          for s in seconds do
4              for p in partitions do
5                  ct = T_{t+1}[d][m][s][p] ∩ T_t[d][m][s][p]
6                  Insert ct in CT
7                  for n in [p − 1, p + 1] do
8                      if T_t[d][m][s][n] and T_t[d][m][s][p] has peoples who are less
                       than 5mt apart then
9                          end_case_t = findpairs( T_t[d][m][s][n], T_t[d][m][s][p])
10                     if T_{t+1}[d][m][s][n] and T_{t+1}[d][m][s][p] has peoples who are
                       less than 5mt apart then
11                         end_case_{t+1} = findpairs(
                           T_{t+1}[d][m][s][n], T_{t+1}[d][m][s][p])
12                 Insert end_case_t ∩ end_case_{t+1} in CT
13 return CT
```

3.4 Dynamic Case

The dynamic case is when users move together in close proximity (within a radius of d m). Due to the movement, their lat/log are continuously changing, resulting in their mapping to two different (non-corresponding) leaf nodes in the M-way like tree. To find out where users in close contact map to the leaf require performing the intersection of each leaf node against all nodes in the latter tree. This is computationally intractable, requiring $O(n^2)$ intersection. Therefore, we solve this issue via an approximation. First, we discuss when users are walking on foot (movement in a car can be handled similarly). On avg, a pedestrian speed is 1.4 m/s. That means, in 5 mins, they walk around 420 m which maps to 14" in lat/long. Therefore, we perform the intersection of each leaf against all leaves in the latter tree which are 14" left and right of the corresponding leaf as shown in the Fig. 7 (left/right is considered since users can move left side or right side of the current latitude. Similarly, up/down movement might happen for longitude). In either case (static/dynamic), once we find users who made contact along the latitude and longitude, we need to compute the distance AB (the actual contact direction) as discussed in Sect. 3.2. To build the final contact list, we proceed as follows. Firstly, we create a set of pairwise contacts from the contact list found by the intersection operation as mentioned previously. This step is repeated for the longitude contact list as well. Secondly, the two sets are intersected to generate the final contact list, i.e., the list of users who actually made a contact event along the direction AB. This operation cost $O(n)$ (due to sets being a hashmap) where the hidden constant is the length of the longest contact list, which is bounded above by a constant since a user can not meet more than a certain number of users in t (=5) mins. From the contact list, we can easily create a **contact vector** of each user and store in an adjacency matrix as discussed in the next section. The entire process of the dynamic case is described step by step in Algorithm 2.

Algorithm 2: Dynamic Case. [] denotes indexing in an array/dictionary

```
   Input: T_t - M-way tree at time t, T_{t+1} - M-way tree at time t + Δt
   Output: CT - Contact Tracing Vector
 1 q = 14                      # 420 meter = 14"
 2 for d in degrees do
 3     for m in minutes do
 4         for s in seconds do
 5             for p in partitions do
 6                 p1 ← []
 7                 p2 ← []
 8                 for n in [p − q, p + q] do
 9                     ct = T_{t+1}[d][m][s][n] ∩ T_t[d][m][s][p]
10                     Insert ct in CT
11                     for n1 in [n − 1, n + 1] do
12                         end_case_t = find peoples who are less than 5mt apart in
                             T_t[d][m][s][n1]
13                         end_case_{t+1} = find peoples who are less than 5mt apart
                             in T_{t+1}[d][m][s][n1]
14                     Insert end_case_t in p1
15                     Insert end_case_{t+1} in p2
16                 Insert p1 ∩ p2 in CT
17 return CT
```

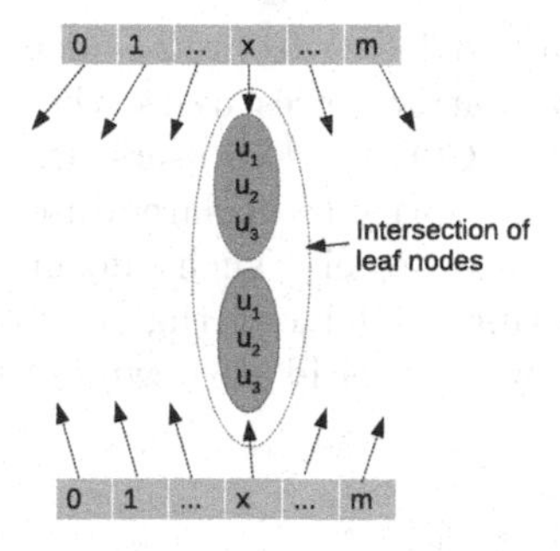

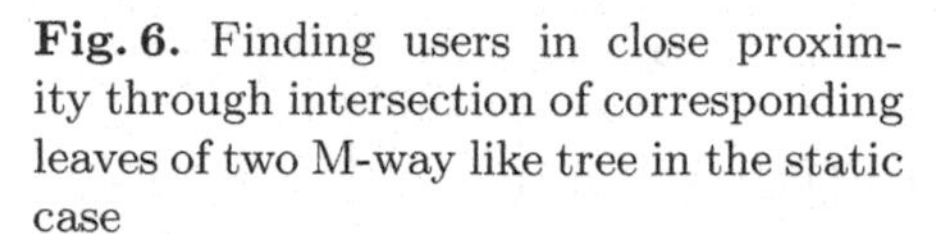

Fig. 6. Finding users in close proximity through intersection of corresponding leaves of two M-way like tree in the static case

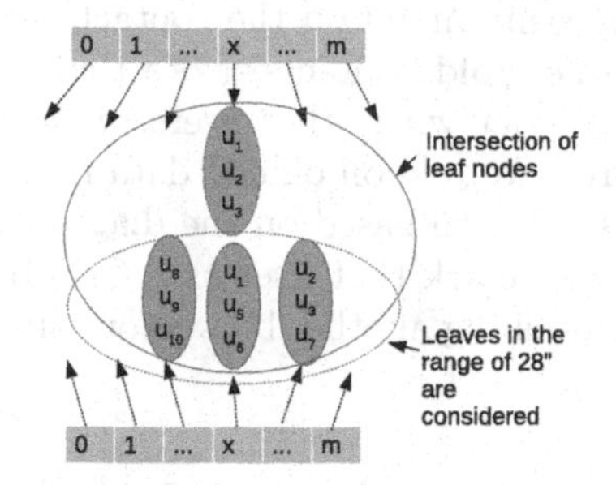

Fig. 7. Finding users in close proximity through intersection of corresponding leaves of two M-way like tree in the dynamic case

4 Potential Hotspot Detection

Our methodology can be used to detect *potential* hotspot areas in a city. Note that there is no unanimous definition of what constitutes a hotspot in epidemiology. As per [9], hotspot can be defined in three ways: (a) as areas of disease elevated occurrence or risk, (b) as areas of frequent disease emergence or reemergence, and (c) an area of elevated transmission efficiency. We take the third definition that is based on the assumption that large gatherings can stimulate the rate of transmission. Therefore, we identify areas in a city where large gatherings are happening along with COVID +ve cases and predict those areas as potential hotspots. The potential hotspot information can help safeguard users and discourage them from freely moving in those areas. Note that our method does declare gatherings as *potential hotspot* if there are no +ve cases in that gathering (cluster). Our assumption is that either COVID +ve user is surrounded by normal users who are not quarantined, and in some cases where COVID +ve user might also be moving, say coming back from a testing center. So, hotspot may contain COVID +ve users who are moving.

In order to locate areas of a potential hotspot, we proceed as follows. Contact list of each user obtained from the contact tracing methodology presented in Sect. 3 from the last 14 d is stored in the form of an adjacency matrix as shown in Eq. 2.

$$\begin{array}{c} \\ u_1 \\ u_2 \\ \vdots \\ u_n \end{array} \begin{array}{c} \begin{array}{cccc} u_1 & u_2 & \ldots & u_n \end{array} \\ \begin{bmatrix} 1 & 1 & \ldots & 1 \\ 0 & 0 & \ldots & 1 \\ \vdots & \vdots & \ddots & \vdots \\ 1 & 0 & \ldots & 1 \end{bmatrix} \end{array} \tag{2}$$

(In actual implementation, it can be implemented using a hashmap where keys are user ids and values are queue/list which stores the ids of users in the order the contact happens). We call each row as a **contact vector** for that user. We can easily maintain the contact vector by adding /deleting from the queue/list the new/old contacts. Note that the adjacency matrix stores users who came in contact with other users. It does not store the COVID +ve users. In other words, we rely on official data from government authority to declare a user as a COVID +ve based on the diagnostic report. Once we receive such information, we can mark that user as a COVID +ve by setting a bit along that row or by maintaining another bit vector which is indexed by the user id as shown in Fig. 8.

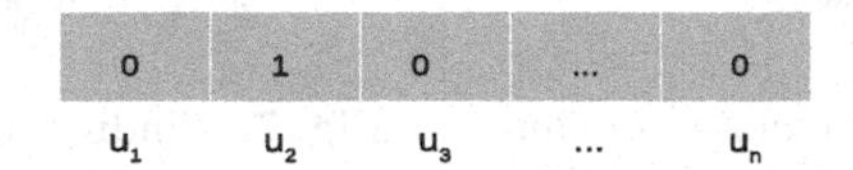

Fig. 8. Bit vector to store COVID +ve users

To detect a *potential hotspot* area, we provide a reference GPS location. This can be a city location as searched by users interested in knowing hotspot areas. The reference GPS is mapped to the lat/long tree created in that period. Assuming a 10 km radius from the reference point though, this can be dynamically adjusted according to the city area. Firstly, all the users are identified who are within the 10 km radius from the reference point. The 10 km of distance maps to around 5.39 min in lat/long. Therefore, all the users in the leaves, which are 5.39' left and right of the reference point, are identified. The user ids of such users are mapped to the bit vector (Fig. 8) to find out users who are COVID +ve within our search area (10 km, for example). Using the adjacency matrix, we also identify users who came in contact with users who are found COVID +ve in the search area. Such users might form *susceptible users*. Further, user ids of the susceptible users are cross-tallied in the search area (since users who came in contact with the COVID +ve user in the search area t days ago may have moved to a different city). Finally, we have built a set of COVID +ve users and suspected users. We can mark COVID +ve and susceptible users on the city map, and if the number of COVID +ve users exceeds a threshold, it can be marked as a *potential hotspot city*.

5 Safe Route Recommendation

Based on the potential hotspot areas, we can recommend a safe route for travel. Users currently use google map/Apple Maps as a primary navigation medium for traveling from one place to another. However, these maps have limited information about whether a route is safe for travel. For example, unless someone marks specifically that a particular route is blocked or traffic is slow, current navigation systems have limited information to tell users in advance that route is not safe. This situation was observed during the recent exodus of people from red zones such as Mumbai/New Delhi. People were following google maps to travel to their hometowns. However, the map took them to other hotspot cities

Algorithm 3: Safe Route Recommendation Algorithm

Input: G - Graph, S - Source, D - Target, $HOTSPOT$ - list of hotspot nodes.
Output: Path from S to T

```
1  for each vertex V in G do
2      distance[V] ← ∞
3      last[V] ← None
4      if V! = S then
5          add V to Priority Queue Q(Using Min-Heap)
6  while Q is Not Empty do
7      U ← Extract min From Q
8      for each unvisited neighbour V of U do
9          if V not in HOTSPOT then
10             temporary_Distance ← distance[U] + edge_Weight(U, V)
11             if temporary_Distance < distance[V] then
12                 distance[V] ← temporary_Distance
13                 last[V] ← U
14 if distance[D] is ∞ then
15     return NoPath
16 return distance[D]
```

on the way. As a result, they have to take a long detour once they reached the hotspot city on the way because entry into the city was not allowed. In such situations and many others, such as avoiding potential gatherings that may be a potential hotspot, our approach can recommend a safe route, thereby avoiding hotspot areas in advance so that users can plan their travel accordingly.

Suppose a user wants to travel from source location s to target location t as shown in Fig. 9. If the user follows Google map, it might show the *green* route since it is the shortest path. However, it may take the user to the hotspot city since it does not have *potential hotspots* information in *real-time*. Following our previous approach, we have the information of *potential hotspots* based on the previous section's discussion. Therefore, we remove the nodes representing the hotspot (red node in the Fig. 9) as well as edges incident to it and run the Dijkstra's algorithm [4,16] on the remaining subgraph. The algorithm produces the shortest path in the subgraph (marked in blue color). This path does not fall in red zones/hotspot areas and thus safe for travel. We verify our approach on a toy dataset and show its efficacy in the experiment section. The safe route recommendation algorithm is described in Algorithm 3.

In the above algorithm, G is the input graph, S and D are the sources and destination nodes (cities), and $HOTSPOT$ is the list of nodes marked as a hotspot thus not safe for travel.

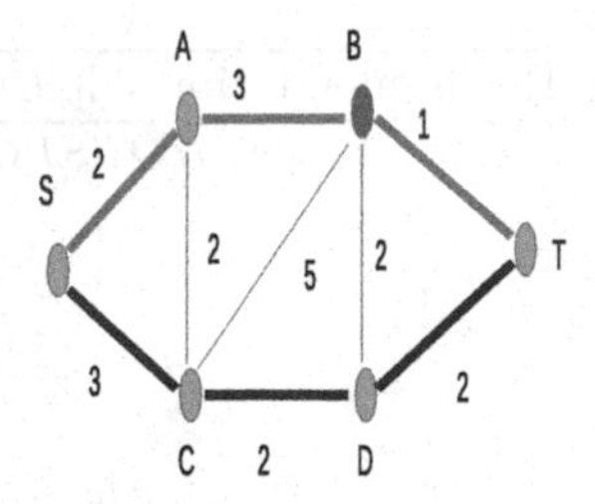

Fig. 9. Safe route recommendation. The shortest route might not be safe. Google Maps recommend the Green route since it is the shortest. Our approach recommends the Blue route though it may not be the shortest, it avoids the potential hotspot areas.

6 Complexity Analysis

In this section, time and space complexity analysis of various operations is presented.

Time Complexity of the Mapping a GPS coordinate to a tree leaf is $O(h)$ where h is the height of the tree. In our case, $h = 4$, so mapping GPS coordinates to tree leaf is a constant time operation $O(1)$. Further, append operation of the users into the list attached to the leaf node will have a time complexity of $O(1)$.

After mapping, we need to find a new cluster of users missed in the mapping process as discussed in Sect. 3.2. Such users can be put into new clusters by sliding a window of size 2, moving over adjacent pairs of clusters, and computing Euclidean/Haversine distance for each pair of users from each cluster. Time taken by this operation is $O(ck\ell)$, where k and ℓ are the sizes of two clusters and $c(= 90 * 60 * 60 * 6)$ is a constant. As discussed previously, the size of the clusters can not grow unbounded in the worst case due to population dynamics. Usually, k and ℓ are in the range of 1 to 100 since within a circle of 5 m radius; we assume no more than 100 users can stay. Further, the operation of finding missed users can be computed efficiently using parallelization (presented in the empirical section).

Time Complexity of the Intersection of leaf node (clusters) in the static case is more straightforward. We need to perform the intersection of $O(c)$ corresponding leaf nodes in two trees obtained at time intervals of $t/2$ mins. Here c is as defined previously. Intersection operation in the dynamic case is involved and incurs extra overhead due to one leaf being intersected with 168 (=28" $\times$ 6) leaf nodes in the other tree for the pedestrian case. Consequently, time cost goes up by a factor of 168 in the dynamic case compared to the static case. Further, note that the intersection operation in both the cases (static and dynamic) can be efficiently parallelized since one leaf node can be intersected with other nodes in parallel. Such parallelization tricks are used in the empirical section to reduce the runtime cost of the intersection operation.

Space Complexity: Total number of internal nodes for mapping latitudes is $90 + 90 * 60 + 90 * 60 * 60 + 90 * 60 * 60 * m = 329490 + 32400m$. Where m is the number of partitions of a second. Take $m = 6$ for $d = 5$,we have total internal

Table 1. Run-time comparison in the static and dynamic cases. All time in seconds. Values in blue and black denote run-time in static and dynamic case respectively. Cells shaded in yellow color show serial run-time whereas cells without shading indicate parallel run-time. Best viewed in color.

Time to Map GPS to Tree				Intersection Tree Time	
Before(at T=t)		After(at T=t+Δt)			
Latitude	Longitude	Latitude	Longitude	Latitude	Longitude
20.18	20.43	20.67	20.96	49.98	46.66
21.99	22.09	23.45	23.67	150.89	161.13
30.43	32.98	35.87	34.88	113.67	122.32
31.9	33.88	37.98	34.98	587.32	632.49
46.76	48.87	43.42	49.44	496.75	513.23
50.32	47.64	48,98	51.32	1208.98	1267.22
55.43	51.4	59.32	24.55	576.23	612.43
65.12	68.32	66.23	71.42	2198.32	2223.64
68.98	72.7	74.89	79.84	786.67	796.87
74.99	76.65	79.43	80.1	3983.78	4073.27
19.89	19.12	20.22	22.43	8.32	7.39
22.43	18.9	21.32	22.94	22.69	29.32
30.21	33.41	33.78	36.98	32.43	40.98
32.76	36.74	38.01	39.63	78.19	102.23
48.64	46.75	50.43	49.31	60.98	73.92
50.76	53.21	48.98	54.67	365.42	398.71
59.45	62.32	58.43	56.49	86.15	92.43
61.23	65.43	66.32	68.32	598.41	631.42
78.23	71.96	75.63	70.54	122.34	134.43
85.78	79.12	83.45	81.12	945.32	990.87

nodes as 2273490. Assume each integer takes 8 bytes. The space complexity of an empty tree is approx 18.18 MB. Further, we are just storing user ids, which for 10M users takes 80 MB. In total, one tree after mapping 10M GPS coordinates will take 98.18 MB of space. If we also store lat/long for 10M users, it will further take 160 MB. Thus in the worst case, one tree will occupy approximately 260 MB. (During execution, we store two trees every 2.5 mins interval. To generate contact vectors of 10M users takes around 17 mins as given in the experimental section; we maintain a total of 16 trees before we can flush the trees and reuse the space. Hence our approach takes roughly 4 GB of space for generating contact vectors compared to 55 GB in [10].)

In our implementation, we take latitude $\simeq$ $(7°, 37°)$ and longitude $\simeq$ $(68°, 97°)$ for India for which empty tree takes $\approx$39 MB when representing internal nodes as four-level dictionary keys and leaves as the empty list in the python programming language. We store app IDs or some other useful global identifier such as IMEI number into leaf nodes to track each user. Thus space complexity for storing user id into leaf will be $O(n)$ where n is the number of user ids. Therefore, the dominating part will be used for storing user ids, and it scales linearly with the users.

One crucial point to consider is that the leaf node is implemented as a list or array which has the cost of $O(1)$ for append operation. However, a growing list beyond specified size incurs extra overhead. However, we must mention that *any* list does not hold more than a constant number of users since each leaf maps GPS range of 5 m and in a circle of 5 m radius, we assume that not more than 100 users can stand.

7 Empirical Demonstration

In this section, we show the working of the proposed methodology for contact tracing, hotspot detection, and safe route recommendation on simulated data.

7.1 Contact Tracing Experiment

To show the scalability of the contact tracing approach, uniformly at random GPS coordinates in the range latitude $\simeq (7°, 37°)$ and longitude $\simeq (68°, 97°)$ are generated for 2/4/6/8/10M users. We will map these GPS coordinates to the M-way tree like data structure as discussed and compute the contact vectors. Our goal is to estimate the time it takes to generate the contact lists. The serial and parallel versions for static and dynamic cases are implemented in C++ with openMP [3] (for multi-threading). The experiments are conducted on a personal computer with Intel(R) Core(TM) *i5-7500U* CPU and 8.0 *GB* memory (RAM) running Ubuntu operating system. The results are shown in the table 1. We have several observation from the results in the table. Firstly, mapping GPS to tree takes at most 80 secs for 10M users using serial implementation, whereas at most 86 secs in the parallel implementation. A little increase in time could be due to thread scheduling overhead in the latter case. In other words, mapping GPS to trees via parallel implementation is not *recommended*, and serial implementation suffices to practical cases. Secondly, compared to mapping GPS coordinates to trees, performing intersection of leaves is time-consuming. Thirdly, intersection time using parallel implementation achieves dramatic speedup. For example, parallel implementation in the dynamic case for 10M users gains 4 × speedup. Finally, generating the final contact vectors for 10M users takes around 16.5 mins (990.87+85.78 secs). In other words, if we want to track contact events every 5 mins, we need to maintain four sets of lat/long trees before we can flush the data in the trees and reuse them again, thereby reducing the memory footprint.

Baselines One of the baselines used to compare the runtime of the contact tracing approach is by performing pairwise comparison (the naive solution).

Table 2. The runtime comparison of our approach with the baseline. The experiment is conducted over 2M users. The number of threads uses in the parallel implementation is set to 10.

	Implementation	CPU time(s)
Naive approach	Serial	1135714.0
	Parallel	127124.59
Ojagh et al. [13]	Serial	428
Our approach	Serial	**185.80**
	Parallel	**52.26**

Another baseline we use it from [13] in which the author perform contact tracing in an indoor setting using a graph-based approach. The result from the serial and parallel implementation of the baselines are shown in Table 2. It is obvious that our approach beats the naive solution by a significant order of magnitude as well as **2.3**× faster tha [13] and favors practical implementation.

7.2 Hotspot Detection Experiment

To verify the approach discussed in Sect. 4 for hotspot detection, we first test it on 20 users and mark them on google map as shown in Fig. 10. All the users, the COVID +ve patients, and the suspected users (users who came in contact with COVID +ve users but still having no symptoms) are plotted on the city map. The figure also shows that at time t_0, users ids 1, 3, 9, and 14 are COVID +ve. The Fig. 10(b) shows that user 6 who came in contact with user 9 (a COVID +ve), has been suspected and marked with the yellow color.

Similarly, user 16 becomes susceptible at time t_2 and so on. Indeed, the figure shows how the infected/suspected users *may* be moving in the city to provide real-time information of danger zones, benefiting normal users to plan accordingly. If the number of infected/suspected users in the map relative to the reference point exceeds a certain threshold, then the area around the reference point can be marked as a potential hotspot area. Alternatively, any area comprising a suspected COVID-19 positive patient can be declared as a *potential hotspot*.

One concern related to privacy is that we are marking users at the individual level. To avoid such a thing, we can draw a circle of larger radius such as 10 or 20 m to conceal the COVID +ve users' identity. Alternatively, we can group nearby users and form large clusters to make the identity indistinguishable at the individual level. Additionally, the user's data remains with a single entity (government entity likely) hence the privacy is not compromised due to data sharing (assuming the government is truthful).

To show the scalability of the proposed approach, the plot of runtime with respect to the number of users considered in a hotspot region is shown in Fig. 12. The graph clearly shows that the runtime is following a *near*-linear trend. Therefore, our approach to detect hotspots in real-time may be useful.

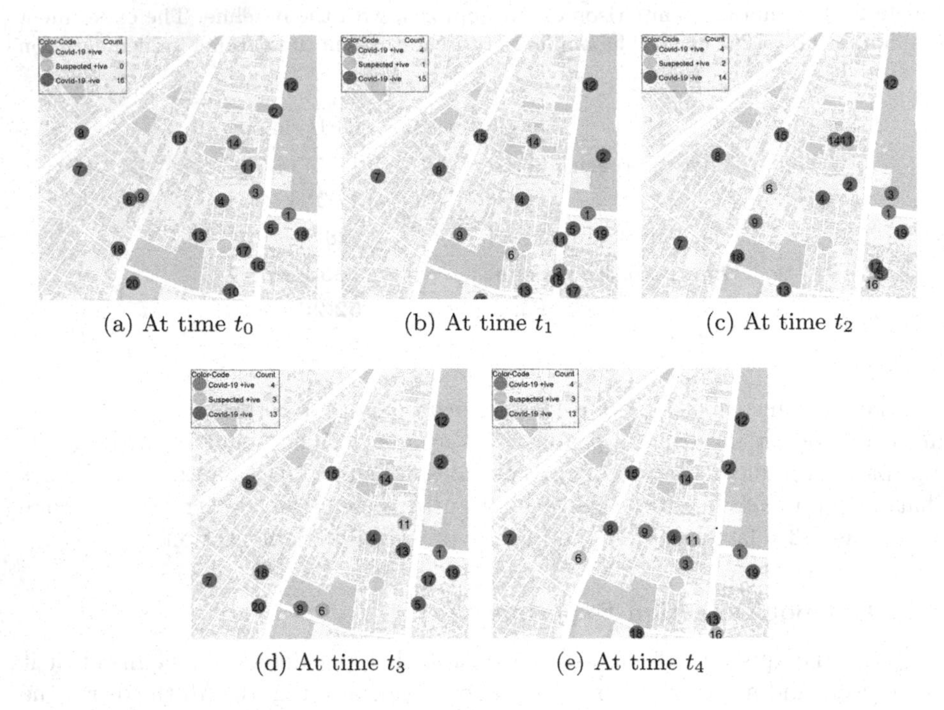

(a) At time t_0 (b) At time t_1 (c) At time t_2

(d) At time t_3 (e) At time t_4

Fig. 10. Hotspot information at 5 consecutive time periods. Best viewed with zooming the image and in color.

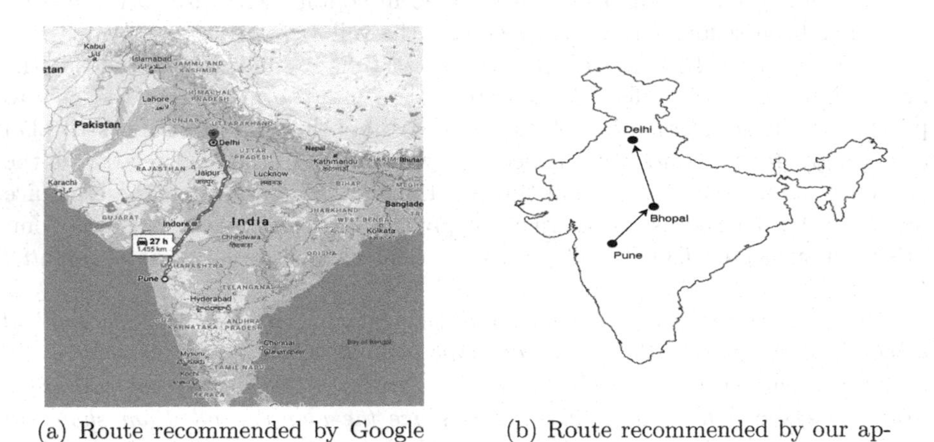

(a) Route recommended by Google map

(b) Route recommended by our approach

Fig. 11. (a) Route recommended by Google map (b) Route recommended by our approach. We can see that Google map recommends a route from Pune to New Delhi, which passes through Indore city, a hotspot area and not allowed for travel. On the other hand, our approach recommends a route that avoids the hotspot city.

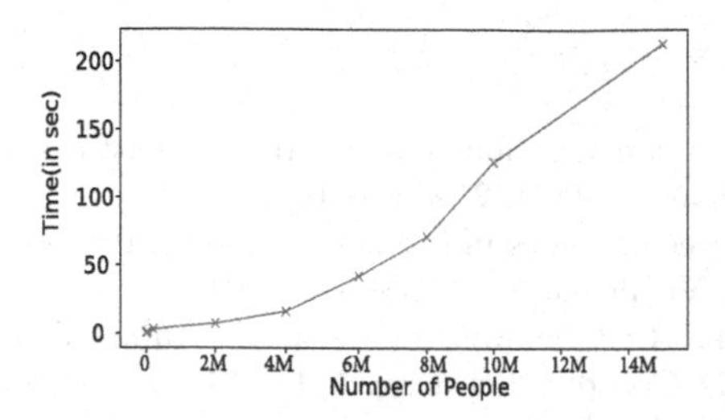

Fig. 12. The runtime of the hotspot detection algorithm wrt different number of users

7.3 Safe Route Recommendation Experiment

The evaluation of the safe route recommendation approach is done in the following way: 1) The proposed method is compared with an existing route Recommendation system, that is, the Google Map, in terms of the total cost of the path. The total distance from the source to destination constitutes the total cost and whether it is a Hotspot Zone free path or not. The efficacy of the proposed approach is analyzed on a real World dataset, which is extracted from Google Map. For the experiment, we have considered major cities in India (Indore, Bhopal, Chennai, Mumbai, Delhi, Bangalore, Lucknow, Hyderabad, Pune, Kolkata).

We have compared our proposed approach with the current Google map recommendation, considering Indore as a Hotspot City, Source as Pune, and Destination as New Delhi. Google maps recommend the path which passes through the hotspot city(Indore) as shown in Fig. 11(a) with a total cost of 1455 Km whereas, the path displayed by the proposed algorithm is an alternate route where the intermediate city is Bhopal(non-hotspot city) instead of Indore with a total cost of 1540 Km which is displayed in Fig. 11(b). Indeed, the path recommended by our approach is not optimal. Still, since our objective is to avoid a hotspot city, our approach is suitable for choosing a safe route rather than an optimal path.

8 Conclusion and Future Work

The proposed approach for contact tracing seems plausible based on the initial experiments on simulated data. Safe route recommendation and potential hotspot information further add new features to our method which is absent in the apps available in the market. Our approach scales well for large data and hence can be used for deploying over big cities. We are planning to release the app in the future after verification on the real-user study.

References

1. Ascione, G.: On the construction of some deterministic and stochastic non-local sir models. Mathematics **8**(12), 2103 (2020)
2. Christaki, E.: New technologies in predicting, preventing and controlling emerging infectious diseases. Virulence **6**(6), 558–565 (2015)
3. Dagum, L., Menon, R.: Openmp: an industry standard api for shared-memory programming. IEEE Comput. Sci. Eng. **5**(1), 46–55 (1998)
4. Dijkstra, E.W., et al.: A note on two problems in connexion with graphs. Numerische mathematik **1**(1), 269–271 (1959)
5. Farrahi, K., Emonet, R., Cebrian, M.: Epidemic contact tracing via communication traces. PloS one **9**(5), e95133 (2014)
6. Ferretti, L., et al.: Quantifying sars-cov-2 transmission suggests epidemic control with digital contact tracing. Science **368**(6491) (2020)
7. Hellewell, J., et al.: Feasibility of controlling covid-19 outbreaks by isolation of cases and contacts. Lancet Global Health **8**(4), e488-e496(2020)
8. Huerta, R., Tsimring, L.S.: Contact tracing and epidemics control in social networks. Physic. Rev. E **66**(5), 056115 (2002)
9. Lessler, J., Azman, A.S., McKay, H.S., Moore, S.M.: What is a hotspot anyway?, Am. J. Tropical Med. Hygiene **96**(6), 1270–1273 (2017)
10. Mahapatra, G., Pradhan, P., Chattaraj, R., Banerjee, S.: Dynamic graph streaming algorithm for digital contact tracing (2020)
11. Martin, T., Karopoulos, G., Ramos, J.L.H., Kambourakis, G., Fovino, I.N.: Demystifying COVID-19 digital contact tracing: a survey on frameworks and mobile apps. CoRR abs/2007.11687 (2020). https://arxiv.org/abs/2007.11687
12. Müller, J., Kretzschmar, M., Dietz, K.: Contact tracing in stochastic and deterministic epidemic models. Math. Biosci. **164**(1), 39–64 (2000)
13. Ojagh, S., Saeedi, S., Liang, S.H.: A person-to-person and person-to-place covid-19 contact tracing system based on ogc indoorgml. ISPRS Int. J. Geo-Inf. **10**(1), 2 (2021)
14. Sanampudi, A.: Indore navigation mobile application using indore positioning system (ips). Int. J. Basic Sci. Appl. Comput. (IJBSAC) **2** (2020)
15. Team, P.P.: Pan-european privacy-preserving proximity tracing (2020)
16. Vaughan, A.: The problems with contact-tracing apps. New Sci. **246**(3279), 9 (2020)
17. Wiki, O.: Precision of coordinates – openstreetmap wiki, (2019). wiki.openstreetmap.org/w/index.php/title/Precision/of/coordinates/oldid/1872063. Accessed 20 May 2020
18. Yoneki, E.: Fluphone study: virtual disease spread using haggle. In: Proceedings of the 6th ACM Workshop on Challenged Networks, pp. 65–66 (2011)

Current Trends in Learning from Data Streams

João Gama[1,3(✉)], Bruno Veloso[1,2,3], Ehsan Aminian[1,3], and Rita P. Ribeiro[1,3]

[1] LIAAD - INESC TEC, Porto, Portugal
jgama@inesctec.pt
[2] University Portucalense, Porto, Portugal
[3] University of Porto, Porto, Portugal

Abstract. This article presents our recent work on the topic of learning from data streams. We focus on emerging topics, including fraud detection, learning from rare cases, and hyper-parameter tuning for streaming data.

Keywords: Data streams · Fraud detection · Rare cases · Hyper-parameter tuning

1 Introduction

Learning from data streams is an hot-topic in machine learning and data mining. This article presents our recent work on the topic of learning from data streams. It is organized into three main sections. The first one, push up the need of forgetting older data to reinforce the focus on the most recent data. It is based on the work presented in [8]. The second topic refers to the problem of learning from imbalanced regression streams. In this setting, we are interested in predicting points in the fringe of the distribution. It is based on the work presented in [2]. The third topic, discuss the topic of hyper-parameter tuning in the context of data stream mining. It is based on the work presented in [7].

2 The Importance of Forgetting

The high asymmetry of international termination rates with regard to domestic ones, where international calls have higher charges applied by the operator where the call terminates, is fertile ground for the appearance of fraud in Telecommunications. There are several types of fraud that exploit this type of differential, being the Interconnect Bypass Fraud one of the most expressive [1,6].

In this type of fraud, one of several intermediaries responsible for delivering the calls forwards the traffic over a low-cost IP connection, reintroducing the call in the destination network already as a local call, using VOIP Gateways. This way, the entity that sent the traffic is charged the amount corresponding to the delivery of international traffic. However, once it is illegally delivered as national traffic, it will not have to pay the international termination fee, appropriating this amount.

S. N. Srirama et al. (Eds.): BDA 2021, LNCS 13147, pp. 183–193, 2021.
https://doi.org/10.1007/978-3-030-93620-4_14

Traditionally, the telecom operators analyze the calls of these Gateways to detect the fraud patterns and, once identified, have their SIM cards blocked. The constant evolution in terms of technology adopted on these gateways allows them to work like real SIM farms capable of manipulating identifiers, simulating standard call patterns similar to the ones of regular users, and even being mounted on vehicles to complicate the detection using location information.

The interconnect bypass fraud detection algorithms typically consume a stream S of events, where S contains information about the origin number $A-Number$, the destination number $B-Number$, the associated timestamp, and the status of the call (accomplished or not). The expected output of this type of algorithm is a set of potential fraudulent $A-Numbers$ that require validation by the telecom operator. This process is not fully automated to avoid block legit $A-Numbers$ and get penalties. In the interconnect bypass fraud, we can observe three different types of abnormal behaviors: (i) the burst of calls, which are $A-Numbers$ that produce enormous quantities of $\#calls$ (above the $\#calls$ of all $A-Numbers$) during a specific time window W. The size of this time window is typically small; (ii) the repetitions, which are the repetition of some pattern ($\#calls$) produced by a $A-Number$ during consecutive time windows W; and (iii) the mirror behaviors, which are two distinct $A-Numbers$ (typically these $A-Numbers$ are from the same country) that produces the same pattern of calls ($\#calls$) during a time window W.

Algorithm 1. The `Lossy Counting` Algorithm.

```
1: procedure LossyCounting(S: A Sequence of Examples; ε: Error margin; α: fast forgetting parameter)
2:     n ← 0; Δ ← 0; T ← 0;
3:     for example e ∈ S do
4:         n ← n + 1
5:         if e is monitored then
6:             Increment Count_e
7:         else
8:             T ← T ∪ {e, 1 + Δ}
9:         end if
10:        if ⌈n/ε⌉ ≠ Δ then
11:            Δ ← n/ε
12:        end if
13:        for all j ∈ T do
14:            if Count_j < δ then
15:                T ← T\{j}
16:            end if
17:        end for
18:    end for
19: end procedure
```

Algorithm 2. The `Lossy Counting` with Fast Forgetting Algorithm.

```
1: procedure LossyCounting(S: A Sequence of Examples; ε: Error margin; α: fast forgetting parameter)
2:     n ← 0; Δ ← 0; T ← 0;
3:     for example e ∈ S do
4:         n ← n + 1
5:         if e is monitored then
6:             Increment Count_e
7:         else
8:             T ← T ∪ {e, 1 + Δ}
9:         end if
10:        if ⌈n/ε⌉ ≠ Δ then
11:            Δ ← n/ε
12:        end if
13:        for all j ∈ T do
14:            Count_j ← α * Count_j
15:            if Count_j < δ then
16:                T ← T\{j}
17:            end if
18:        end for
19:    end for
20: end procedure
```

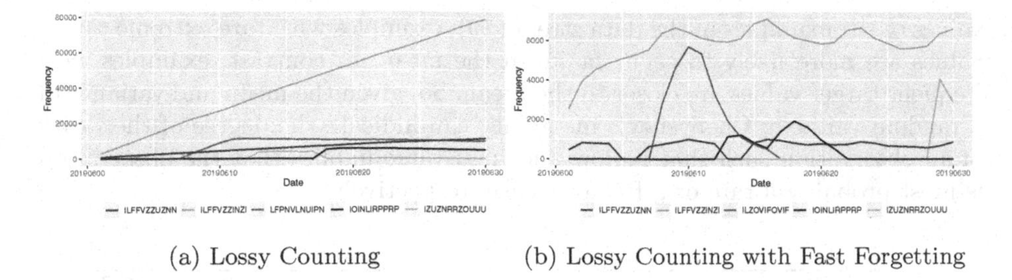

(a) Lossy Counting (b) Lossy Counting with Fast Forgetting

Fig. 1. Fraud detection on number of calls

3 Learning Rare Cases

Few approaches in the area of learning from imbalanced data streams discuss the task of regression. In this study, we employ the Chebyshev's inequality as an heuristic to disclose the type of incoming cases (i.e. frequent or rare). We discuss two methods for learning regression models from imbalanced data streams [2]. Both methods use Chebyshev's inequality to train learning models over a relatively balanced data stream once the incoming data stream is imbalanced. The mentioned inequality derived from Markov inequality is used to bound the tail probabilities of a random variable like Y. It guarantees that in any probability distribution, 'nearly all' values are close to the mean. More precisely, no more than $\frac{1}{t^2}$ of the distribution's values can be more than t times standard deviations away from the mean. Although conservative, the inequality can be applied to completely arbitrary distributions (unknown except for mean and variance). Let Y be a random variable with finite expected value $\overline{y}$ and finite non-zero variance σ^2. Then for any real number $t > 0$, we have:

$$\Pr(|y - \overline{y}| \geq t\sigma) \leq \frac{1}{t^2} \tag{1}$$

Only the case $t > 1$ is useful in the above inequality. In cases $t < 1$, the right-hand side is greater than one, and thus the statement will be "always true" as the probability of any event cannot be greater than one. Another "always true" case of inequality is when $t = 1$. In this case, the inequality changes to a statement saying that the probability of something is less than or equal to one, which is "always true".

For $t = \frac{|y-\overline{y}|}{\sigma}$ and $t > 1$, we define *frequency* score of the observation $\langle x, y \rangle$ as:

$$P(|\,\overline{y} - y\,| \geq t) = \frac{1}{\left(\frac{|y-\overline{y}|}{\sigma}\right)^2} \tag{2}$$

The above definition states that the probability of observing y far from its mean is small, and it decreases as we get farther away from the mean. In an imbalanced data streams regression scenario, considering the mean of target values of the examples in the data stream ($\overline{y}$), examples with rare extreme target values are more likely to occur far from the mean. In contrast, examples with frequent target values are closer to the mean. So, given the mean and variance of a random variable, Chebyshev's inequality can indicate the degree of the rarity of an observation such that its low and high value implies that the observation is most probably a rare or a frequent case, respectively.

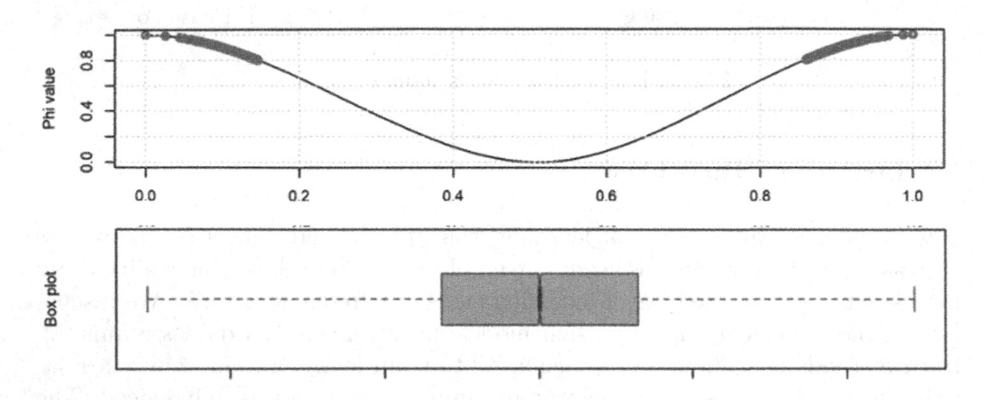

Fig. 2. Data-points relevance and the box plot for the target variable of Fried data set.

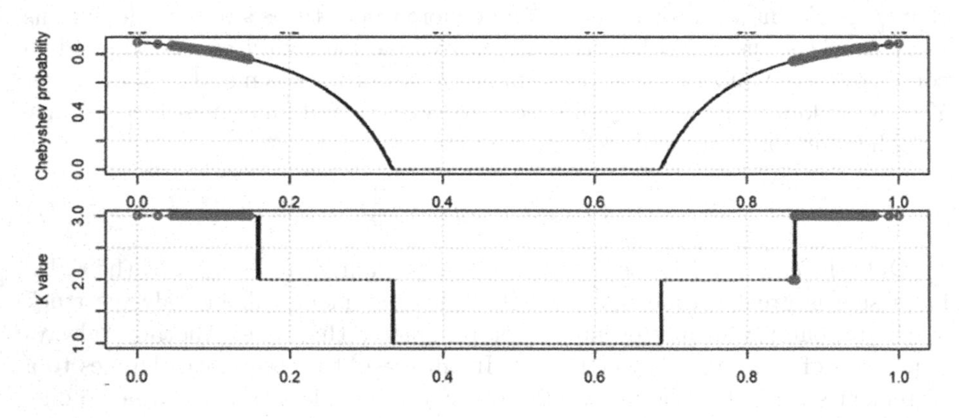

Fig. 3. Chebyshev probability used by the under-sample approach (top panel), and the K-value used in the over sample approach (bottom panel) for the target variable of Fried data set.

Figure 3 shows the probability values calculated from Eq. 2 for Fried data set, described in [5], along with the box plot of the target variable. As can be seen from the figures and as we expected, Chebyshev's probability value for examples near the mean is close to one. It decreases as we get far from the mean until it gets close to zero, for example, at the farthest distance to the mean. Accordingly, interpretation of the output value of Eq. 2 for an example as its frequency score makes sense. Moreover, it meets the imbalance regression problem definition, w.r.t. rare extreme values of the target variable.

Having equipped with the heuristic to discover if an example is rare or frequent, the next step is to use such knowledge in training a regression model.To do that, ChebyUS and ChebyOS are the two methods proposed in this paper. They are described in detail in the next subsections.

3.1 ChebyUS: Chebyshev-Based Under-Sampling

The proposed under-sampling method is presented in Algorithm 3. This algorithm selects an incoming example for training the model if a randomly generated number in $[0, 1]$ is greater or equal to its Chebyshev's probability which is calculated as:

$$P(\mid y - \overline{y} \mid \geq t) = \begin{cases} \frac{\sigma^2}{|y-\overline{y}|^2}, & t > 1 \\ 1, & t \leq 1 \end{cases} \tag{3}$$

If the example is not selected, it is assumed that the example is probably a frequent case. Still, it receives a second chance for being selected if the number of frequent cases selected so far is less than that of rare cases. If so, the example is selected with a second chance probability (input parameter sp).

The descriptive statistics (μ and σ^2) of the target variable of examples can be computed through incremental methods [4]. The greater the number of examples, n, we have, the more accurate the estimation will be. For the first examples, mean and variance are not accurate, and therefore, Chebyshev's probability will not be accurate enough. But as more examples are received, those statistics (i.e. mean and variance) and, consequently, Chebyshev's probability are getting more stable and accurate.

At the end of the model's training phase, we expect the model to have been trained over approximately the same portion of frequent and rare cases.

3.2 ChebyOS: Chebyshev-Based Over-Sampling

Another way of making a balanced data stream is to over-sample rare cases of the incoming imbalanced data stream. Since those rare cases in data streams can be discovered by their Chebyshev's probability, they can be easily over-sampled by replication. Algorithm 4 describes our over-sampling proposed method.

For each example, a t value can be calculated by Eq. 4 which yields the result in $[0 \quad +\infty)$.

$$t = \frac{|y - \overline{y}|}{\sigma} \tag{4}$$

While t value is small for examples near the mean, it would be larger for ones farther from the mean and has its largest value for examples located in the farthest distance to the mean (i.e. extreme values). We limit the function in Eq. 4 to produce only natural numbers as follows:

Algorithm 3. ChebyUS: Chebyshev-based Under-Sampling Algorithm

```
 1: procedure CHEBYUS(S: Data stream, sp: second chance probability)
 2:     H ← CreateEmptyModel()
 3:     i ← 0
 4:     RCounter ← 0                                   ▷ rare cases counter
 5:     FCounter ← 0                               ▷ frequent cases counter
 6:     while true do
 7:         ⟨x, y⟩ ← GetExample()
 8:         ȳ, σ ← UpdateStatistics(y)
 9:         p ← ComputeProbability(y, ȳ, σ)                   ▷ Equation 3
10:         if RandomNumber ≥ p then
11:             H ← TrainModel(H, ⟨x, y⟩)
12:             RCounter ← RCounter + 1
13:         else
14:             if FCounter ≤ RCounterANDRandomNumber ≤ sp then
15:                 H ← TrainModel(H, ⟨x, y⟩)
16:                 FCounter ← FCounter + 1
17:             end if
18:         end if
19:         i ← i + 1
20:     end while
21: end procedure
```

Algorithm 4. ChebyOS: Chebyshev-based Over-Sampling Algorithm

```
 1: procedure CHEBYOS(S: Data stream)
 2:     H ← CreateEmptyModel()
 3:     i ← 0
 4:     while true do
 5:         ⟨x, y⟩ ← GetExample()
 6:         ȳ, σ ← UpdateStatistics(y)
 7:         k ← ComputeK(y, ȳ, σ)                             ▷ Equation 5
 8:         for i ← 1 to k do
 9:             H ← TrainModel(H, ⟨x, y⟩)
10:         end for
11:         i ← i + 1
12:     end while
13: end procedure
```

$$K = \left\lceil \frac{|y - \overline{y}|}{\sigma} \right\rceil \tag{5}$$

K is expected to have greater numbers for rare cases. In our proposed over-sampling method, we use K value computed for each incoming example and present that example exactly K times to the learner.

Examples that are not as far from the mean as the variance, are most probably frequent cases. They contribute only once in the learner's training process while the others contribute more times.

3.3 Experimental Evaluation

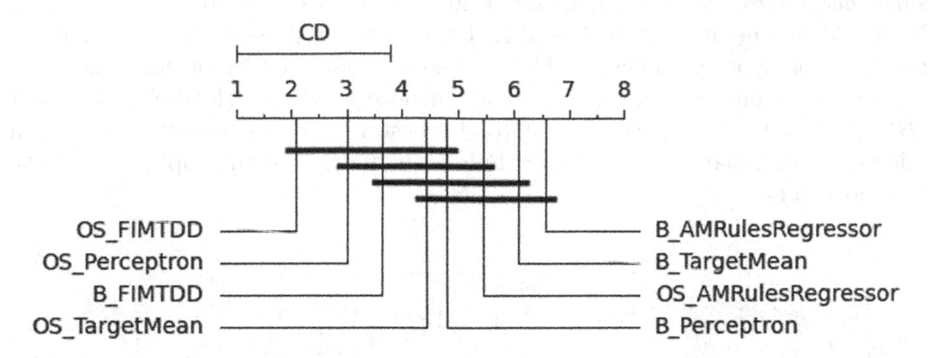

Fig. 4. Critical Difference diagrams considering both extreme rare cases ($thr_{\phi} = 0.8$), for four regression algorithms with no sampling (Baseline) and with the Chebyshev-based Over-Sampling (ChebyOS) strategy.

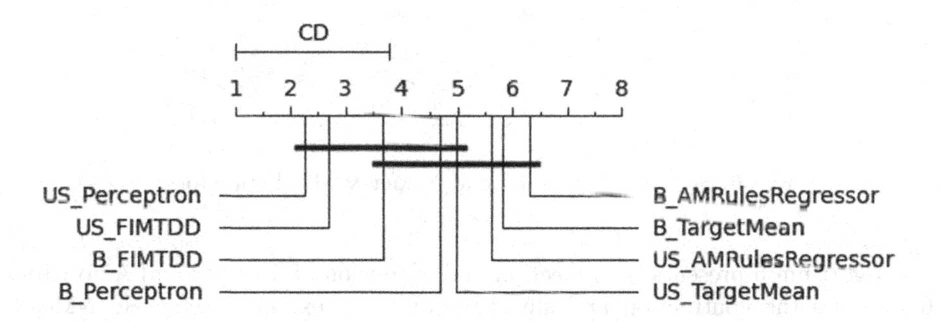

Fig. 5. Critical Difference diagrams considering both extreme rare cases ($thr_{\phi} = 0.8$), for four regression algorithms with no sampling (Baseline) and with the Chebyshev-based Under-Sampling (ChebyUS) strategy.

Figures 4 and 5 presents the critical difference diagrams [3] for four regression algorithms with no sampling (Baseline) and with the proposed sampling strategies: Chebyshev-based Under-Sampling (ChebyUS) andChebyshev-based Over-Sampling (ChebyOS).

4 Learning to Learn: Hyperparameter Tunning

This algorithm is a simplex search algorithm for multidimensional unconstrained optimization without derivatives. The vertexes of the simplex, which define a convex hull shape, are iteratively updated in order to sequentially discard the vertex associated with the largest cost function value.

The Nelder-Mead algorithm relies on four simple operations: *reflection*, *shrinkage*, *contraction* and *expansion*. Figure 6 illustrates the four corresponding Nelder-Mead operators R, S, C and E. Each vertice represents a model containing a set of hyper-parameters. The vertexes (models under optimisation) are ordered and named according to the root mean square error (RMSE) value: best (B), good (G), which is the closest to the best vertex, and worst (W). M is a mid vertex (auxiliary model). Algorithms 5 and 6 describe the application of the four operators.

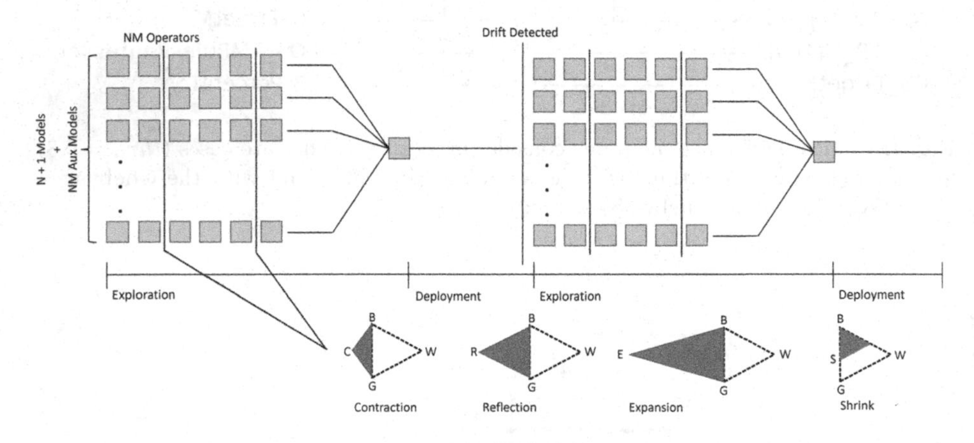

Fig. 6. SPT working modes and Nelder & Mead operators.

Algorithm 5 presents the reflection and extension of a vertex and Algorithm 6 presents the contraction and shrinkage of a vertex. For each Nelder-Mead operation, it is necessary to compute an additional set of vertexes (midpoint M, reflection R, expansion E, contraction C and shrinkage S) and verify if the calculated vertexes belong to the search space. First, Algorithm 5 computes the midpoint (M) of the best face of the shape as well as the reflection point (R). After this initial step, it determines whether to reflect or expand based on the set of predetermined heuristics (lines 3, 4 and 8).

Algorithm 5. Nelder-Mead - reflect (a) or expand operators (d).

```
1: M = (B + G)/2
2: R = 2M − W
3: if f(R) < f(G) then
4:     if f(B) < f(R) then
5:         W = R
6:     else
7:         E = 2R − M
8:         if f(E) < f(B) then
9:             W = E
10:        else
11:            W = R
12:        end if
13:    end if
14: end if
```

Algorithm 6. Nelder-Mead - contract (c) or shrink (b) operators.

```
1: M = (B + G)/2
2: R = 2M − W
3: if f(R) ≥ f(G) then
4:     if f(R) < f(W) then
5:         W = R
6:     else
7:         C = (W + M)/2
8:         if f(C) < f(W) then
9:             W = C
10:        else
11:            S = (B + W)/2
12:            if f(S) < f(W) then
13:                W = S
14:            end if
15:            if f(M) < f(G) then
16:                G = M
17:            end if
18:        end if
19:    end if
20: end if
```

Algorithm 6 calculates the contraction point (C) of the worst face of the shape – the midpoint between the worst vertex (W) and the midpoint M – and shrinkage point (S) – the midpoint between the best (B) and the worst (W) vertexes. Then, it determines whether to contract or shrink based on the set of predetermined heuristics (lines 3, 4, 8, 12 and 15).

The goal, in the case of data stream regression, is to optimise the learning rate, the learning rate decay and the split confidence hyper-paramcters. These hyper-parameters are constrained to values between 0 and 1. The violation of this constraint results in the adoption of the nearest lower or upper bound.

4.1 Dynamic Sample Size

The dynamic samplc size, which is based on the RMSE metric, attempts to identify significant changes in the streamed data. Whenever such a change is detected, the Nelder-Mead compares the performance of the $n+1$ models under analysis to choose the most promising model. The sample size S_{size} is given by Eq. 6 where σ represents the standard deviation of the RMSE and M the desired error margin. We use $M = 95\%$.

$$S_{size} = \frac{4\sigma^2}{M^2} \tag{6}$$

However, to avoid using small samples, that imply error estimations with large variance, we defined a lower bound of 30 samples.

4.2 Stream-Based Implementation

The adaptation of the Nelder-Mead algorithm to on-line scenarios relies extensively on parallel processing. The main thread launches the $n+1$ model threads and starts a continuous event processing loop. This loop dispatches the incoming events to the model threads and, whenever it reaches the sample size interval, assesses the running models and calculates the new sample size. The model assessment involves the ordering of the $n + 1$ models by RMSE value and the application of the Nelder-Mead algorithm to substitute the worst model. The Nelder-Mead parallel implementation creates a dedicated thread per Nelder-Mead operator, totalling seven threads. Each Nelder-Mead operator thread generates a new model and calculates the incremental RMSE using the instances of the last sample size interval. The worst model is substituted by the Nelder-Mead operator thread model with lowest RMSE.

4.3 Experimental Evaluation

Figure 7 presents the critical difference diagram [3] of three hyper-parameter tuning algorithms: SPT, Grid search, default parameter values on four benchmark classification datasets. The diagram clearly illustrate the good performance of SPT.

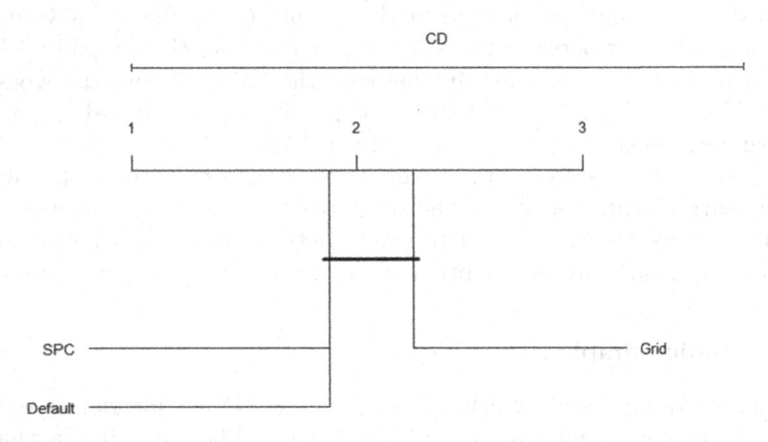

Fig. 7. Critical Difference Diagram comparing Self hyper-parameter tuning, Grid hyper-parameter tuning, and default parameters in 4 classification problems.

5 Conclusions

This paper reviews our recent work in learning from data streams. The two first works present different approaches to deal with imbalanced data: from applied research in fraud detection to basic research on using Chebyshev inequality to guide under-sampling and over-sampling. The last work presents a streaming optimization method to find the minimum of a function and its application in finding the hyper-parameter values that minimize the error. We believe that the three works reported here will have an impact on the work of other researchers.

Acknowledgements. This work was supported by the CHIST-ERA grant CHIST-ERA-19-XAI-012, and project CHIST-ERA/0004/2019 funded by FCT.

References

1. Ali, M., Azad, M., Centeno, M., Hao, F., Van Moorsel, A.: Consumer-facing technology fraud: economics, attack methods and potential solutions. Future Gener. Comput. Syst. **100**, 408–427 (2019)
2. Aminian, E., Ribeiro, R.P., Gama, J.: Chebyshev approaches for imbalanced data streams regression models. Data Mining Knowledge Discovery, pp. 1–78 (2021). https://doi.org/10.1007/s10618-021-00793-1
3. Demšar, J.: Statistical comparisons of classifiers over multiple data sets. J. Mach. Learn. Res. **7**, 1–30 (2006)
4. Finch, T.: Incremental calculation of weighted mean and variance. Technical Report, University of Cambridge Computing Service, Cambridge, UK (2009)
5. Friedman, J.: Multivariate adaptive regression splines. Ann. Statist. **19**(1), 1–67 (1991)
6. Laleh, N., Abdollahi Azgomi, M.: A taxonomy of frauds and fraud detection techniques. In: Prasad, S.K., Routray, S., Khurana, R., Sahni, S. (eds.) Information Systems, Technology and Management. ICISTM 2009. Communications in Computer and Information Science, vol. 31. Springer, Heidelberg (2009). https://doi.org/10.1007/978-3-642-00405-6_28
7. Veloso, B., Gama, J., Malheiro, B., Vinagre, J.: Hyperparameter self-tuning for data streams. Inf. Fusion **76**, 75–86 (2021)
8. Veloso, B., Tabassum, S., Martins, C., Espanha, R., Azevedo, R., Gama, J.: Interconnect bypass fraud detection: a case study. Ann. Telecommun., **75**(9), 583–596 (2020). https://doi.org/10.1007/s12243-020-00808-w

References

[illegible]

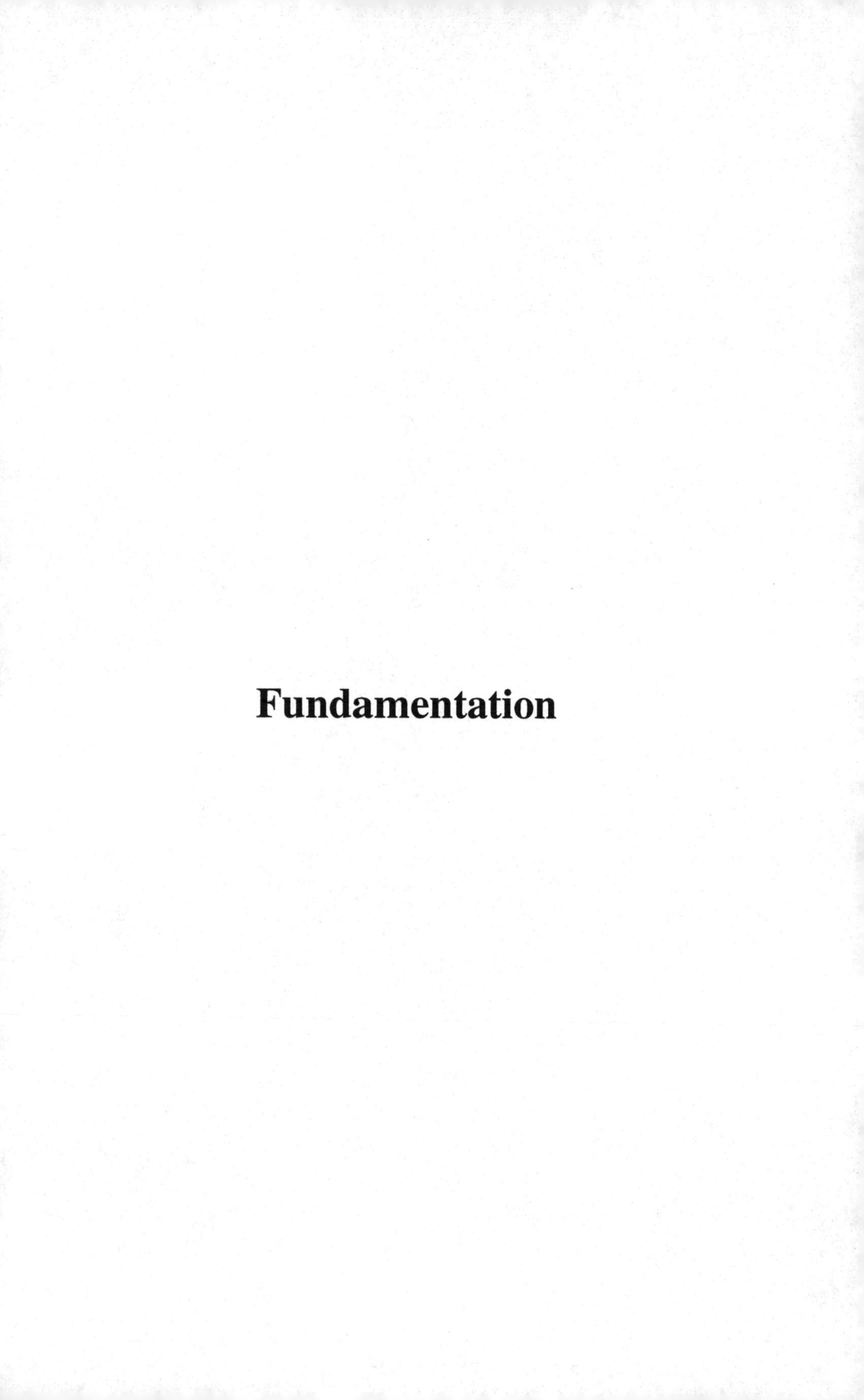

Fundamentation

Diagnostic Code Group Prediction by Integrating Structured and Unstructured Clinical Data

Akshara Prabhakar[1(✉)], Shidharth Srinivasan[1], Gokul S. Krishnan[1,2], and Sowmya S. Kamath[1]

[1] Healthcare Analytics and Language Engineering (HALE) Lab, Department of Information Technology, National Institute of Technology Karnataka, Surathkal 575025, India
{aksharap.181it132,sowmyakamath}@nitk.edu.in

[2] Robert Bosch Centre for Data Science and Artificial Intelligence, Indian Institute of Technology, Madras, India

Abstract. Diagnostic coding is a process by which written, verbal and other patient-case related documentation are used for enabling disease prediction, accurate documentation, and insurance settlements. It is a prevalently manual process even in countries that have successfully adopted Electronic Health Record (EHR) systems. The problem is exacerbated in developing countries where widespread adoption of EHR systems is still not at par with Western counterparts. EHRs contain a wealth of patient information embedded in numerical, text, and image formats. A disease prediction model that exploits all this information, enabling accurate and faster diagnosis would be quite beneficial. We address this challenging task by proposing mixed ensemble models consisting of boosting and deep learning architectures for the task of diagnostic code group prediction. The models are trained on a dataset created by integrating features from structured (lab test reports) as well as unstructured (clinical text) data. We analyze the proposed model's performance on MIMIC-III, an open dataset of clinical data using standard multi-label metrics. Empirical evaluations underscored the significant performance of our approach for this task, compared to state-of-the-art works which rely on a single data source. Our novelty lies in effectively integrating relevant information from both data sources thereby ensuring larger ICD-9 code coverage, handling the inherent class imbalance, and adopting a novel approach to form the ensemble models.

Keywords: Clinical decision support systems · Healthcare informatics · Disease prediction

---

A. Prabhakar and S. Srinivasan—Equal contribution.
G. S. Krishnan—Author contributed to this work as part of Ph.D. research in HALE Lab, NITK.

S. N. Srirama et al. (Eds.): BDA 2021, LNCS 13147, pp. 197–210, 2021.
https://doi.org/10.1007/978-3-030-93620-4_15

1 Introduction

Electronic Health Records (EHR) represent a consolidated digital portfolio of a patient's medical history, which doctors can access at any time and share with other healthcare professionals. It encompasses vital information such as past medications, immunizations administered, progress notes, laboratory data, and clinical notes. The use of EHR systems has increased dramatically in hospitals due to their operational efficiency and use in secondary healthcare analytics like disease diagnosis, ICU mortality prediction, and clinical decision making. Clinical Decision Support System (CDSS) technology acts as a backbone and builds upon the foundations of EHR, and supports health-related decision making by aiding clinicians to incorporate its useful suggestions along with their knowledge.

ICD-9 (International Classification of Diseases, 9th edition) is a hierarchical taxonomy maintained by the World Health Organization (WHO) that assigns unique diagnostic codes for various medical conditions of patients. Insurance companies rely hugely on EHRs and use ICD classification to settle dues and reimbursements during patient discharge. Currently, medical coders assign appropriate ICD-9 codes after reviewing a patient's record using their domain knowledge in the medical field. However, the overwhelming rate of patient data generation makes manual coding a very cumbersome, error-prone (over-coding/under-coding), and expensive task [1,2]. As predicting unique ICD-9 codes has been found to be poor performing [3], researchers made attempts to capture the categories of diseases as a step prior to coding and therefore ICD-9 group prediction was adopted [4–6]. Hence, we focus on this problem and group codes into higher-order categories to reduce the feature set.

Important patient-related data can be found in lab test reports, which are structured in nature, and clinical notes written during their admission tenure, which are in the form of unstructured text. Utilizing only one of these would result in some observations or indicative symptoms being omitted, thus causing incorrect diagnostic coding. These aspects make automatic coding a challenging task indeed and there is a necessity for developing effective disease coding or grouping models based on integrated structured and unstructured clinical data sources. The major contributions of this paper are:

1. We present mixed ensemble models for ICD-9 code group prediction, which integrate relevant properties of both structured lab events and unstructured text data.
2. A novel ensembling mechanism based on correlation is proposed to effectively combine various individual models.
3. We perform extensive analysis on the predictions of ICD-9 code groups on the MIMIC-III dataset using standard multi-label metrics like AUROC and Hamming Loss, which can be used as a baseline for further research.

2 Related Work

Computerized ICD-9 coding, a task that has been actively explored over the past decade, has seen a significant volume of research. Previous works aiming to predict ICD-9 codes have either used only the structured clinical lab test data [5,7], or only the clinical notes texts, specifically from the discharge summary category [1–3,8,9]. Larkey & Croft [10] assigned ICD-9 codes using discharge summaries by combining kNNs, relevance feedback, and Bayesian classifiers. A single code characterized each patient visit, but in reality, there could be many. Several works proposed RNN based models and considered this as a multi-label classification task. Lipton et al. [7] utilized LSTMs with target replication and dropout to predict 128 diagnoses from irregularly sampled structured EHR data. In [1], a simple LSTM model was fed with Glove word embeddings of the discharge summary notes to predict codes.

Prakash et al. [11] utilized a condensed version of memory networks [12] with effective use of several memory hops, instead of LSTMs, and Wikipedia as the knowledge source. This, however, increases the number of model parameters and training time. Due to several unique ICD-9 codes which are very granular, most studies have reduced this task; by focusing on a small subset of codes, e.g., considering top-10/top-50 codes and categories [2] or the most commonly occurring 50/100 labels [11], or on a specific outcome such as mortality rate. Purushotham et al. [5] benchmarked ICD-9 group prediction on extensive healthcare data using a Super Learner model. This used only the structured patient data from clinical lab tests and predicted the ICD code group(s) for a patient, ignoring the large volume of data that could be obtained from the easily available clinical notes in text format. TAGS [6] utilized nursing notes and adopted vector space and topic modeling approaches to capture text semantics aiding in diagnosis. An initial attempt to predict ICD-9 codes based on structured and unstructured data was presented in [13].

We build upon this and integrate structured and unstructured data considering a more diverse cohort, and apply various types of learning algorithms to effectively extract and integrate features from these diverse sources and use a novel correlation-driven approach to form ensemble models which produce better results overall.

3 Materials and Methods

We investigate the use of various predictive models for ICD-9 coding integrating both structured and text data. Figure 1 illustrates the processes of our proposed methodology. We separately pre-process structured and text data tables and then join the two, based on hospital admission (*hadm_id*) as key.

3.1 Data Preparation and Preprocessing

MIMIC-III [14] is a database containing clinical data relating to 61,532 critical care patients who were admitted to Beth Israel Deaconess Medical Center, New York, USA. For structured data analysis, we used the contents of the *admissions*, *patients*, *labevents*, and *diagnosis_icd* tables which provided us with statistical details regarding the ICU stay and tests undertaken. For unstructured data

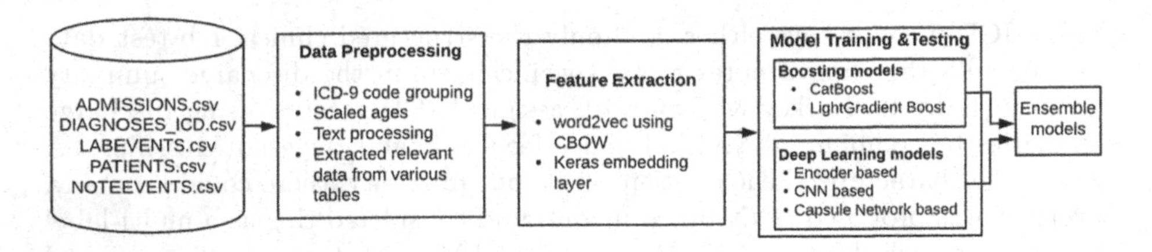

Fig. 1. Proposed methodology

analysis, we used the *discharge summary* and *radiology* notes columns of the *noteevents* table.

The *admissions* table contains information about patients' admission such as admission/discharge times and patient demographics. Every hospital visit is assigned a unique *hadm_id*. The *patients* table provides the details about a specific patient represented by a *subject_id*. Laboratory based test observations are found in the *labevents* table, which has the clinical test values for about 480 tests. *diagnosis_icd* gives the mapping between every admission and corresponding manually assigned medical ICD-9 codes. The clinical notes associated with every admission can be found in the *noteevents* table which contains several categories of notes such as radiology, discharge summary, physiology, and nursing.

Structured Data Preprocessing. The *admissions* and *patients* tables were joined on the *subject_id* key and the resulting table was joined with the *labevents* table on *hadm_id* key. Patient age was calculated as the difference between *admit_time* from the *admissions* table and *DOB* value from the *patients* table. In the dataset, the date of birth of patients above 89 years were adjusted to recondite their true age, so the ages were scaled appropriately. The age and gender distribution of the patients considered is shown in Fig. 2. The ICD-9 disease codes present in the *diagnosis_icd* table were aggregated into 20 ICD-9 disease groups, similar to existing literature [5,6]. Therefore, 20 code groups were reviewed with binary values: 0, indicating absence and 1, for presence of the ailment. The groups considered are shown in Table 1.

For every *hadm_id*, the presence/absence of the 20 disease groups was determined, which was the training target. In some cases, the same clinical tests have been taken multiple times during the admission tenure. In such a scenario, the test values were sorted according to *chart_time* in ascending order, and the values corresponding to the earliest test were considered as it reflects the patients' initial condition during admission time. These 480 different clinical tests, which are continuous variables from the *labevents* table were added to the prepared cohort with values for the tests undertaken by the patients, and others as 0. We drop records for which *hadm_id* is missing (indicates that the patient was out-patient i.e. not admitted or possible data error), and each clinical admission is treated as a unique case.

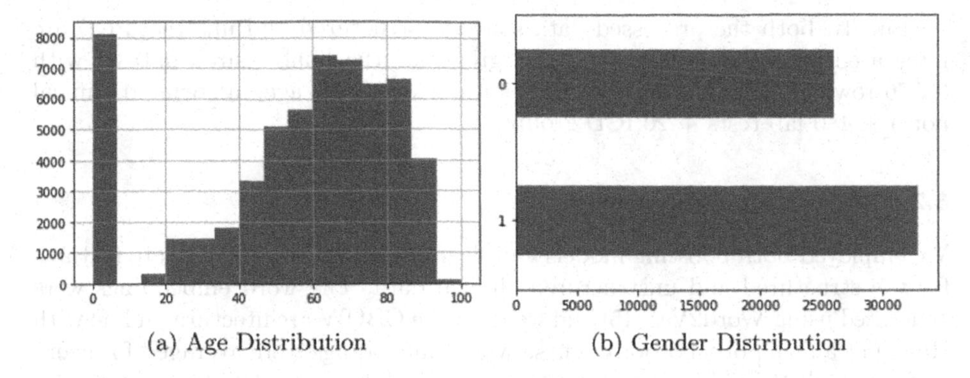

(a) Age Distribution (b) Gender Distribution

Fig. 2. Patient metadata statistics after Structured Data Preprocessing. The bump seen at 0 in (a) is due to the inclusion of newborns/neonates too in our study. In (b) classes 0 and 1 represent female and male patients respectively

Table 1. ICD-9 Code groups - *Label Statistics w.r.t Study data*

Group	Range	Occurrence	Brief description
1	001 – 139	102	*Infectious and parasitic Diseases*
2	140 – 239	502	*Neoplasms*
3	240 – 279	251	*Endocrine, nutritional, metabolic, immunity disorders*
4	280 – 289	108	*Blood and blood-forming organs' diseases*
5	290 – 319	257	*Mental disorders*
6	320 – 389	565	*Diseases of the Nervous system and sense organs*
7	390 – 459	452	*Diseases of the Circulatory system*
8	460 – 519	206	*Diseases of the Respiratory system*
9	520 – 579	406	*Diseases of the Digestive system*
10	580 – 629	260	*Diseases of the Genitourinary system*
11	630 – 677	203	*Related to pregnancy, childbirth & Puerperium*
12	680 – 709	159	*Diseases of skin and subcutaneous tissue*
13	710 – 739	374	*Musculoskeletal System and Connective Tissue*
14	740 – 759	262	*Congenital Anomalies*
15	760 – 779	206	*Conditions originating in Perinatal period*
16	780 – 796	264	*Symptoms & Non-specific abnormal findings*
17	797 – 799	18	*Ill-defined/unknown causes of morbidity & mortality*
18	800 – 999	1396	*Injury and Poisoning*
19	E Codes	502	*External causes of injury*
20	V Codes	491	*Supplementary, factors influencing health status*

Unstructured Data Preprocessing. We performed standard preprocessing on the error-free notes (having 0 in the *iserror* column) to clean the text corpus: tokenization, removal of stopwords, stemming, and lemmatization. The processed free-text notes were then concatenated by grouping based on *hadm_id*, to get a condensed note for each admission.

Finally, both the processed data sources (structured and unstructured) are integrated using the *hadm_id* as key, giving us the final desired dataset with 58976 rows $\times$ 504 columns [4 (admission days, gender, age, aggregated clinical note) + 480 lab tests + 20 ICD groups].

3.2 Feature Engineering

We employed both boosting models and deep neural models to generate features for the structured and unstructured clinical data. The word embeddings were generated using Word2Vec [15] and we used the CBOW architecture, in line with Huang et al.'s [2] observations. These word embeddings were averaged for every *hadm_id* giving equal importance to each word of the text related to a particular *hadm_id*. For feature extraction in deep neural models, the Keras embedding layer was used. This layer takes one-hot encoded data as input and generates embeddings, guided by the loss function during training of the model, which are more suited towards the specific task being trained, rather than just based on contextual similarity/word occurrences as in pre-trained embedding models. Following tokenization, the input sequence was padded, keeping max length as the average length of all sequences.

3.3 Disease Group Prediction Models

As a patient can suffer from multiple diseases, we model our task as a binary classification of multiple labels. The models have not been fine-tuned as our emphasis was on exploring different architectures.

Boosting Models. Boosting is a standard ensemble method where weak learners are trained sequentially, each attempting to fit on its predecessors' residual error. CatBoost [16] performs well on textual, image data and is based on the gradient boosting machine learning algorithm. The structured data was passed through a pipeline consisting of OneVsRestClassifier and LGBM Classifier [17] to get the predicted probabilities for each ICD code group. In the case of unstructured clinical data, the word embeddings generated were sent to a similar pipeline as above, consisting of OneVsRestClassifier and CatBoost[16] /LGBM Classifier. Our dataset is highly imbalanced, with over 90% of the label entries being 0, indicate the absence of disease. We used class weights to overcome sparsity issue. Let a, b be class labels (here 0 and 1) and P_a, P_b their occurrence, then the weights used are:

$$class\_weight_a = P_b/(P_b + P_a)$$
$$class\_weight_b = P_a/(P_b + P_a)$$

respectively.

Deep Neural Models. We also experimented with three different deep neural architectures – Encoder model, convolutional neural model and capsule networks. The details of the experiments are discussed in detail below.

a. Encoder Architecture: After the embedding layer, the vectors are passed to an encoder to utilize multi-headed self-attention [18] which jointly attends to information from different sub-spaces. Since common medical terms are coded similarly, they are given more importance due to attention. Positional encoding was not used as we are more concerned with medical terms indicating a particular disease's presence instead of sentence structure.

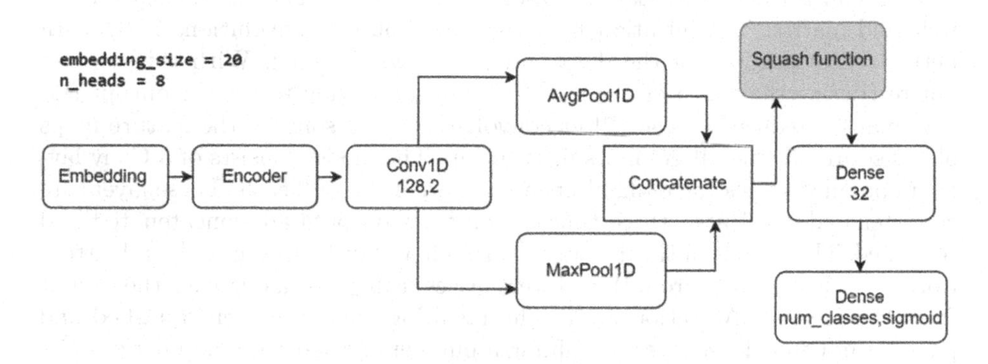

Fig. 3. Architecture 1: encoder model

The batch normalization layers were replaced with squash functions, and the number of units in the feed-forward layer was reduced. The encoder's output, taken from the hidden layer, was fed into a Conv layer with 128 filters and a kernel size of 2. This output is separately passed through MaxPool and AvgPool layers, which are later concatenated to combine all the words' average weights and the most weighted word in every sliding window. After concatenation, the output is squashed and passed to a feed-forward layer with 32 units followed by the output layer, which contains the number of units equal to the number of classes and uses a sigmoid activation function instead of a softmax function, as the aim is to find probabilities of each class occurrence rather than the most confident one. The motivation behind this was to leverage the word encodings generated by a transformer based encoder. By concatenating outputs from different pooling functions we try get more features from existing features. We found the squash activation function [19] to work better than ReLU.

b. Convolutional Neural Network (CNN): After passing through the embedding layer, the input is squashed by the squashing function, [19] which is an activation function used to keep the variance in the range (0,1) (Eq. 1). We used it here instead of other well-known activation functions, as applying it on the output reduces the variance, where $\mathbf{s}_j$ and $\mathbf{v}_j$ are the input and output vectors of capsule

j. $\mathbf{v}_j$ is passed to a Conv layer with 64 filters and kernel size of 2. Next, the output is squashed and passed through grouped convolutions [20], and an attention layer that uses a weighted average to calculate the importance of every word of the given input as shown in Eq. 2

$$\mathbf{v}_j = \frac{||\mathbf{s}_j||^2}{1 + ||\mathbf{s}_j||^2} \frac{\mathbf{s}_j}{||\mathbf{s}_j||} \tag{1}$$

$$e_t = h_t w_a; \qquad a_t = \frac{exp(e_t)}{\sum_{i=1}^{T} exp(e_i)}; \qquad v = \sum_{i=1}^{T} a_i h_i \tag{2}$$

where h_t is the hidden state representation at time t, and w_a is the attention weighted matrix. The intuition behind using grouped convolutions is to learn better representations of the data and parallelize learning. With this, we are not restricting the width of our model, and computation to get the output feature maps is reduced as each filter convolves only on some of the feature maps obtained from kernel filters in its filter group. The model consists of 3 Conv layers in parallel, consisting of 32 filters and a kernel size of 2 each. These layers are concatenated, replicated three times; then their outputs are concatenated and squashed. The output of this is passed through a MaxPool, AvgPool, and Attention layer. The outputs from these layers are concatenated along with the output from the previous Attention layer. The resulting output is then squashed and passed through a dense layer containing a number of classes as a number of units and passed through a sigmoid activation function to give each class's prediction weights. Since our data is highly imbalanced, conventional deep networks tend to overfit easily by generalizing all the predicted labels as the majority class.

c. Capsule Network: When CNNs and pooling layers such as max-pooling and average-pooling are used for text classification tasks, many useful features are lost as max-pooling only retains the feature with the highest activation, and average pooling represents the input vector at each position equally. However, in Capsule Networks, [19] dynamic routing chooses to preserve not only one but all features that are useful, as long as they are "agreed" among layers. This was the key factor for using a capsule network based model, as medical notes often have multiple medical terms signifying the presence or absence of certain diseases. As these medical terms are common they are given more weightage during the routing compared to other text.

The architecture used, depicted in Fig. 5, begins with a convolution layer that extracts n-gram features at various sentence locations using different filters. Next is the primary capsule layer. We used softmax in our routing algorithm as we have a multi-label classification task on our hands. Capsule Layer takes in output from Keras embedding layer, which transforms each word in our corpus to a 20-dimensional (20 - number of class labels) vector. A convolutional layer follows this with kernel size (9,1) and stride length of 1. Every layer so forth is a capsule layer consisting of 10 capsules, each instantiated with 16-dimensional parameters, and the length indicates the probability of the capsules' existence. These layers are

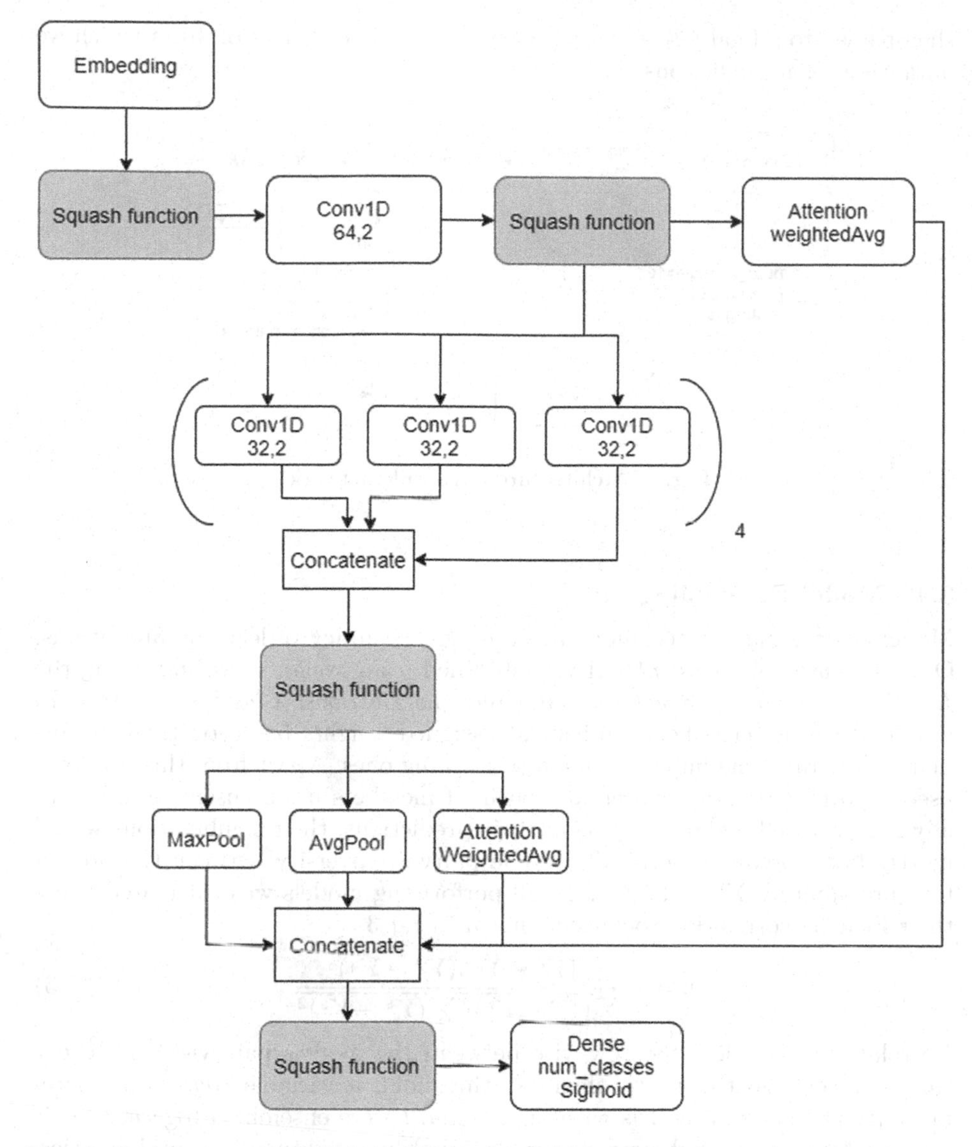

Fig. 4. Architecture 2: CNN based Deep Neural Model, bracket and number denote number of parallel blocks concatenated.

connected via transformation matrices, with every connection multiplied by a routing coefficient, which is computed dynamically by the routing mechanism. In our architecture, we use consecutive capsule layers after the embedding layer. We use an *AttentionWeightedAverage* layer which is an attention layer that uses a weighted average to calculate the importance of every word, shown in Eq. 2. The capsule layers' output is passed to both *AttentionWeightedAverage* and *Flatten* layers simultaneously. The outputs of these layers are concatenated and then passed to a final feed-forward layer with sigmoid activation from which we obtain the final predictions.

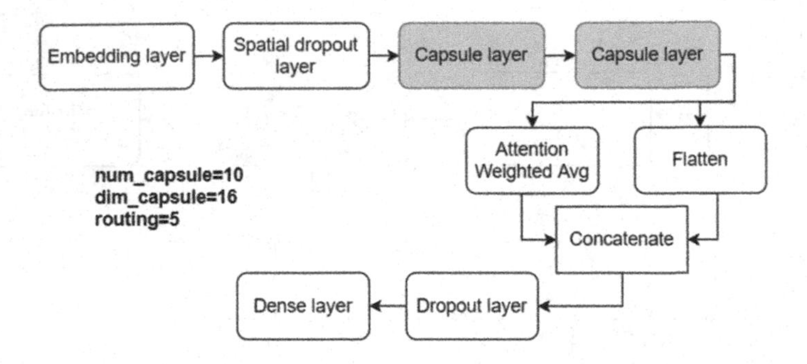

Fig. 5. Architecture 3: capsule network

3.4 Model Ensembling

Model ensembling is a technique to boost the accuracy of learning algorithms. Our first ensemble, *Ensemble A* was obtained using weighted voting, using the *LGBM Classifier on structured data* and the *CatBoost Classifier on text.* In weighted voting, classifiers are instead assigned weights based on their performance to reduce the impact of poor-performing ones. Apart from this, we have used a novel correlation-driven approach. If members in an ensemble individually perform well and are diverse in their predictions, their combinations would mostly lower prediction errors [21]. The pairwise diversity between the prediction probabilities Y^r and Y^s of 2 well-performing models was calculated using their Pearson correlation coefficient shown in Eq. 3,

$$r = \frac{\sum(Y_i^r - \bar{Y^r})(Y_i^s - \bar{Y^s})}{\sqrt{\sum(Y_i^r - \bar{Y^r})^2 \sum(Y_i^s - \bar{Y^s})^2}} \tag{3}$$

Correlation was calculated row-wise between the models pair-wise, and every time the coefficient was less than the threshold, a variable *count* was incremented. The pairs of models with higher *count* were ensembled together (as it indicated diverse, though accurate predictions) by weighting their probabilities. The weights used were the log-reduced values of the calculated *count*. The final model, *Ensemble B* was built using - *LGBM Classifier on structured data, CNN, Encoder* and *Capsule Network.*

4 Results and Analysis

4.1 Baseline Models and Experimental Setup

To the best of our knowledge, no prior work has considered an integrated approach similar to ours, so a direct comparison is not plausible. However, we compare with some existing benchmark works, Super Learner and MMDL [5] and

TAGS [6], which use a single data source. There are significant differences in our data modeling, as well as the final cohort used for prediction:

1. We take every admission (including re-admissions) as an independent instance without any data leak, rather than taking just the first admission of a patient. This results in a larger number of entries (58,976), which indirectly helps to simulate more patients having different target labels each time.
2. Integrating structured and text data ensures better ICD-9 code coverage. Unlike some other works that consider only adults (>15 years), all patients are taken, so our model is capable of predicting neonatal and congenital diseases as well.

We used Sklearn, Pandas, Gensim, and Keras and experimented on NVIDIA M40 processor for the experiments. The optimizer used was *Adam* and the loss function was *binary_crossentropy*. Models were trained for 1000 epochs, the learning rate was set as 0.1, embedding size as 1000, minimum word frequency as 3, and window size as 10 for Word2Vec. For our Pearson correlation approach, we set the threshold to 0.7. All hyper-parameters were empirically determined. We use standard scoring metrics: AUROC Score, a label-based metric, is a plot showing the trade-off between true positive and false-positive rates for various thresholds and tells how well the model distinguishes among the classes. Hamming loss, an example-based metric that represents the fraction of misclassified labels, is considered a better metric for multi-label prediction tasks [22,23].

4.2 Results

Table 2 shows the individual models' performance on structured and unstructured data independently and results of ensembling these, selected as described in Sect. 3.4 along with existing baseline models. Ensemble A has a lower hamming loss as it consists of a combination of boosting models, where we could prevent the class imbalance by assigning class weights (refer Sect. 3), and hence the results are more accurate patient-wise. Ensemble B consists of a combination of boosting models as well as various deep learning based models, where the latter was used to capture features from the text. However, as deep learning models fall prey to class imbalance, we get a higher AUROC score, but the hamming loss increases compared to Ensemble A. Additionally, even though we consider a larger number of hospital admissions and more patient types, our results are comparable to existing works, whose cohort is less diverse.

4.3 Discussions

Achieving good results with regards to both Hamming loss and AUROC scores is essential, as in multi-label classification, no misclassification is a hard wrong or right prediction. Having a low hamming loss ensures a more accurate prediction of labels overall as it indicates a row-wise error (at patient-level), and optimizing this metric implies more patients are correctly diagnosed. In AUROC, the

Table 2. Comparison of proposed models with existing works. **Explained in Sect.* 3.4. *(–) denotes unreported value*

Sl. no.	Model	Hamming Loss	AUROC
Baselines			
1	FFN (Structured only, on 38425 entries) [5]	–	0.717
2	RNN (Structured only, on 38425 entries) [5]	–	0.724
3	Super Learner (Structured only, on 38425 entries) [5]	–	0.758
4	MMDL (Structured only, on 38425 entries) [5]	–	0.777
5	TAGS (Unstructured only, on 6532 entries) [6]	–	**0.787**
I. Performance on structured data			
1	*OneVsRestClassifier + CatBoostClassifier*	0.193	0.686
2	*OneVsRestClassifier + XGBClassifier*	0.173	0.647
3	*OneVsRestClassifier + LGBMClassifier*	0.169	0.672
II. Performance on unstructured data			
1	*Word2Vec + LGBMClassifier*	0.182	0.664
2	*Word2Vec + CatBoostClassifier*	0.181	0.667
3	*Word2Vec + XGBClassifier*	0.182	0.656
4	*Architecture 1: Encoder*	0.172	0.670
5	*Architecture 2: CNN*	0.187	0.651
6	*Architecture 3: Capsule Network*	0.176	0.659
III. Performance of Ensembles on both structured and unstructured data			
1	**Ensemble A* (on 58976 entries)**	**0.154**	0.748
2	**Ensemble B* (on 58976 entries)**	0.172	**0.768**

average of all labels' AUROC scores is taken, which may be biased. For instance, some labels may exist for which the model yields very accurate results, while for some, the results may be abysmal. So being able to separate the various labels (ICD-9 code groups) and predicting the codes correctly for every admission is vital, which is why we have tried to optimize both the metrics. Most previous works [5,6,8,24] have not assessed their models based on hamming loss but have achieved good results for AUROC. Our ensemble models are able to achieve comparable AUROC scores and good hamming loss scores too.

A major problem we face is dealing with the huge class imbalance in the multilabel classification task. A patient rarely has more than 3 diseases at a time, and the class imbalance is more apparent in some classes representing rare disease groups. Deep learning models tend to usually pick this up and learn the labels in majority and generalize it over all classes. This does give good accuracy scores overall; however, it is incorrect. Hence we are in a way forced to look towards combining results from other models, as every model captures different sets of features and ensembling these will help overcome the bias towards the majority label, as we are able to capture a wider range of features, which any single model might have missed. This also reflects in our results, where we get far better results when we ensemble different models.

5 Conclusion and Future Work

In this work, an approach that integrates structured and text data features for ICD-9 code group prediction is presented. To combat the extensive code combinations, we aggregated them into groups and used frequency-based weights to handle class imbalance. Pearson correlation was utilized to select models and the weightage to be given to them in our ensembles. Our study serves as the benchmark for better ICD-9 code coverage (combining structured + unstructured data), patient coverage (every admission instance considered, and neonates also included), and evaluating relevant metrics like AUROC score and hamming loss. There is scope to improve performance by better data modeling, exploiting pretrained word embeddings, and appropriate fine-tuning of hyperparameters. We also observe an undeniable trade-off between the AUROC and hamming loss in our models, and optimizing both would be future work.

References

1. Ayyar, S., Don, O., Iv, W.: Tagging patient notes with icd-9 codes. In: Proceedings of the 29th Conference on Neural Information Processing Systems, pp. 1–8 (2016)
2. Huang, J., Osorio, C., Sy, L.W.: An empirical evaluation of deep learning for icd-9 code assignment using mimic-iii clinical notes. Comput. Methods Programs Biomed. **177**, 141–153 (2019)
3. Perotte, A., et al.: Diagnosis code assignment: models and evaluation metrics. J. Am. Med. Inf. Assoc. JAMIA **21** (2013)
4. Choi, E., Bahadori, M.T., Schuetz, A., Stewart, W.F., Sun, J.: Doctor ai: predicting clinical events via recurrent neural networks. JMLR Workshop and Conf. Proc. **56**, 301–318 (2016)
5. Purushotham, S., Meng, C., Che, Z., Liu, Y.: Benchmarking deep learning models on large healthcare datasets. J. Biomed. Inf. **83** (2018)
6. Gangavarapu, T., Jayasimha, A., Krishnan, G.S., S., S.K.: Predicting icd-9 code groups with fuzzy similarity based supervised multi-label classification of unstructured clinical nursing notes. Knowl. Based Syst. **190**, 105321 (2020)
7. Lipton, Z.C., Kale, D.C., Elkan, C., Wetzel, R.: Learning to diagnose with LSTM recurrent neural networks. In: 4th International Conference on Learning Representations, ICLR 2016, San Juan, Puerto Rico(2016)
8. Xie, P., Xing, E.: A neural architecture for automated ICD coding. In: Proceedings of the 56th Annual Meeting of the ACL. ACL, pp. 1066-1076 (2018)
9. Krishnan, G.S., Kamath S.S.: Ontology-driven text feature modeling for disease prediction using unstructured radiological notes. Computación y Sistemas **23**(3) (2019)
10. Larkey, L.S., Croft, W.B.: Combining classifiers in text categorization. In: Proceedings of the 19th Annual International ACM SIGIR Conference on Research and Development in Information Retrieval. ACM, pp. 289-297 (1996)
11. Prakash, A., et al.: Condensed memory networks for clinical diagnostic inferencing. In: Thirty-First AAAI Conference on Artificial Intelligence (2017)
12. Sukhbaatar, S., Szlam, A., Weston, J., Fergus, R.: End-to-end memory networks. In: Proceedings of the 28th International Conference on Neural Information Processing Systems. Vol. 2, pp. 2440–2448. NIPS'15, MIT Press, Cambridge, MA, USA (2015)

13. Akshara, P., Shidharth, S., Krishnan, G.S., Kamath, S.: Integrating structured and unstructured patient data for icd9 disease code group prediction. In: 8th ACM IKDD CODS and 26th COMAD, p. 436. CODS COMAD 2021, Association for Computing Machinery, New York, NY, USA (2021)
14. Johnson, A.E., et al.: Mimic-iii, a freely accessible critical care database. Sci. Data **3**(1), 1–9 (2016)
15. Mikolov, T., Chen, K., Corrado, G., Dean, J.: Efficient estimation of word representations in vector space (2013)
16. Prokhorenkova, L., Gusev, G., Vorobev, A., Dorogush, A.V., Gulin, A.: Catboost: unbiased boosting with categorical features. In: Proceedings of the 32nd International Conference on Neural Information Processing Systems, pp. 6639–6649. NIPS'18, Curran Associates Inc., Red Hook, NY, USA (2017)
17. Ke, G., et al.: Lightgbm: a highly efficient gradient boosting decision tree. In: Proceedings of the 31st International Conference on Neural Information Processing Systems, p. 3149–3157. NIPS'17, Curran Associates Inc., Red Hook, NY, USA (2017)
18. Vaswani, A., et al.: Attention is All You Need, pp. 6000–6010. NIPS'17, Curran Associates Inc., Red Hook, NY, USA (2017)
19. Sabour, S., Frosst, N., Hinton, G.E.: Dynamic routing between capsules. In: Proceedings of the 31st International Conference on Neural Information Processing Systems, pp. 3859–3869. NIPS'17, Curran Associates Inc., Red Hook, NY, USA (2017)
20. Krizhevsky, A., Sutskever, I., Hinton, G.E.: Imagenet classification with deep convolutional neural networks. In: Pereira, F., Burges, C.J.C., Bottou, L., Weinberger, K.Q. (eds.) Advances in Neural Information Processing Systems. vol. 25, pp. 1097–1105. Curran Associates, Inc. (2012)
21. Sluban, B., Lavrac, N.: Relating ensemble diversity and performance: a study in class noise detection. Neurocomputing **160**, 120–131 (2015)
22. Wu, X.-Z., Zhou, Z.-H.: A unified view of multi-label performance measures. In: Proceedings of the 34th International Conference on Machine Learning. Vol. 70, pp. 3780–3788. ICML'17, JMLR.org, Sydney, NSW, Australia (2017)
23. Zhang, M., Zhou, Z.: A review on multi-label learning algorithms. IEEE Trans. Knowl. Data Eng. **26**(8), 1819–1837 (2014)
24. Shickel, B., Tighe, P.J., Bihorac, A., Rashidi, P.: Deep ehr: a survey of recent advances in deep learning techniques for electronic health record (ehr) analysis. IEEE J. Biomed. Health Inf. **22**(5), 1589–1604 (2018)

SCIMAT: Dataset of Problems in Science and Mathematics

Snehith Kumar Chatakonda, Neeraj Kollepara, and Pawan Kumar(✉)

CSTAR, International Institute of Information Technology, Hyderabad, India
{snehith.kumar,neeraj.kollapara}@students.iiit.ac.in,
pawan.kumar@iiit.ac.in
http://faculty.iiit.ac.in/~pawan.kumar

Abstract. Datasets play an important role in driving innovation in algorithms and architectures for supervised deep learning tasks. Numerous datasets exist for images, language translation, etc. One of the interesting challenge problems for deep learning is to solve high school problems in mathematics and sciences. To this end, a comprehensive set of dataset containing hundreds of millions of samples, and the generation modules is required that can propel research for these problems. In this paper, a large set of datasets covering mathematics and science problems is proposed, and the dataset generation codes are proposed. Test results on the proposed datasets for character-to-character transformer architecture show promising results with test accuracy above 95%, however, for some datasets it shows test accuracy of below 30%. Dataset will be available at: www.github.com/misterpawan/scimat2.

Keyword: Question-Answering, Mathematics, Science, Transformer

1 Introduction

We all wonder how a child learns to solve problems in science and mathematics. It is one of the most difficult tasks for most of us to gradually acquire. We also acknowledge that most of us acquire language skills first, i.e., we learn how to read, write, and speak in our language. This is followed by systematic development of skills in counting, obtaining familiarity with names of objects, learning rules and laws of science, which includes physics, chemistry, biology, commerce, etc. We also acknowledge that language skills help us in deciphering the question in our language followed by extracting the problem statement in a form that we can put them into known laws or methods, and arrive at a final solution. Hence, we assume that problem solving in science and mathematics requires a decent understanding of natural language as a first step.

There have been remarkable advances in last 10 years in tasks such as language translation, speech recognition and translation, together classified as natural language processing or speech processing. Deep neural networks such as Recurrent Neural Networks (RNN), Long Short Term Memory (LSTM) [9], and a simplified variant of LSTM, the gated recurrent unit (GRU) [6] have been a

S. N. Srirama et al. (Eds.): BDA 2021, LNCS 13147, pp. 211–226, 2021.
https://doi.org/10.1007/978-3-030-93620-4_16

method of choice until recently. More recently, attention based methods such as Transformers [13] was proposed. The related architectures such as BERT [7] and very recently, in the year 2020, a large GPT-3 [3] model with 175 Billion parameters was proposed. Availability of datasets propels research towards discovering powerful neural architectures thereby improving the state-of-the-art supervised learning algorithms.

In general, deep learning architectures such as CNN have been extremely impactful in other wide variety of tasks such as those related to vision, speech, etc. Given the powerful representation ability of the modern architectures, we wonder whether they are powerful enough to do other related tasks where labelled data can be generated. One such problem is science and mathematical problem solving. We however, realize that problems in science and mathematics present us with a hierarchy of difficulty levels. To revisit some of the problems in mathematics, we were taught root finding, polynomial divisions, differential equations, integration and differentiation, matrices, statistics, etc. On the other hand, in high school science, we were taught equations of motion, force, sound, work, power, energy, chemical equations, thermodynamics, etc. These questions are written in a natural language, and we need to decipher the problem, understand the rules or laws, then formulate equations and eventually solve it.

We discuss about related work in this area in Sect. 2 of this paper, to facilitate further research in this area, we have the following contributions.

1. **Comprehensive labelled datasets for high school science and mathematics problems:** The dataset includes wide variety of question answer-data pairs in plain .txt file format. For each type of problems, there are roughly 0.5 to 4 million samples. We discuss about different types of problems available in the dataset in detail in Sect. 3 of this paper.
2. **Codes for generating these datasets:** For the datasets, corresponding dataset generation codes in python are provided. The codes can be easily modified to generate different variations of datasets and encoding formats. An example for a generation code is explained in Sect. 4.3 of this paper.
3. **Results with character to character encoding:** For these new datasets, we show results with char2char encoding for transformer model [13]. For some of these problems, we achieve a test accuracy of 95%–100%, for some, we achieve accuracy between 75%–95%, and we also show results for some where the accuracy is very poor. We discuss more about char2char transformer model and results obtained on it in Sect. 4 of this paper.

2 Related Work

The powerful representation and function approximation ability of neural networks led to its usage in specific problems in science and mathematics. Plenty of science and engineering problems involve modelling the problem by differential equations. For example, Navier-Stokes problems model fluid flow problems, Schodinger equation is used to model the wave function of quantum mechanical system, Maxwell equations is considered a mathematical model to model electricity and magnetism. All these models are described by partial differential equations. In [12], neural networks were used to solve such partial differential equations described above. In the context of dynamical systems, an interesting application of neural architectures were used to replicate chaotic attractors. For puzzle type problems, neural networks, in particular, LSTM were used [4], and for mathematics problems, datasets and some test accuracy results were shown in [11]. However, the dataset provided were limited in scope.

Since 1963, for solving word problems in math, various semantic rules and models have been proposed starting from [1,2,8] in the NLP community. For example, some word problems in [10] are as follows: *Anne has nine flowers. Kathy has five flowers less than Anne. How many flowers does Kathy have?* or *Dennis has eleven apples. Fred has six apples. How many apples does Fred have less than Dennis?*, Kintsch and Greeno's (1985) theory of comprehension and solution for arithmetic word problems was used in this paper, classical approaches with semantic rules were also used in these papers. Also, there are datasets for logical reasoning and English comprehension. For example, logical reasoning question answer dataset is proposed in [14]. Problems involving single supporting fact, two supporting fact, counting, path finding, size reasoning, etc. can be considered as various type of reasoning. For example, a sample question for size reasoning is as follows: *The football fits in the suitcase. The suitcase fits in the cupboard. The box is smaller than the football. Will the box fit in the suitcase? Will the cupboard fit in the box?*. The answers for above two questions is *yes* and *no* respectively. This dataset is a part of bAbI project[1] of facebook research.

3 Datasets

3.1 Existing DeepMind Datasets

Datasets for some mathematics problems were proposed in [11]. These datasets were divided into train-easy, train-medium, train-hard on the difficulty levels. The dataset contains 1.5 million questions and answers. The dataset corresponds to the following topics:

1. **Algebra:** This dataset contains problems for solving system of linear equations, finding LCM, GCD, and prime factorization of integers.
2. **Polynomials:** This dataset includes problems on polynomial addition, polynomial expansions, etc.
3. **Arithmetic:** In this dataset, the problems are on basic addition, division, multiplication, etc.
4. **Comparisons:** This dataset contains problems for sorting, finding kth biggest number, etc.
5. **Others:** This dataste contains problems for base conversions, finding place values, rounding decimal numbers, etc.

[1] https://github.com/facebookarchive/bAbI-tasks.

3.2 Our New Datasets

New Additional Mathematics Dataset. We propose following extensive set of datasets for mathematics from various topics. Some of the topics are listed below.

1. **Calculus:** Differentiation and Integration (for product and composition). In this topic, the problems involve calculating symbolic differentiation or integration of product and compositions of functions. Note that the architecture needs to learn not only the differentiation of single function, but learn to apply product rule for the product of functions and chain rule for the composition of functions. Example for this type of problems are in item 1 and 2.
2. **Linear Algebra:** In this topic, the problems involve sum of matrices, difference of matrices, product of matrices, rank and trace of matrices, row reduced echelon form of a matrix, determinant of a matrix, etc. As we can see, some of these operations such as rank, row echelon form, and determinant of a matrix are hard tasks for the transformers to learn. We show results for these in experiments section. Example for this type of problems are in items 3 and 4.
3. **Set Operations:** Here the problems involve basic set operations such as set union, set intersection, set difference, etc. Example for this type of problems are in items 5 and 6.
4. **Statistics:** In this topic, the problems involve computing mean, median, variance of a given dataset. A sample question for this type is shown in item 7.
5. **Number Theory:** In this topic, the problems involve computing GCD, LCM, prime factor etc. Example for this type of problems are in items 8, 9 and 10.
6. **Probability:** In this topic, the problems involve computing probabilities based on counting. A sample question for this type is shown in item 11.

We have contributed 1 Million to 3 Million labelled samples for each type of problem. Some of the problems above for example, integration and differentiation require symbolic computation. For this we used `SymPy` to generate the symbolic answers. The sample questions and answers from these datasets developed by us are as follows:

3.3 Sample Question in Mathematics

1. **Question:** Differentiate 137 * x * (cos(x) + sec(x)) with respect to x
   Answer: 137 * x * (−sin(x) + tan(x) * sec(x)) + 137 * cos(x) + 137 * sec(x)
2. **Question:** Integrate 654*x^2*(sin(x)+cos(x)) with respect to x
   Answer: 654*x^2*sin(x) − 654*x^2 * cos(x) + 1308*x*sin(x) + 1308*x*cos(x) − 1308*sin(x) + 1308*cos(x)
3. **Question:** Calculate the Rank of Matrix [[2, 1, 3, 7] , [1, 0, 4, 2] , [3, 1, 7, 9]]
   Answer: 2
4. **Question:** Calculate the Trace of Matrix [[27, 11, 34] , [89, 28, 55] , [19, 17, 49]]
   Answer: 104

5. **Question:** Find the difference of { 1, 3, 5, 7, 9 } and { 1, 4, 7, 8, 9 }
   Answer : $\{3, 5\}$
6. **Question:** What is the union of { 1, 4, 7 } with { 1, 2, 3, 7, 9 }
   Answer : $\{1, 2, 3, 4, 7, 9\}$
7. **Question:** What is the median of the sequence (31, 15, 17, 4, 27)
   Answer : 17
8. **Question:** Calculate the highest common divisor of 250 and 200.
   Answer: 50
9. **Question:** Round 1345678.963852741 to 5 decimal places
   Answer: 1345678.96385
10. **Question:** What is 5 (base 7) in base 9?
    Answer: 5
11. **Question:** What is prob of picking 5 h when five letters picked without replacement from hghhegeghggeh ?
    Answer: 1/1287

Datasets from Science. In the following we show some sample question and answer for our dataset. The problems from high school science can be divided into the following categories.

1. **Acids And Bases:** In this topic, problems are based on PH, POH and Neutralization reactions. A sample question for this type is shown in item 1.
2. **Atomic Structure:** In this topic, problems are based on isotopes and dependencies between the atomic number, atomic weight and number of electrons, protons, neutrons in the atom. Example for this type of problem is in item 2.
3. **Stoichiometry:** In this topic, problems are based on molar mass, number of moles, number of particles, and Avogadro number. Example for this type of problem is in item 3.
4. **Thermodynamics:** In this topic, problems are based on Adiabatic and Isothermal processes, Gibbs Free Energy, relation between Enthalpy, Work done and Internal energy. This also include questions on Refrigerator and Steam Engine. A sample question for this type is shown in item 4.
5. **Units And Dimensions:** In this topic, problems are based on Unit System conversions, Significant Digits, Least Count, Error and Estimated values. Example for this type of problems is in item 5.
6. **Kinematics:** In this topic, problems based on Equations of Motion are included. These involves the relationships between Initial speed/velocity, Final speed/velocity, Time, Acceleration and Distance/Displacement. This section also includes questions on Basics of Vectors, Relative Velocity and Projectile Motion. A sample question for this type is shown in item 6.
7. **Laws of Motion:** In this topic, problems are based on Newtons laws of motion and Force. This also includes questions on Friction, Impulse, Momentum, Collisions, Bodies in Contact. Examples for this type of problems are in items 7 and 8.
8. **Work Power Energy:** In this topic, problems are based on Work, Energy and Power. This includes problems on notion of Work, Potential Energy (PE), Kinetic Energy (KE) and Conservation of Energy. Example for this type of problems is in item 9.

9. **Rotatory Motion:** In this topic, problems are based on Rotatory Kinematics, Moment of Inertia, Angular Momentum, Rotational Kinetic Energy, Center of Gravity. A sample question for this type is shown in item 10.
10. **Gravitation:** In this topic, problems are based on Gravitational Force between two objects using Universal law of Gravitation, Motion of Objects under influence of Gravitational Force, Gravitational Field and Potential. Example for this type of problems is in item 11.
11. **Electricity:** In this topic, problems are based on Electrical Current, Electrical Circuit, Electrical Potential, Ohm's law, Resistance of a System of Resistors, Capacitance of a System of Capacitors, Dipole, Flux, Electric Charges and Fields, Alternating Current, LCR circuit. Examples for this type of problems are in items 12 and 13.
12. **Moving Charges and Magnetism:** In this topic, problems are based on Magnetic Fields, Force and Torque due to Magnetic Fields, Magnetic Moment, Magnetic Effects due to Moving Charges and Sensitivities. Examples for this type of problems are in items 14 and 15.
13. **Electro Magnetic Induction:** In this topic, problems are based on induced EMF, self and mutual inductance. Example for this type of problems is in item 16.
14. **Alternating Current and Electro Magnetic Waves:** In this topic, problems are based on angular frequency, alternating current in LCR circuits, Basics of EM Waves. A sample question for this type is shown in item 17.
15. **Ray Optics and Optical Instruments:** In this topic, problems are based on Magnification, Mirror Formula, Lens Formula, Lensmakers Formula, Myopia-Hypermetropia, Refractive Index, Power of Lens, Apparent and Real Depth, Critical Angle, Prism, Telescope. These includes questions on different types of mirrors/lens for different positions of objects and images. Examples for this type of problems are in items 18 and 19.
16. **Wave Optics and Dual Nature of Matter:** In this topic, problems are based on Young's double slit experiment, brewster angle, fresnel distance. A sample question for this type is shown in item 20.
17. **Mechanical Properties of Solids and Liquids:** In this topic, problems are based on Surface Tension, Viscosity, Angle of Contact, Bernouli Equation, Laminar Flow, Pressure, Relative density, Bulk Modulus, Young Modulus, Shear Modulus and their relations. Examples for this type of problems are in items 21 and 22.
18. **Thermal Properties of Matter, Kinetic theory of Gases:** In this topic, problems are based on average Thermal Energy, Mean Free Path, Vrms, Vavg, Grahams law, Ideal Gas Equation, Linear, Areal and Volume expansions, relation between Temperature Scales, Specific Heats. Example for this type of problems is in item 23.
19. **Sound:** In this topic, problems are based on Speed of Sound and its relationship to Wavelength and Frequency, and whether a Sound would be Audible or not to humans. A sample question for this type is shown in item 24.
20. **Waves And Oscillations :** In this topic, problems are based on Simple Harmonic Motion, Springs, Damping, Transverse and Longitudinal Waves, Phase Difference, Wave Equation, Superposition of Waves, Resonance and Beats. A sample question for this type is shown in item 25.
21. **SemiConductors and Communication Systems:** In this topic problems are based on Amplifiers, p-n junction, Modulation Index. Examples for this type of problems are in items 26 and 27.

Some of the problems above for example, vector differentiation require symbolic computation. For this we used SymPy to generate the symbolic answers. And some of the science datasets requires element names along with its properties like atomic number, etc. For this we used mendeleev. The sample questions and answers from these datasets developed by us are as follows:

3.4 Sample Questions in Science

1. **Question:** Calculate the pH of 0.97512 N of HCl acid.
   Answer: 0.011
2. **Question:** If an atom X have an atomic mass of 301 u and 97 protons in its nucleus. How many neutrons does it have?
   Answer: 204
3. **Question:** If one mole of Nitrogen atoms weigh 14.0 g, what is the mass (in grams) of 4097 atoms of Nitrogen?
   Answer: 9.5e-20 g
4. **Question:** If a monoatomic gas of 4 mol at 109 atm and volume 11 lit is adiabatically changed to pressure 141 atm, then what will be its volume.
   Answer : 9.4 lit
5. **Question:** Two gold pieces of masses are 9.48 g and 49500.0042 g respectively. What is the sum of the masses of the pieces to correct significant figures?
   Answer: 4.95e+04 g
6. **Question:** A body is dropped from a height of 6065 m with an initial velocity of 75 m/s. With what velocity will it strike the ground ?
   Answer: 352.8 m/s
7. **Question:** A force is applied on a body of mass 42 kg placed on a smooth surface resulting in acceleration of 45 m/s2, then what is the force applied ?
   Answer: 1890 N
8. **Question:** Two objects are moving along the same line and direction with velocities of 2 m/s and 33 m/s respectively. They collide and after the collision their velocities are 12 m/s and 25 m/s respectively. If the mass of second object is 10 g, then determine the mass of first object.
   Answer: 8.0 g
9. **Question:** If a force of 1i + −3j + 9k N acts on a body and displaces it 47 m in direction of x, 3 m in direction of y and 15 m in direction of z, then what is the work done by the force on the body.
   Answer: 173 J
10. **Question:** A solid cylinder and a hollow cylinder rolls down two different inclined planes of the heights 13 and 11 and with angles of inclination of 52 and 23, which will reach the ground faster ?
    Answer: solid cylinder
11. **Question:** What is the gravitational field at a point P at a distance of 86 m from the center of the uniform solid sphere of radius 99 m and mass 11 kg.
    Answer: −6.50e-14 J/kg
12. **Question:** What is the lowest total resistance that can be secured by combinations of three coils of resistance 188, 154, 33?
    Answer: 23.7 Ω
13. **Question:** An Electric heater of resistance 13 Ω draws 6 A current from the service mains in 3069 s. Calculate the rate at which heat is developed in the heater?
    Answer: 468 W

14. **Question:** Two moving coil meters M1 and M2 have the following particulars R1 = 3 Ω, N1 = 2, A1 = 0.0003 m^2, B1 = 10 T, K1 = 2 and R2 = 6 Ω, N2 = 3, A1 = 0.0004 m^2, B1 = 5 T, K2 = 6. Determine the ratio of voltage sensitivities of M1 and M2.
Answer: ratio of current sensitivity of M1 to M2 is 6.0
15. **Question:** A square coil of side 21 cm consists of 20 turns and carries a current of 7 A. The coil is suspended vertically and the normal to the plane of the coil makes an angle of 49 degress with the direction of a uniform horizontal magnetic field of magnitude 9 T. What is the magnitude of the torque experienced by the coil.
Answer: 41.9 N-m
16. **Question:** A rectangular wire loop of sides 8 cm and 15 cm with a small cut is moving out of a region of uniform magnetic field of magnitude 11.1 T directed normal to the loop. What is the emf developed across the cut if the velocity of the loop is 31 cm/s in a direction normal to the two different sides of the loop. For how long does the induced voltage last in each case.
Answer: 2.75e-01 V for 2.58e-01 s if velocity is perpendicular to 8 cm side, 5.16e-01 V for 4.84e-01 s if velocity is perpendicular to 15 cm side
17. **Question:** A series LCR circuit is connected to a variable frequency 230 V source with L = 187 H, C = 114 muF, R = 145 Ω. Determine the source frequency which drives the circuit in resonance?
Answer: 6.8 rad/s
18. **Question:** An object is placed at 437 cm in front of a concave mirror of focal length 921 cm. Find at what distance image is formed and its nature.
Answer: 831.6 cm and virtual
19. **Question:** A person with a hypermetropic eye cannot see objects within 49864 mm distinctly. What should be the type of the corrective lens used to restore proper vision?
Answer: convex lens, 3.98 diatrope
20. **Question:** In a Youngs double-slit experiment, the slits are separated by 0.34 mm and the screen is placed 8.3 m away. The distance between the central bright fringe and the 2 bright fringe is measured to be 10 cm. Determine the wavelength of the light used for the experiment.
Answer: 2.05e-06 m
21. **Question:** A piece of copper having a rectangular cross-section of 66 mm x 46 mm is pulled in tension with 11400 N force, producing only elastic deformation. Calculate the resulting strain? Shear modulus of elasticity of copper is 42 x 10^9 N/m^2.
Answer: 89.4 x 10^(−6)
22. **Question:** What is the largest average velocity of blood flow in an artery of radius 804 μm and length 1989 mm if flow must remain laminar, what is the corresponding flow rate.
Answer: velocity is 2.45 ms^{-1}, corresponding flow rate is 4.97e-06 m^3s^{-1}
23. **Question:** On an unknown scale temperature of the melting point and boiling point of the water are 70 °C and 221 °C respectively, If temperature on the Fahrenheit scale is 142 °C F, then what is the temperature shown on that scale.
Answer: 162.3
24. **Question:** A sound wave travels at a speed of 267440.2 m/s, if it's wavelength is 38 m, will the sound wave be audible?
Answer: audible
25. **Question:** A spring balance has a scale that reads from 0 to 175 kg. The length of the scale is 15 cm. A body suspended from this balance, when displaced and

released, oscillates with a period of 610 ms. What is the weight of the body?
Answer: 1.06e+03 N

26. **Question:** Two amplifiers are connected one after the other in series (cascaded). The first amplifier has a voltage gain of 42 and the second has the voltage gain of 25. If the input signal is 0.103 V, calculate the output ac signal.
Answer: 1.08e+02 V
27. **Question:** For an amplitude modulated wave, the maximum amplitude is found to be 9.78 V while the minimum amplitude is found to be 7.12 V. Determine the modulation index.
Answer: 0.16

4 Experimental Results and Analysis

4.1 Transformer Architecture and Char2Char Encoding

As mentioned before, the transformer architecture is already considered state-of-the-art for language translation. The seq2seq architecture using characters as sequence has been used to solve some mathematics problems [11]. For the new datasets proposed in this paper, we use the transformer architecture to check the accuracy for the proposed dataset. We use the standard transformer described in [13] with our own specifications as follows. We use an encoder which is composed of stack of $N = 4$ identical layers. The embedding size $d_{\text{model}} = 128$, attention heads $h = 8$. The inner layer size of feed forward network used in each layers of encoder stack $d_{ff} = 512$. We minimize the sum of log probabilities of the correct tokens via the Adam optimizer with adaptive learning rate.

4.2 Computational Resources Used

The code is written in Python and PyTorch is used. The models are trained on dual Intel Xeon E5-2640 v4 processors, providing 40 virtual cores per node, 128 GB of 2400MT/s DDR4 ECC RAM and four Nvidia GeForce GTX 1080 Ti GPUs, providing 14336 CUDA cores.

4.3 Dataset Organization and Generation

The dataset is called `SCIMAT`. It contains two folders `mathematics` and `science`. Both the folders contains folders with chapter names, each of these folders contain folders with name same as problem name. For example, science folder contains a folder named `WorkPowerEnergy`, which in turn contains a folder named `KE`, which contains problems on kinetic energy. Each of the problem folder contains generator code written in Python, which on execution generates corresponding datasets `questions.txt` and `answers.txt` with as many samples as required. Typically, as mentioned before, we generated 0.5 Million to 4 Million samples per problem. The dataset generation time for most of the datasets is very fast and would happen in few seconds. But some datasets like differentiation, integration, etc. which uses SymPy for computation would typically take few minutes based on number of samples.

In any of the chapter, first we identify some of the important problem types, for example in Work Power Energy (WPE) chapter, we have identified problem types such as notion of work, KE, conservation of energy, etc. as stated above in Sect. 3, then for each of the problem types, we find some questions from basic class 9 to class 12 problems (based on chapter), modify them slightly if needed and create some variants of that problem type. For example we have found KE problem type in WPE and created some variants of that problem type.

Now Let's discuss how a generator file is written with the help of KE example. In Listing 1, at line numbers 4, 5 and 6 we show 3 sample questions(variants) for a problem type are being generated. The calculations for computing KE is done in lines 13–14, calculation for computing mass is done in lines 16–17, and finally calculation for computing velocity is done in lines 19–20. For each of these types random values for mass, velocity or energy is assumed to create different samples. Total number of samples to be created is indicated in line 11, and the files to write the questions and answers are in line numbers 9 and 10 respectively. This generator file can be used to create more samples with different range of values for mass, velocity or KE. Moreover we can use this file to generate different wordings for the same questions.

```
1 import random
2 import math
3
4 # What is the kinetic energy of an object of mass m kg moving
     with a velocity of v m/s ?
5 # What is the mass of an object with kinetic energy KE J moving
     with a velocity of v m/s ?
6 # What is the velocity of an object of mass m kg having kinetic
     energy KE J ?
7 # KE = 1/2*m*v*v
8
9 qns = open('./questions.txt', 'w')
10 ans = open('./answers.txt','w')
11 no_of_samples = 2000000
12
13 def calculation_KE(m, v):
14     return round(1/2 * m * v * v , 1)
15
16 def calculation_m(KE, v):
17     return round((2*KE) / (v * v), 1)
18
19 def calculation_v(KE, m):
20     return round(math.sqrt((2*KE) / m) , 1)
21
22 def type1():
23     m = random.randint(1,1000)
24     v = random.randint(1,1000)
25     KE = str(calculation_KE(m,v)) + " joule\n"
26     q = "What is the kinetic energy of an object of mass " +
     str(m)
27     + " kg moving with a velocity of " + str(v) + " m/s ?\n"
28     return q,KE
29
30 def type2():
31     KE = random.randint(499000,500000)
32     v = random.randint(1,1000)
33     m = str(calculation_m(KE,v)) + " kg\n"
```

```
34        q = "What is the mass of a object with kinetic energy of "
          + str(KE)
35        + " J moving with a velocity of " + str(v) + " m/s ?\n"
36        return q,m
37
38    def type3():
39        KE = random.randint(498000,500000)
40        m = random.randint(1,1000)
41        v = str(calculation_v(KE,m)) + " m/s\n"
42        q = "What is the velocity of an object of mass " + str(m)
43        + " kg having kinetic energy " + str(KE) + " J ?\n"
44        return q,v
45
46    for i in range(no_of_samples):
47        types = random.randint(0,3)
48        if types == 0:
49            ques,answer = type1()
50        if types == 1:
51            ques,answer = type2()
52        if types == 2 or types == 3:
53            ques,answer = type3()
54        qns.write(ques)
55        ans.write(answer)
56    qns.close()
57    ans.close()
```

Listing 1.1. Example of Dataset Generation for Kinetic Energy Problems

4.4 Evaluation Criterion and Splitting of Train and Test

We have used binary scoring for test samples. That is, if the predicted answer matches completely with the expected answer character by character (or) word by word, then we assign a score of 1, else it is given a score of 0. Percentage accuracy, i.e., the number of test samples on which the predicted answer matches with true answer is used as the metric for evaluating the performance of our model.

Test Train Split: The number of test samples used were 200, and the training sample size varied from 6000 to 120000 for our dataset. We use lesser number of samples on char2char transformer because we observe that char2char transformer produces similar kinds of results when trained on relatively lesser number of samples as well. The datasets for which the train or test accuracy were not high, we tried to increase the samples with the hope of obtaining better accuracy.

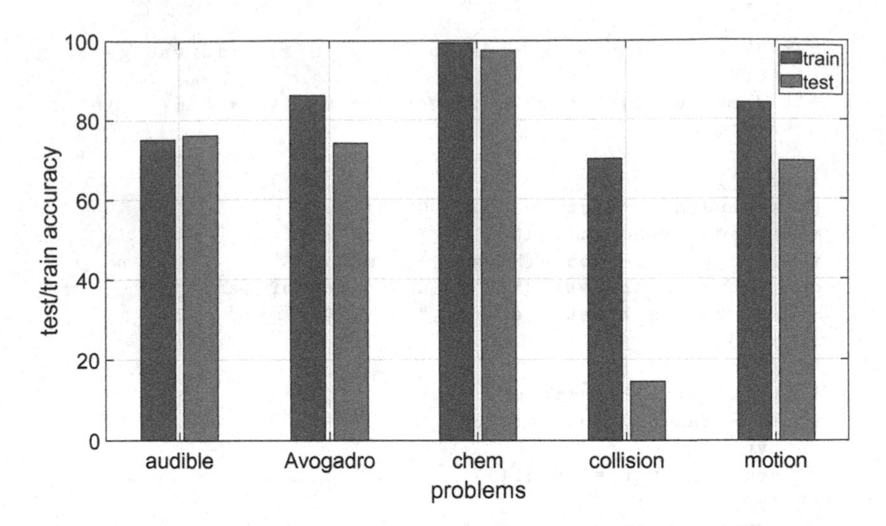

Fig. 1. Comparison of train and test accuracy for some of the science datasets.

4.5 Comparison of Train and Test Accuracy

Using char2char encoding we use the transformer architecture as described above for the generated datasets. In Fig. 1, we show the train and test accuracy for selected problems. We observe that in some cases such as the problem named "collision" (Q8 of sample questions) in the figure, train accuracy is 70.0%, whereas, the test accuracy is mere 14.5%. We believe that this is a case of overfitting and further analysis is needed to understand the reason for this behaviour. The datasets which involves calculation of squares and square roots (for example, Q6 in sample questions) gave less test accuracies. On the other hand, for example, in equation of motion ($v = u + a * t$) we got accuracy of 96%. However, it was able to perform well on datasets which includes log and trigonometric functions, for example for questions Q1, Q10 and Q15 in sample questions.

4.6 Discussion of Test Accuracy for Generated Datasets

In Table 1, we train the transformer architecture for the mathematics datasets proposed in [11]. This verifies that we are able to reproduce the results for char2char encoding as mentioned in the paper. We observe that the test accuracy is greater than 95% for seven problems, and there are four datasets with test accuracy between 75% and 95%. The remaining seven datasets have the accuracy below 75%. Some of the problems such as list prime factors, LCM, GCD have low accuracy, perhaps it has to do with the fact that network needs to learn the prime factors, which is known to be a hard task.

In Table 2, we use the same untrained transformer architecture, and retrain our newly proposed datasets. Once again we observe that there are about twelve datasets, where the accuracy is greater than 95%, there is one dataset with

Table 1. Performance comparison of our optimized model with the state-of-the-art model used in [11]. The Char2Char is denoted by C2C below.

Type of problem	C2C Accuracy
Simple addition, subtraction	99%
Polynomial addition (multiple terms)	96%
Base conversion	7%
Function evaluation	60%
Arithmetic multiplication	75%
Arithmetic division	86%
Remainder	39%
Unit conversion	91%
Rounding decimals	100%
Place value	100%
Sequence predicting	92%
LCM	58%
GCD	61%
Sorting	98%
Polynomial differentiation	99%
Kth biggest	99%
Prime check	59%
List Prime factor	10%

Table 2. Comparison of our Model trained on new datasets with Char2Char transformer. The Char2Char is denoted by C2C below.

Type of problem	C2C Accuracy
Differentiation of sum of Basic Trigonometric functions	99%
Integration of sum of Basic Trigonometric functions	100%
Matrix Addition	49%
Matrix Subtraction	74%
Matrix Transpose	100%
Determinant of Matrix	32%
Matrix Multiplication	32%
Trace of a Matrix	100%
Product of Matrix with a Scalar number	100%
Row Reduced Echelon form of a Matrix	76%
Rank of a Matrix	100%
Mean of a sequence	95%
Variance of a sequence	39.5%
Median of a sequence	99%
Union of Sets	100%
Intersection of Sets	97.5%
Sets Difference	100%
Symmetric Difference between sets	100%

accuracy between 75% and 95%, and there are five datasets with accuracy below 75%. Hence, we can claim that for majority of datasets, the network seems to be effective. Perhaps, a variation in the dataset with more number of samples for the datasets where the network fails needs to be explored as future research work.

Table 3. Accuracy for science datasets with Char2Char transformer. The Char2Char is denoted by C2C below. We have also listed the reference questions.

Type of problem	C2C Accuracy
pH (Q1)	99.8%
Mass of atoms using Avogadro number (Q3)	74%
Adiabatic (Q4)	76.3%
Operations with Significant Digits (Q5)	80.4%
Equation of motion, ($v^2 - u^2 = 2as$ vertically projected body) (Q6)	8.5%
Force, mass, acceleration (Q7)	45.5%
Collision and conservation of momentum (Q8)	14.5%
Solid, Hollow Cylinders rolling down on inclined planes (Q10)	98.1%
Gravitational Field (Q11)	94.2%
Series/Parallel combination of Resistance (Q12)	88.9%
Power, Current, Resistance (Q13)	99.3%
Current Sensitivity (Q14)	98.1%
Torque due to Magnetic Field (Q15)	84.5%
LCR circuit (Q17)	91.3%
Mirror formula for concave mirror (Q18)	79.6%
Blood Pressure (Q22)	99.9%
Relation Between different temperature scales (Q23)	61.6%
Is the sound audible? (Q24)	76%
Modulation index (Q27)	89.6%
Kinetic Energy	73.5%
Force between wires	75.2%
Equation of motion, ($v = u + at$ vertically projected body)	96%
Work, mass, velocity	10%
Sound wave through different materials	71%

In Table 3, we show preliminary results for some of the datasets for new science problem dataset. In parenthesis we indicate the question number from the list of questions we mentioned in the Sect. 3.4. There are 6 datasets with accuracy greater than 95%, and there are 10 datasets with accuracy between 70%

Table 4. Accuracy for science datasets compared on Char2Char transformer, Word2Word transformer, Word2Word performer. The Char2Char transformer is denoted by C2C, Word2Word transformer is denoted by W2W Tran., Word2Word performer is denoted by W2W Per. below. We have also listed the reference questions. The best results are shown in bold.

Comparison between various architectures and encodings			
Type of problem	C2C	W2W Tran.	W2W Per.
pH (Q1)	**99.8%**	97.3%	84.7%
Operations with Significant Digits (Q5)	**80.4%**	72.9%	67.4%
Equation of motion, ($v^2 - u^2 = 2as$) (Q6)	8.5%	**12.8%**	12.2%
Series/Parallel combination of Resistance (Q21)	**88.9%**	32.5%	45.7%
Kinetic Energy	**73.5%**	72.4%	71.8%

and 95%. There are 8 datasets with accuracy below 70%. In particular, we obtain very poor test accuracy for problems involving equations of motion, collision, and for work, mass, velocity type problems. More investigation is needed, and more varied samples using the dataset generator may help improve accuracies for these problems.

Finally in Table 4, we show comparison between 3 different approaches. Again, in paranthesis we indicate the question number from the list of questions we mentioned in the Sect. 3.4. The first one is a transformer with char2char encoding, Second is a transformer with word2word encoding and finally we have a performer [5] architecture with word2word encoding. In this table we observed that in four of the datasets char2char transformer performed better than the other two. We observe that in "equation of motion problem", all the models give poor accuracy, however char2char transformer is the worst. For the problem "series/parallel combination of resistance", we observe that char2char had significantly better accuracy compared to the other two models.

5 Conclusion

Labelled datasets drive research and innovation in machine learning research. Being able to solve problems in mathematics and sciences is arguably a more difficult task than many natural language processing tasks, because apart from understanding the intricacies of natural languages in the questions, one needs to understand the method, law, or rules to solve the problem. A comprehensive set of datasets for problem solving in mathematics and sciences have been missing except for a small dataset released by deep mind [11]. This paper contributes hundreds of millions of labelled question-answer samples in raw text format for high school problems in mathematics and sciences. This raw text format can be further processed into any desired encoding for further research. We have shown some preliminary results on the test accuracy of various transformer architectures with char2char and word2word encodings on these new datasets.

We found that for some problems the accuracy was above 95%, and for some it was below 70%. For problems where the accuracy is very low, further research on encoding or on architecture should be focused on for these challenging problems. In future, we plan to keep augmenting the datasets with more samples.

References

1. Bobrow, D.G.: Natural language input for a computer problem solving system (1964)
2. Briars, D.J., Larkin, J.H.: An integrated model of skill in solving elementary word problems (1984)
3. Brown, T., et al.: Language models are few-shot learners. In: Larochelle, H., Ranzato, M., Hadsell, R., Balcan, M.F., Lin, H. (eds.) Advances in Neural Information Processing Systems, vol. 33, pp. 1877–1901. Curran Associates, Inc. (2020). https://proceedings.neurips.cc/paper/2020/file/1457c0d6bfcb4967418bfb8ac142f64a-Paper.pdf
4. Cheng, S., Chung, N.: Simple mathematical word problems solving with deep learning. Technical report, Stanford University
5. Choromanski, K., et al.: Rethinking attention with performers (2021)
6. Chung, J., Gülçehre, Ç., Cho, K., Bengio, Y.: Empirical evaluation of gated recurrent neural networks on sequence modeling. ArXiv abs/1412.3555 (2014)
7. Devlin, J., Chang, M., Lee, K., Toutanova, K.: BERT: pre-training of deep bidirectional transformers for language understanding. CoRR abs/1810.04805 (2018). http://arxiv.org/abs/1810.04805
8. Feigenbaum, E., Feldman, J.: Computers and Thought (1963)
9. Hochreiter, S., Schmidhuber, J.: Long short-term memory. Neural Comput. **9**(8), 1735–1780 (1997). https://doi.org/10.1162/neco.1997.9.8.1735
10. Fletcher, C.R.: Understanding and solving arithmetic word problems: a computer simulation. Behav. Res. Methods Instrum. Comput. **17**, 565–571 (1985). https://doi.org/10.3758/BF03207654
11. Saxton, D., Grefenstette, E., Hill, F., Kohli, P.: Analysing mathematical reasoning abilities of neural models. ArXiv abs/1904.01557 (2019)
12. Sirignano, J., Spiliopoulos, K.: DGM: a deep learning algorithm for solving partial differential equations. J. Comput. Phys. **375**, 1339–1364 (2018). https://doi.org/10.1016/j.jcp.2018.08.029
13. Vaswani, A., et al.: Attention is all you need. In: Guyon, I., et al. (eds.) Advances in Neural Information Processing Systems, vol. 30, pp. 5998–6008. Curran Associates, Inc. (2017). https://proceedings.neurips.cc/paper/2017/file/3f5ee243547dee91fbd053c1c4a845aa-Paper.pdf
14. Weston, J., et al.: Towards AI-complete question answering: a set of prerequisite toy tasks (2015)

Rank-Based Prefetching and Multi-level Caching Algorithms to Improve the Efficiency of Read Operations in Distributed File Systems

Anusha Nalajala[1](✉), T. Ragunathan[1], Rathnamma Gopisetty[2], and Vignesh Garrapally[3]

[1] SRM University, Andhra Pradesh, India
{anusha_nalajala,ragunathan_t}@srmap.edu.in
[2] Gitam University, Hyderabad, India
[3] Kakatiya Institute of Technology and Sciences, Warangal, India

Abstract. In the era of big data, web-based applications deployed in cloud computing systems have to store and process large data generated by the users of such applications. Distributed file systems are used as the back end storage component in the cloud computing systems and they are used for storing large data efficiently. Improving the read performance of the distributed file system is the important research problem as most of the web-based applications deployed in the cloud computing systems carry out read operations more frequently. Prefetching and caching are the two important techniques used for improving the performance of the read operations in the distributed file system. In this paper, we have proposed novel rank-based prefetching, multi-level caching and rank-based replacement algorithms for the effective caching process. Our simulation results reveal that the proposed algorithms improve the performance of the read operations carried out in the distributed file systems better than the algorithms proposed in the literature.

Keywords: Distributed file system · Rank · Prefetching · Multi-level caching · Cache replacement policies

1 Introduction

In the emerging Big data scenario, web-based applications are deployed in cloud computing systems. Distributed file systems (DFSs) are used as the back end storage component in the cloud computing systems and they are used for storing and processing large data efficiently. Following are the popular DFSs used in the cloud computing systems for storing large data: GFS [5], HDFS [21], Blobseer [12]. Most of the users of the web-based applications very frequently issue read requests and less frequently issue write requests. Hence, improving the performance of the read operations carried out in the distributed file system is the important research problem.

S. N. Srirama et al. (Eds.): BDA 2021, LNCS 13147, pp. 227–243, 2021.
https://doi.org/10.1007/978-3-030-93620-4_17

Prefetching and caching are the two important techniques discussed in the literature for improving the performance of the read operations carried out in the DFS. Prefetching techniques are used to fetch the data in advance before it is requested by the user. Client side caching techniques fill the local caches of the client nodes by some useful data which can be accessed by the client application programs running locally. These techniques help in improving the performance of the read requests issued by the users of the web applications deployed in the cloud.

Currently, computer systems are available with solid state drives (SSDs) which are more faster than the hard disk drives (HDDs) [16]. The SSD devices are used for improving the performance of file read and write operations. In this paper, we have considered a rack organized DFS in which a master node (MN), a global node (GN) and a set of data nodes (DNs) are present. We have proposed a novel rank-based prefetching technique which fetches the frequently accessed file blocks (FAFBs) by considering the frequency and the recent timestamps of the accessed file blocks of the DNs. We have computed the average read access time and hit ratios of the proposed rank-based prefetching and multi-level caching algorithms. We have compared the performance of our algorithms with the simple support based (SB) [18] algorithm proposed in the literature and found that our algorithms perform better than the SB algorithm.

The remainder of the paper is organised as follows: the literature related to prefetching and client side caching techniques is discussed in Sect. 2. The proposed algorithms are presented in Sect. 3. Simulation results of are covered in Sect. 4. The last section covers the conclusion and future work.

2 Related Work

In this section, we discuss about various prefetching and client side caching techniques presented in the literature.

The authors in this paper [19] discussed about the prefetching based metadata management in advanced multitenant hadoop where graphs drawn from past access requests in order to predict future access requests. The downside with this paper is that they used an additional storage for handling the native hadoop metadata and requires an authorized ones to access the information about metadata without the intervention of hadoop master servers.

In [4], prefetching and caching techniques are proposed to improve the performance of NoSQL databases which allows to define certain rules for prefetching to do simulations and execution plans in certain cases. A multi-tiered and distributed I/O buffering system is introduced in [11] which is considered as a middle ware layer. A server push data prefetching is done which performs low when server is heavily loaded.

Smeta [2] is a data correlation directed prefetching method which fetches data correlations instilled in a file using syntax analysis. The correlated files whose frequency is not high may not be missed by using this method. However, if a file is requested, all its correlated files are also prefetched and cached in the DFS client where the size of the cache may be a problem.

In [13–15] the authors proposed initiative data prefetching prediction algorithms where prefetching is done at storage servers by analysing past disk I/O accesses. As prefteching is done at server side there will be an extra burden on the client to find the address of the client and the I/O requests here are still served from the disk.

Support-based prefetching and hierarchical collaborative caching techniques are proposed to improve the efficiency of read operations in DFS by fetching the popular file blocks and caching them in an hierarchical manner where multi level memory devices are not used to fasten the operations [6,7].

In [3] prefetching is done statically and dynamically by predicting file access patterns by using machine learning techniques. Medium of storage is at the disk level for storing the prefetched files. Here entire file is prefetched and if the size of thc file is big memory may be wasted unnecessarily.

CAPre [23] code analysis based prefetching is done on objects of static code by analysing object oriented applications. Prefetching is done at application's compile time and objects can be prefetched in parallel from various nodes at the same time. Its the burden of application to generate hints for prefetching.

Hycache [25] is a user level file system embedded between distributed file systems and client file systems. The performance will be degraded because more context switching is done due to the small block sizes in the client file system.

Hyper cache [10] isused for reducing overhead of I/O virtualisation for caching the popular blocks on virtual machines. However migrating cache components to another operating system is difficult due to heavy guest traffic.

Most of the prefetching and client side caching techniques proposed in literature mainly based on prefetching the files, fileblocks and fileblock access patterns without considering multi level memories for caching on client side. In this paper, we have proposed Rank-based prefetching technique which includes timestamp rank along with the support rank for the file block. We have used multilevel memories in each node for caching the prefetched file blocks which are more popular. By using this way of prefetching and caching the I/O requests to be served from the disk will be reduced which in increases the performnace of the DFS.

3 Proposed Algorithms

In this section, first, we discuss the architecture of the DFS and how to calculate ranks for file blocks. After this, we discuss the proposed support rank based prefetching and access timestamp rank based prefetching algorithms.

3.1 Architecture

We considered a rack organization with 10 racks, 10 data nodes (DNs) in each rack as shown in the Fig. 1. We have considered that a master node (MN) and a global node (GN) are also present in the racks. Each data node consists of three tier storage i.e., hard disk (disk), solid state drive (ssd) cache and main memory. Please note that main memory is used to implement the local cache. The MN consists of meta data of files and file blocks present in the data nodes. The GN is a global cache directory with local cache and ssd caches which are accessible by all the applications running in the DNs. The DN is the place where all the user application programs will be executed and the DN also maintains the meta data about file blocks requested by the user application programs.

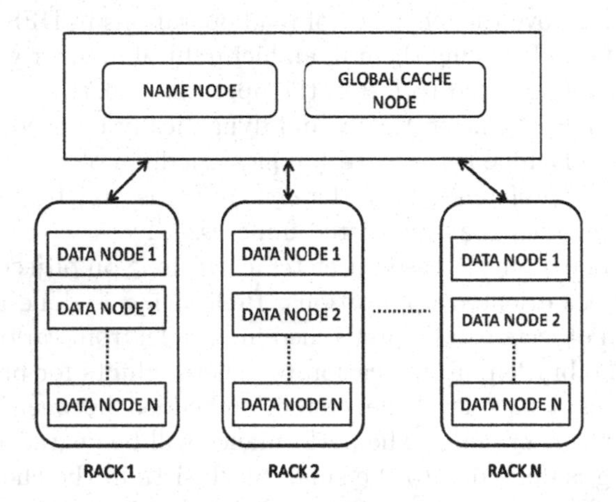

Fig. 1. Architecture of DFS

First, from the log, FAFBs are prefetched based on support, support rank and access timestamp rank. Then we have filled the prefetched FAFBs based on their support, support rank and access timestamp rank in the multi-level caches of the nodes considered in the Fig. 1. After filling the caches, read and write requests are generated and these requests are served based on the procedures presented in the next sections. Finally, we have calculated the average read access time and cache hit ratios using the proposed prefetching algorithms and compared them with existing simple support based algorithm that are described in the next section.

3.2 Rank-Based Prefetching

As discussed in Sect. 3.1, logs are maintained in all DNs to record the information about the files and file blocks requested by the user applications running locally. The frequently accessed file blocks (FAFB) are extracted based on two types of ranks namely support rank and access timestamp rank. First, from the log, support value is calculated for all files and the files whose support value is greater than the given threshold value are considered as frequently accessed files (FAFs). In the same way support value is calculated for all blocks of FAFs and whose support value is above the given threshold are considered as FAFBs. Then a rank value is assigned to each block in such a way that a block with highest support value has highest rank and a block with least support value will have the lowest rank. Next, the support rank and access time stamp rank are calculated and assigned to all FAFBs which are presented in the next sections.

Support Rank. We have considered a log with n entries of file and file blocks requested by the user applications in all the DNs. An example of a session in the log is shown in Fig. 2. Each entry in the log is considered as a session and in each session there will be rack id, data node id, user application id, file name followed by the number of blocks requested in that particular file along with access time stamp. From these sessions of the log, support and support rank are calculated for all the FAFBs.

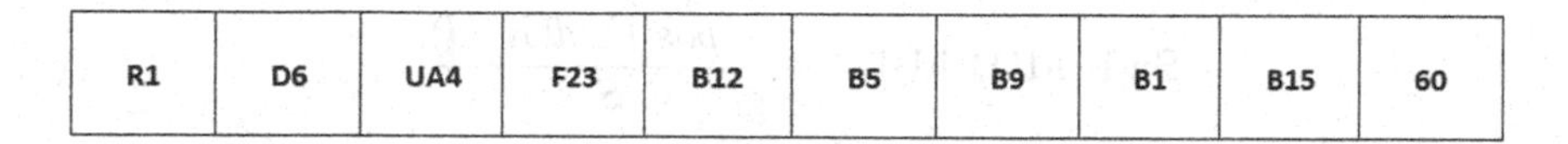

R1
D6
UA4
F23
B12
B5
B9
B1
B15
60

Fig. 2. A session in the log.

From the log, first we have calculated the local support for all the files in the logs of all DNs. For example, local support for a file with respect to a single data node DN_id(lets say the name of the file is F_id) is calculated as show in Eq. 1.

$$\textbf{Local support(F\_id)} = \frac{Frequency\ of\ F\_id\ in\ the\ log\ of\ DN\_id}{log\ size} \tag{1}$$

After calculating support values for all the files in the log of DN_id, these are sorted in descending order based on their support values. Now a local threshold value (*lf_th*) is fixed to extract the more frequently files. A file whose support value is above the given *lf_th* is considered as local FAF. All these files along with support values are stored in *lfaf_list*. In the same way **Global support(F_id)** for all the files are computed with respect to the entries in all DNs as shown in Eq. 2 and a global threshold value (*gf_th*) is fixed. The files whose global support value is greater than *gf_th* are considered as global FAFs and are stored in *gfaf_list*.

$$\textbf{Global support(F\_id)} = \frac{Frequency\ of\ F\_id\ in\ the\ log\ of\ all\ DNs}{log\ size} \tag{2}$$

For all blocks of the files present in *lfaf_list* and *gfaf_list*, local and global support values are calculated. First, we will see how the local support for a block B_id in a file F_id with respect to data node DN_id is computed as shown in Eq. 3 and then calculation of the global support in shown in Eq. 4.

$$\textbf{Local support}[\textbf{F\_id}(\textbf{B\_id})] = \frac{Frequency\ of\ the\ B\_id\ in\ DN\_id}{Frequency\ of\ F\_id\ in\ the\ log\ of\ DN\_id} \quad (3)$$

$$\textbf{Global support}[\textbf{F\_id}(\textbf{B\_id})] = \frac{Frequency\ of\ the\ B\_id\ in\ all\ DNs}{Frequency\ of\ F\_id\ in\ the\ log\ of\ all\ DNs} \quad (4)$$

Next, a local threshold (*lfb_th*) is fixed and the blocks whose local support value is greater than *lfb_th* are stored in a list called *lfafb_list*. In the same way a global threshold (*gfb_th*) is fixed and the blocks whose global support is above the fixed threshold are stored in *gfafb_list*. Now for all the blocks in *lfafb_list* and *gfafb_list*, rank is assigned as discussed in 3.2. Then the support rank (Sr) for a file block [F_id(B_id)] is computed as shown in Eq. 5.

$$\textbf{Sr}[\textbf{F\_id}(\textbf{B\_id})] = \frac{S - pos[F\_id(B\_id)]}{S} \quad (5)$$

The pos[F_id(B_id)] is the position of the file block in the list and S is the size of the list.

The procedure for prefetching FAFBs locally and globally based on support rank are presented in Algorithms 1 and 2.

Algorithm 1. Procedure for local prefetching

---

for each file F_id in the log of data node DN_id **do**
  Calculate **Local support(F_id)**
  if Local support(F_id) $\geq$ *lf_th* **then**
    add F_id to *lfaf_list*
  end if
end for
for each block B_id of F_id in *lfaf_list* **do**
  Calculate textbfLocal support[F_id(B_id)]
  if Local support[F_id(B_id)] $\geq$ *lfb_th* **then**
    add F_id(B_id) to *lfafb_list*
  end if
end for
for each block B_id of F_id in *lfafb_list* **do**
  Assign rank for B_id of F_id in *lfafb_list*
  Calculate **Sr[F_id(B_id)]**
end for

After calculating the support rank for all the file blocks in *lfafb_list*, this list is sorted in descending order based on their support rank values and named as *sr_lfafb_list*.

Algorithm 2. Procedure for global prefetching

---

for each file Fi_id in the log with respect all data nodes **do**
  Calculate **Global support(F_id)**
  if Global support(F_id) $\geq$ *gf_th* **then**
    add F_id to *gfaf_list*
  end if
end for
for each block B_id of F_id in *gfaf_list* **do**
  Calculate **Global support[F_id(B_id)]**
  if Global support[F_id(B_id)] $\geq$ *gfb_th* **then**
    add F_id(B_id) to *gfafb_list*
  end if
end for
for each block B_id of F_id in *gfafb_list* **do**
  Assign rank for B_id of F_id in *gfafb_list*
  Calculate **Sr[F_id(B_id)]**
end for

In this way, the support rank is assigned to all file blocks in *gfafb_list* and are sorted in descending order based on their support rank values and are stored in a new list called *sr_gfafb_list*.

Access Timestamp Rank. From the log, the time stamp rank assigned to all file blocks in *lfafb_list* and *gfafb_list* are extracted and ordered in such a way that recently accessed file block is placed at the top of the lists. Now the access time stamp rank (Rk) for a file block [F_id(B_id)] is computed as defined in Eq. 6.

$$\mathbf{Rk[F\_id(B\_id)]} = \frac{N - pos[F\_id(B\_id)]}{N} \tag{6}$$

where pos[F_id(B_id)] is the position of the file block in the list and the list size is denoted by N.

The local and global prfetching procedure for extracting FAFBs are presented in Algorithm 3.

After calculating the access timestamp rank values for all the blocks in *lfafb_list* and *gfafb_list*, the blocks are sorted in descending order according to their ranks and are added to new lists *rk_lfafb_list* and *rk_gfafb_list*.

3.3 Multi-level Caching

In this section, we will discuss about the procedure for caching the prefetched FAFBs which are stored in *sr_lfafb_list* and *sr_gfafb_list* in the local cache and

Algorithm 3. Prefetching based access timestamp rank

```
for each block B_id of F_id in lfafb_list do
    Order B_id of F_id according to the timestamp
    Calculate Rk[F_id(B_id)]
end for
for each block B_id of F_id in gfafb_list do
    Order B_id of F_id according to the timestamp
    Calculate Rk[F_id(B_id)]
end for
```

SSD caches of the DNs and the GN according to their respective sizes. The access timestamp rank prefetching algorithm also follows the same procedure.

Caching in Local DNs. In this section we discuss about the procedure how the prefetched FAFBs are cached in the multi-level memories of the local DNs. Here the local cache, SSD cache of DN_i are filled with the FAFBs whose rank value is greater than the local prefetch threshold (lprefetch_threshold) according to their respective sizes. The procedure for caching the FAFBs in DN_i from the *sr_lfafb_list* is mentioned in Algorithm 4.

Algorithm 4. Procedure for caching in DN_i

```
for each DN_id do
    while (local cache of DN_id is not full) && (ssd cache of DN_id is not full) do
        for each block F_id(B_id) in sr_lfafb_list do
            if Sr[F_id(B_id)] ≥ lprfetch_threshold then
                add F_id(B_id) to local cache of DN_id
            end if
            if local cache of DN_id is full then
                add F_id(B_id) to ssd cache of DN_id
            end if
        end for
    end while
end for
```

Caching in Global Node. The procedure for caching the prefetched blocks whose new rank value in the *sr_gfafb_list* is above the global prefetching threshold (gprefetch_threshold) in the local cache and ssd cache of the global node is discussed in Algorithm 5.

Algorithm 5. Procedure for caching in GN

```
while (local cache of GN is not full) && (SSD cache of GN is not full) do
  for each block F_id(B_id) in sr_gfafb_list do
    if Sr[F_id(B_id)] ≥ gprfetch_threshold then
      add F_id(B_id) to local cache of GN
    end if
    if local cache of GN is full then
      add F_id(B_id) to SSD cache of GN
    end if
  end for
end while
```

3.4 Reading from the DFS

Default Read Operation. The default read procedure for reading a file block from DFS whenever a file block is requested in a particular data node is the DFS client program running the DN contacts MN to get the metadata about the requested file block. The MN in response sends the addresses of the addresses of the data nodes where the file block is residing. Then the DFS client program identifies the nearest DN to read the file block and delivers it to the user application.

Proposed Read Operation. An user application (UA) program running in a DN_id issues a read request for the file block F_id(B_id). The procedure for serving the read request is explained in Algorithm 6.

Algorithm 6. Proposed read algorithm

```
if F_id(B_id) is available in local cache of DN_id then
  UA reads F_id(B_id) from local cache
else if F_id(B_id) is available in local cache of GN then
  F_id(B_id) is read from local cache of GN and delivered to UA
  copy F_id(B_id) to local cache of DN_id
  if local cache of DN_id is full then
    Follow least rank replacement policy
  end if
else if F_id(B_id) is available in SSD cache of DN_id then
  F_id(B_id) is read from SSD cache of DN_id and delivered to UA
else if F_id(B_id) is available in SSD cache of GN then
  F_id(B_id) is read from SSD cache of GN and delivered to UA
  copy F_id(B_id) to local cache of DN_id
  if local cache of DN_id is full then
    Follow least rank replacement policy
  end if
else
  Follow default read operation
end if
```

Least Rank Replacement Policy. In this section, the Algorithm 7 explains the procedure for replacement if any of the caches of local DN or GN are full. This replacement procedure is also followed by access timestamp rank based prefetching algorithm.

Algorithm 7. Least rank replacement policy

```
Remove the F_id(B_id) with least support rank value
calculate the support rank of F_id(B_id)
if Sr[F_id(B_id)] ≥ least support rank value in local cache of GN then
   add F_id(B_id) to local cache of GN
   if local cache of GN is full then
      Remove the F_id(B_id) with least support rank value
      add F_id(B_id) to ssd cache of GN
      if ssd cache of GN is full then
         Remove the F_id(B_id) with least support rank value
      end if
   end if
else
   add F_id(B_id) to ssd cache of DN_id
   if ssd cache of DN_id is full then
      Remove the F_id(B_id) with least support rank value
   end if
end if
```

3.5 Writing to DFS

The procedure writing a file block F_id(B_id) issued by an user application which is getting executed in a particular DN_id is structured in Algorithm 8.

Algorithm 8. Write opeartion

```
for each DN_id do
   if F_id(B_id) is present in local cache and ssd cache of DN_id,GN then
      invalidate F_id(B_id)
   else
      write F_id(B_id) in local cache of DN_id
      if local cache of DN_id is full then
         Follow least rank replacement policy
      end if
   end if
end for
```

4 Experimental Results

In this section, first, we discuss about the various parameters considered for the simulation and then the experimental setup for conducting simulation experiments followed by the results obtained from the simulation.

4.1 Parameters

1. Time taken to read a 4 kb file block from main memory is 0.00009 ms [24]
2. Time taken for reading 4 kb block from SSD is 0.0104 ms [8]
3. Time taken to read 4 kb block from disk is 0.0890 ms [20]
4. Communication time needed for transferring 4 kb block from remote to local data node is 0.004 ms [1]
5. Time for reading a data block from remote memory requires 0.00409 ms.
6. The time required to transfer a data block from data node present in another rack is 0.00569 ms.

4.2 Experimental Setup

In this sub section, we discuss about the environment which we have simulated our algorithms. We considered a rack organisation in which 10 racks are present and for each rack 10 data nodes are present. A global node is also maintained which acts as a global directory and a master node to store the metadata of the files.

Regarding the data set, we have generated a log with 1000000 entries using Medisyn [22] technique. We considered 10000 files and 1000 blocks and its frequencies are generated by using zipf distribution by taking the popularity parameter as 0.8 [17]. The block size is fixed as 4 kb. Each entry in the log is considered as a session and in each session 5 to 15 blocks may be accessed. We considered 80% read and 20% write requests in the log [9]. First, we calculated support for finding frequent files and blocks and the top 60% are considered as frequently accessed files and blocks. Then we calculated the support rank and access timestamp rank for the FAFBs and cached them in the local and ssd caches as explained in Sect. 3.3. For each run, we generated 100000 read and write requests and observed the performance of proposed support rank- and access timestamp rank-based prefetching algorithms with the existing support-based algorithm by calculating average read access time and hit ratios.

4.3 Simulation Results

The simulation results obtained for the proposed algorithms support rank based prefetching (SRBP) algorithm, Access timestamp rank based prefetching (ARBP) algorithm and the simple support based (SB) algorithm proposed in the literature are presented in this section. By observing the results we found that the proposed algorithms SRBP and ARBP perform better than the existing

SB algorithm. And the proposed SRBP algorithm achieved highest performance than ARBP algorithm.

Figure 3 shows the average read access time achieved by our proposed and existing algorithms. We observed the performance by varying the local cache size from 100 to 500 blocks, local ssd cache size from 1000 to 5000 blocks and by fixing the size of local cache in GN as 2500 blocks and the size of the ssd cache in GN as 25000 blocks, We found that SRBP algorithm performs better than ARBP algorithm which in turn performs better than existing SB algorithm.

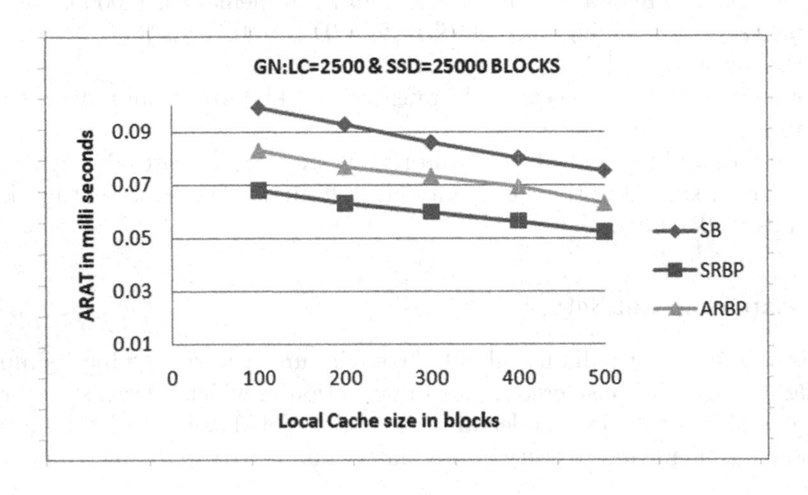

Fig. 3. Average read access time.

The same results are observed in Fig. 4 by varying the global node's local cache size to 5000 blocks and ssd cache size to 50000 blocks.

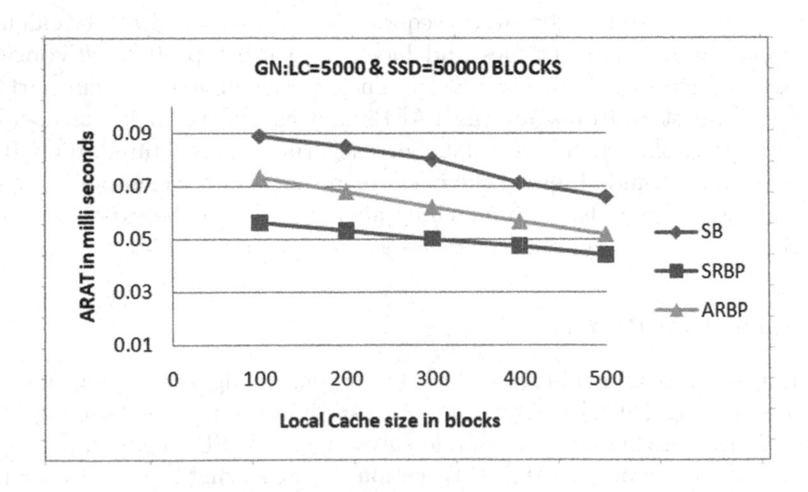

Fig. 4. Average read access time.

The local cache hit ratios of the data nodes obtained from the SRBP, ARBP and SB algorithms by varying the local cache sizes from 100 blocks to 500 blocks and fixing the GNs local cache size to 2500 blocks are depicted in Fig. 5.

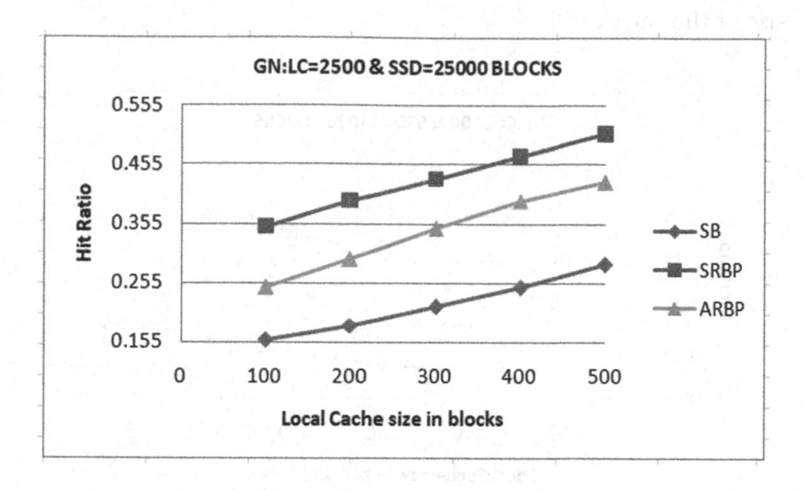

Fig. 5. Local cache hit ratios.

We observed similar trend by varying the GNs local cache size to 5000 blocks and ssd cache size to 50000 blocks as shown in Fig. 6.

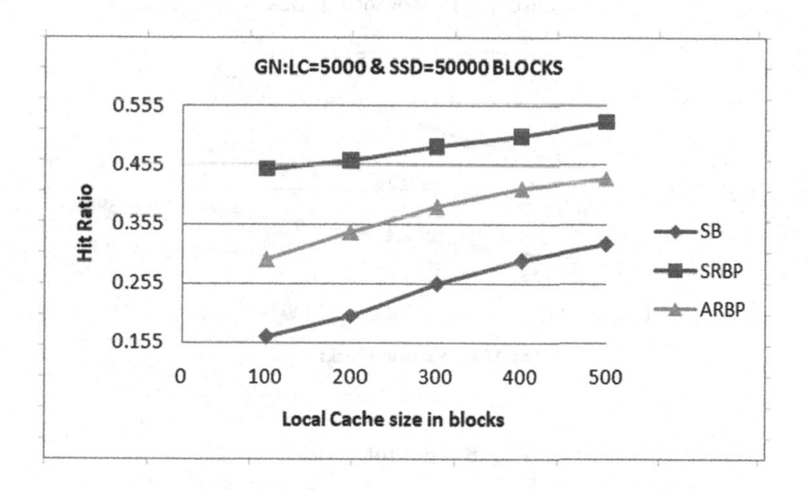

Fig. 6. Local cache hit ratios.

The ssd cache hit ratios of the data nodes are presented in Figs. 7 and 8. Here, we observed the hit ratios by changing ssd cache size from 1000 blocks to 2500 and 5000 blocks and fixing ssd cache size of GN to 25000 and 50000 blocks. We also noticed that ssd cache hit ratios are raised with respect to the increase in the size of the cache.

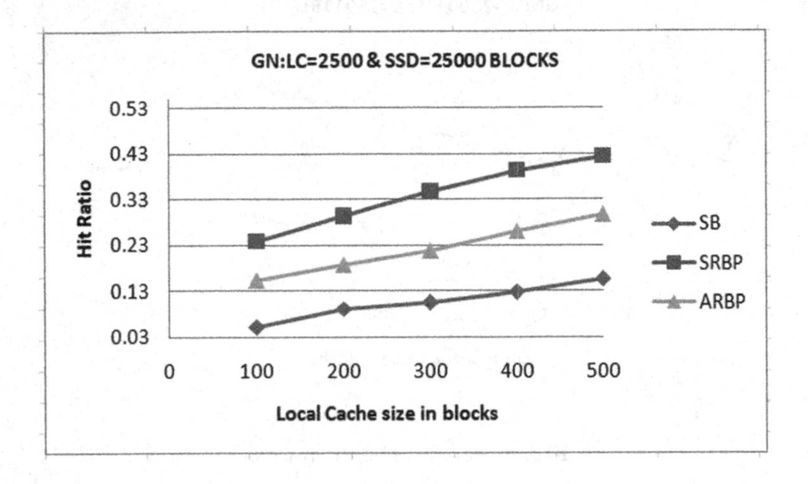

Fig. 7. SSD hit ratios.

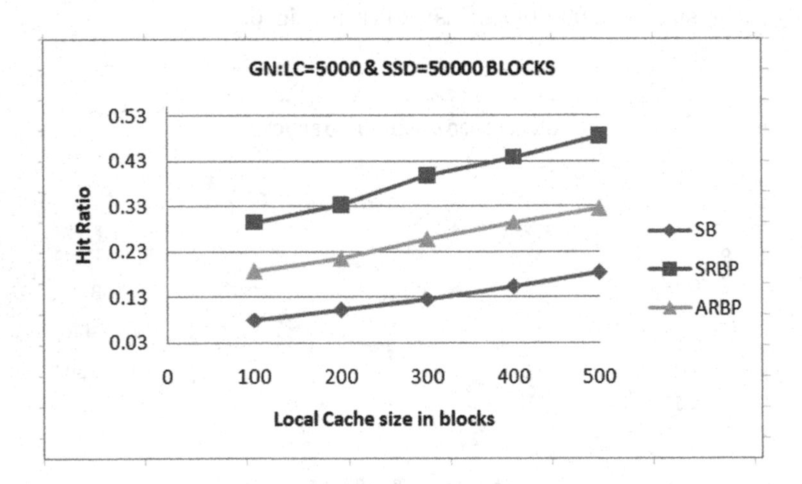

Fig. 8. SSD hit ratios.

Overall, we can conclude by observing the results obtained from the simulation that SRBP algorithm achieved highest performnace than ARBP algorithms which performs better than SB algorithm. The reason for SRBP achieving the highest performance is that the highest ranked block has highest support and could have been accessed very recently. Note that, we have combined support and access time values for calculating the rank and hence this algorithm performs better than the remaining algorithms. Also, we have followed rank-based cache replacement algorithm which removes the file block which has the lowest rank, from the cache. Please note that, this removed block may not be referred by the application programs again. While coming to the ARBP algorithm we prioritised the blocks which are accessed recently by calculating access timestamp rank. In SB algorithm we considered only simple support for the file blocks without considering the timestamp value so the blocks which have highest support value may not be accessed very recently.

5 Conclusion

In this paper, we have proposed support rank-based and access time rank-based algorithms for prefetching the frequently accessed file blocks for improving the performance of read operations carried out in the DFS. We have proposed the multi-level caching algorithms for placing the prefetched file blocks in the data nodes and global node present in the DFS environment. The performance of the proposed algorithms are compared with the simple support-based algorithm discussed in the literature. Our simulation results indicate that the support rank-based algorithm performs better than simple support-based and access time rank-based algorithms by considering the least rank replacement policy. In future, we wish to work on the implementation of the proposed algorithms in Hadoop distributed file system to demonstrate their efficiency.

References

1. Cisco Nexus 5020: Switch performance in market-data and back-office data delivery environments (2010). https://www.cisco.com/c/en/us/products/collateral/switches/nexus-5000-series-switches/white_paper_c11-492751.html
2. Chen, Y., Li, C., Lv, M., Shao, X., Li, Y., Xu, Y.: Explicit data correlations-directed metadata prefetching method in distributed file systems. IEEE Trans. Parallel Distrib. Syst. **30**(12), 2692–2705 (2019)
3. Cherubini, G., Kim, Y., Lantz, M., Venkatesan, V.: Data prefetching for large tiered storage systems. In: 2017 IEEE International Conference on Data Mining (ICDM), pp. 823–828. IEEE (2017)
4. Daniel, G., Sunyé, G., Cabot, J.: Prefetchml: a framework for prefetching and caching models. In: Proceedings of the ACM/IEEE 19th International Conference on Model Driven Engineering Languages and Systems, pp. 318–328 (2016)
5. Ghemawat, S., Gobioff, H., Leung, S.T.: The google file system. In: Proceedings of the Nineteenth ACM Symposium on Operating Systems Principles, pp. 29–43 (2003)

6. Gopisetty, R., Ragunathan, T., Bindu, C.S.: Support-based prefetching technique for hierarchical collaborative caching algorithm to improve the performance of a distributed file system. In: 2015 Seventh International Symposium on Parallel Architectures, Algorithms and Programming (PAAP), pp. 97–103. IEEE (2015)
7. Gopisetty, R., Ragunathan, T., Bindu, C.S.: Improving performance of a distributed file system using hierarchical collaborative global caching algorithm with rank-based replacement technique. Int. J. Commun. Netw. Distrib. Syst. **26**(3), 287–318 (2021)
8. Intel: list of intel ssds (2019). https://en.wikipedia.org/w/index.php?title=List_of_Intel_SSDs&oldid=898338259
9. Intel: resource and design center for development with intel (2019). https://www.intel.com/content/www/us/en/design/resourcedesign-center.html
10. Kim, T., No, J., Piao, Z., Yoo, S.J.: I/o access frequency-aware cache method on kvm/qemu. Cluster Comput. **20**(3), 2143–2155 (2017)
11. Kougkas, A., Devarajan, H., Sun, X.H.: I/o acceleration via multi-tiered data buffering and prefetching. J. Comput. Sci. Technol. **35**(1), 92–120 (2020)
12. Krishna, T.L.S.R., Ragunathan, T., Battula, S.K.: Improving performance of a distributed file system using a speculative semantics-based algorithm. Tsinghua Sci. Technol. **20**(6), 583–593 (2015)
13. Liao, J.: Server-side prefetching in distributed file systems. Concurrency Comput. Pract. Experience **28**(2), 294–310 (2016)
14. Liao, J., Trahay, F., Gerofi, B., Ishikawa, Y.: Prefetching on storage servers through mining access patterns on blocks. IEEE Trans. Parallel Distrib. Syst. **27**(9), 2698–2710 (2015)
15. Liao, J., Trahay, F., Xiao, G., Li, L., Ishikawa, Y.: Performing initiative data prefetching in distributed file systems for cloud computing. IEEE Trans. Cloud Comput. **5**(3), 550–562 (2015)
16. Micheloni, R., Marelli, A., Eshghi, K. (eds.): Inside Solid State Drives (SSDs). SSAM, vol. 37. Springer, Singapore (2018). https://doi.org/10.1007/978-981-13-0599-3
17. Nagaraj, S.: Zipf's law and its role in web caching. In: Web Caching and its Applications. The International Series in Engineering and Computer Science, vol. 772, pp. 165–167. Springer, Boston (2004). https://doi.org/10.1007/1-4020-8050-6_19
18. Nalajala, A., Ragunathan, T., Rajendra, S.H.T., Nikhith, N.V.S., Gopisetty, R.: Improving performance of distributed file system through frequent block access pattern-based prefetching algorithm. In: 2019 10th International Conference on Computing, Communication and Networking Technologies (ICCCNT), pp. 1–7. IEEE (2019)
19. Nguyen, M.C., Won, H., Son, S., Gil, M.S., Moon, Y.S.: Prefetching-based metadata management in advanced multitenant hadoop. J. Supercomput. **75**(2), 533–553 (2019)
20. Seagate: storage reviews (2015). https://www.storagereview.com/seagate_enterprise_performance_10k_hdd_review
21. Shvachko, K., Kuang, H., Radia, S., Chansler, R.: The hadoop distributed file system. In: 2010 IEEE 26th Symposium on Mass Storage Systems and Technologies (MSST), pp. 1–10. IEEE (2010)
22. Tang, W., Fu, Y., Cherkasova, L., Vahdat, A.: Medisyn: a synthetic streaming media service workload generator. In: Proceedings of the 13th International Workshop on Network and Operating Systems Support for Digital Audio and Video, pp. 12–21 (2003)

23. Touma, R., Queralt, A., Cortes, T.: Capre: code-analysis based prefetching for persistent object stores. Future Gener. Comput. Syst. **111**, 491–506 (2020)
24. Vengeance, C.: Corsair vengeance lpx ddr4 3000 c15 2x16gb cmk32gx4m2b3000c15 (2019). https://mram.userbenchmark.com/Compare/CorsairVengeance-LPX-DDR4-3000-C15-2x16GB-vs-Group-/92054vs10
25. Zhao, D., Raicu, I.: Hycache: a user-level caching middleware for distributed file systems. In: 2013 IEEE International Symposium on Parallel & Distributed Processing, Workshops and Phd Forum, pp. 1997–2006. IEEE (2013)

Impact-Driven Discretization of Numerical Factors: Case of Two- and Three-Partitioning

Minakshi Kaushik(✉), Rahul Sharma, Sijo Arakkal Peious, and Dirk Draheim

Information Systems Group, Tallinn University of Technology, Akademeia tee 15a, 12618 Tallinn, Estonia
{minakshi.kaushik,rahul.sharma,sijo.arakkal,dirk.draheim}@taltech.ee

Abstract. Many real-world data sets contain a mix of various types of data, i.e., binary, numerical, and categorical; however, many data mining and machine learning (ML) algorithms work merely with discrete values, e.g., association rule mining. Therefore, the discretization process plays an essential role in data mining and ML. In state-of-the-art data mining and ML, different discretization techniques are used to convert numerical attributes into discrete attributes. However, existing discretization techniques do not reflect best the impact of the independent numerical factor onto the dependent numerical target factor. This paper proposes and compares two novel measures for order-preserving partitioning of numerical factors that we call *Least Squared Ordinate-Directed Impact Measure* and *Least Absolute-Difference Ordinate-Directed Impact Measure*. The main aim of these measures is to optimally reflect the impact of a numerical factor onto another numerical target factor. We implement the proposed measures for two-partitions and three-partitions. We evaluate the performance of the proposed measures by comparison with human-perceived cut-points. We use twelve synthetic data sets and one real-world data set for the evaluation, i.e., school teacher salaries from New Jersey (NJ). As a result, we find that the proposed measures are useful in finding the best cut-points perceived by humans.

Keywords: Discretization · Partitioning · Numerical attributes · Data mining · Machine learning · Association rule mining

1 Introduction

In data mining and machine learning, discretization is an essential data preprocessing step to achieve discretized values from numeric columns. Numeric attributes can be discretized by partitioning the range of numeric attributes into different intervals. In-state of the art, several discretization approaches such as equi-depth, equi-width [3], ID3 [19], etc., are proposed. However, the existing

S. N. Srirama et al. (Eds.): BDA 2021, LNCS 13147, pp. 244–260, 2021.
https://doi.org/10.1007/978-3-030-93620-4_18

discretization methods do not provide optimal results for the discretization process, and they also have some drawbacks like information loss, etc., for mining algorithms.

In this paper, we propose an optimal way to find out intervals or partitions of a numerical attribute that reflect best the impact of one independent numerical attribute on a dependent numerical attribute. We provide a *Least Squared Ordinate-Directed Impact Measure (LSQM)* and *Least Absolute-Difference Ordinate-Directed Impact Measure (LADM)* for order-preserving partitioning of numerical factors. The measures provide a simple way to search the appropriate cut-points for finding the optimal partitions. For best cut-points, order-preserving partitioning on an independent factor is performed and implemented for two-partitions and three-partitions. The order of the independent variable is preserved using the value of data points. Therefore, the value of data points of one partition will always be less than the value of data points of the next partition. The measures' performance is assessed using one real-world data set and twelve synthetic data sets (including two-step and three-step functions). The outcomes are first compared to human perceived cut-points, and then their outcomes are also compared to one another. The following are the key contributions of this article:

1. We develop two measures to find out the partitions which best reflect the impact of one numerical factor on another numerical factor.
2. We evaluate the proposed measures on one real-world data set and twelve synthetic data sets, including two-step functions, three-step functions, compare it with human-perceived cut-points.
3. We provide the comparison of the results of both measures.

In Sect. 2, related work is discussed. In Sect. 3, we discuss the motivation of the proposed measures with an example. We provide methods in Sect. 4. In Sect. 5, we evaluate the proposed measures with a variety of data sets, including one real-world data set, two-step, and three-step data points. We finish with the paper with a conclusion in Sect. 6.

2 Related Work

In the literature, many concepts related to correlation and inter-dependency among variables are discussed in statistical reasoning, e.g. Pearson correlation [17,23], linear regression [15], ANOVA (Analysis of Variance) [8] etc. However, these tools do not find the partition of the numerical variable that reflects best the impact on another variable.

The idea of this research emerged from the research on partial conditionalization [5], association rule mining [20,21] and numerical association rule mining [9,22]. In these papers, the discretization process is discussed as an essential step for numerical association rule mining. We have also presented a tool named Grand report [18] which reports the mean value of a chosen numeric target column concerning all possible combinations of influencing factors. The measures proposed in this paper are important for discretization, which is an essential step in frequent itemset mining, especially for quantitative association rule mining [22] or numerical association rule mining.

There are various discretization processes available in the literature. Researchers and data scientists proposed different algorithms using different methods such as clustering, partitioning. However, these methods mainly focus on discretizing the continuous factor by finding the appropriate cut-points to make suitable intervals, or some of them use distance measures to create clusters. In this paper, our work is related to discretization and provides the partitions of one factor that best describes the impact of one factor on another.

Mehta et al. [14] worked in this direction and proposed a PCA-based unsupervised correlation preserving discretization method, which discretizes continuous attributes in multivariate data sets. The work ensures the use of all attributes simultaneously to decide the cut-points in place of one attribute at a time.

Dougherty et al. [4] reviewed and classified discretization methods along three separate axes; global versus local, supervised versus unsupervised, and static versus dynamic. Dougherty et al. [4] compared binning, unsupervised discretization method to entropy-based and purity-based supervised methods. Global methods, such as binning, partition all the data set attributes into regions, and each attribute is independent of other attributes. The static methods discretize each feature separately, whereas dynamic methods obtain inter-dependencies among features via conducting the search through space.

Liu et al. [12] performed a systematic study of existing discretization methods and proposed a hierarchical framework for discretization methods from the perspective of splitting and merging. The unsupervised static discretization methods such as equal-width and equal-frequency are simple and relevant to our work. The Equal-width discretization algorithm uses the minimum and maximum values of the continuous attribute and then divides the range into equal-width intervals called bins. The equal-frequency algorithm determines an equal number of continuous values and places them in each bin.

Ludl and Widmer [13] present RUDE (Relative Unsupervised Discretization) algorithm for discretizing numerical and categorical attributes. The algorithm combines the aspect of both supervised and unsupervised discretization. The algorithm is implemented in three steps: pre-discretizing, structure projection, and merging split points. The primary step is structure projection which projects the structure of each source attribute onto the target attribute. Then clustering is performed using projected intervals and merges split points if the difference is less than or equal to the user-specified minimum difference.

Recently, H. M. Abachi et al. [1] worked on statistical unsupervised method SUFDA (Statistical Unsupervised Feature Discretization Algorithm). The SUFDA tries to provide discrete intervals with low temporal complexity and good accuracy by decreasing the differential entropy of the normal distribution. Multi-scale and information entropy-based discretization method is also proposed in [24]. In 1988, Eubank [6] and Konno et al. [10] worked on the best piecewise constant approximation of a function f of single variable. Eubank used the population quantile function as a tool to show the best piecewise constant approximation problem. Later Bergerhoff [2] proposed an approach using particle swarm optimization for finding optimal piecewise constant approximations of one-dimensional signals. Our work is different because we are not using signals, and our main focus is on data sets that use several data points for one value of influencing factor.

3 Motivation

A number of discretization methods have been proposed in the state of the art [7,11]; however, they have not considered the type of target attribute, such as binary, categorical, or numerical. We use numerical attributes as both influencing and response factors in the proposed impact-driven discretization method. In general, when one variable influences another, the human brain is trained to notice changes and can easily discern compartments or partitions. However, in a real-world data set, it is difficult for a human to determine the most suitable compartments; for example, both the *Experience* and *Salary* attributes are numerical in the graph shown in Fig. 1. A human cannot easily find the appropriate compartment using this graph. As a result, the proposed measures partition the numerical attribute and determine its impact on a target attribute. This section presents a motivating example explaining why a specific measure is required to locate the suitable compartments.

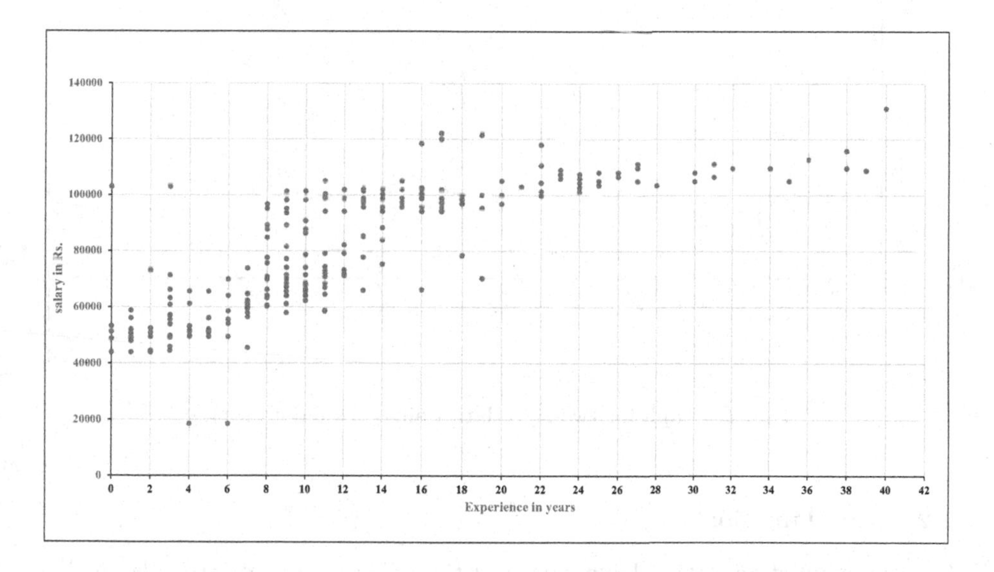

Fig. 1. An example for motivation of real-world data set.

4 Our Approach

The basic idea of our approach is to take one numerical independent variable and one target variable from a data set and discretize the independent variable in such a way as to find the appropriate cut-points, which are not observed easily by humans.

4.1 Key Intuition

We claim to discretize the independent numerical attribute by using order-preserving partitioning and see the impact on the numerical target attribute. In Fig. 2, we provide the graph for two-step data points where *X factor* is the numerical independent attribute, and *Y factor* is the numerical target attribute. It is the extreme case where data points are distributed as a step-function. In this case, humans can easily find out the cut-points without any difficulty. We evaluate the same data set with the proposed measures and compare the results with human perceived partitions.

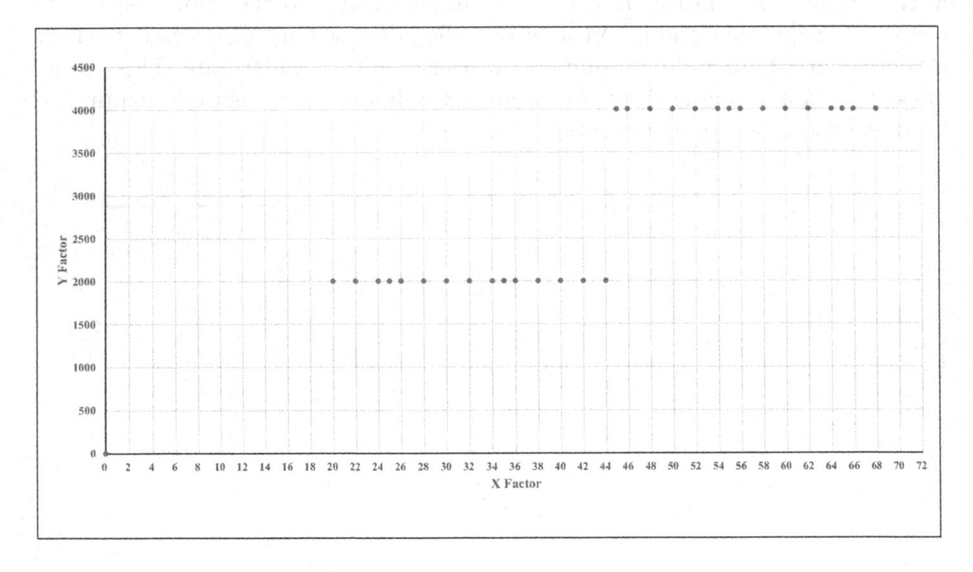

Fig. 2. Graph for two-step data-points (DS5 data set).

4.2 Step Function

In a *Step function f*, the domain is partitioned into several intervals. *f(x)* is constant for each interval, but the constant can be different for each interval. The different constant values for each interval create the jumps between horizontal line segments and develop a staircase which is also known as a step function.

Definition 1. *A Step function f on interval [a,b] is a piece wise constant function which contains many finite pieces. There exist a partition* $P = \{a = x_0, x_1, \ldots, x_n = b\} \in p[a,b]$ *such that* $f(x)$ *for all* $x \in (x_{r-1}, x_r)$ *for each* $r \in \{1,2,3,\ldots,n\}$. *The jump of* x_r *for* $r \in \{0,1,2,\ldots,n\}$ *is defined to be* $f(x_r^+) - f(x_r^-)$.

4.3 Definitions

To compute appropriate cut-points of independent numerical variables, we introduce the following impact measures as per the below definition.

Definition 2 (Least Squared Ordinate-Directed Impact Measure). *Given* $n \geq 2$ *real-valued data points* $(< x_i, y_i >)_{1 \leq i \leq n}$, *we define the* least squared ordinate-directed impact measure *for k-partitions (with* $k-1$ cut-points*) as follows:*

$$\min_{i_0=0<i'_1<\ldots<i'_{k-1}<i'_k=n} \sum_{j=1}^{k} \sum_{i'_{j-1}<i"\leq i'_j} (y_{i"} - \mu_{i'_{j-1}<\phi\leq i'_j})^2 \quad (1)$$

where the average of data values in a partition $\mu_{a<\phi\leq b}$ *between indexes a and b* $(a < b \leq n)$ *is defined as*

$$\mu_{a<\phi\leq b} = \frac{\sum_{a<\phi\leq b} y_\phi}{b-a} \quad (2)$$

In (1), we have that i'_j is the highest element in the *j-th* partition, where *highest element* means the data point with the highest index.

Definition 3 (Least Absolute-Difference Ordinate-Directed Impact Measure). *Given* $n \geq 2$ *real-valued data points* $(<x_i, y_i>)_{1\leq i\leq n}$, *we define the* least absolute-difference ordinate-directed impact measure *for k-partitions (with* $k-1$ cut-points*) as follows:*

$$\min_{i_0=0<i'_1<\ldots<i'_{k-1}<i'_k=n} \sum_{j=1}^{k} \sum_{i'_{j-1}<i"\leq i'_j} |y_{i"} - \mu_{i'_{j-1}<\phi\leq i'_j}| \quad (3)$$

where the average of data values in a partition $\mu_{a<\phi\leq b}$ *between indexes a and b* $(a < b \leq n)$ *is defined as*

$$\mu_{a<\phi\leq b} = \frac{\sum_{a<\phi\leq b} y_\phi}{b-a} \quad (4)$$

Listing 1. Pseudo-code for finding the three-partitions that reflect best the impact of a numerical variable on another numerical variable.

```
FUNCTION finding the first partition(array_d)
  FOR a=0 to (array_d.length-2)
    MEAN_POINT1: mean(array_d[0 to a+1])
    FOR b= 0 to a+1
      ABS_DIFF1: absolute.difference(array_d[b]-MEAN_POINT1)
      C1_SUM: C1_SUM + SQUARE(ABS.DIFF1)
    ENDFOR
    FOR j =(a+1) to (array_d.length-1)
        MEAN_POINT2: mean(array_d[a+1 to j+1])
        FOR i = a+1 to (j+1)
           ABS.DIFF2: absolute.difference(array_d[i]-MEAN_POINT2)
           C2_SUM: C2_SUM + SQUARE(ABS.DIFF2)
        ENDFOR
        MEAN_POINT3: mean(array_d[j+1 to (array_d.length)]
        FOR k = j+1 to (array_d.length)
           ABS_DIFF3: absolute.difference(array_d[k]-MEAN_POINT3)
           C3_SUM = C3_SUM + SQUARE(ABS_DIFF3)
        ENDFOR
        TOTAL_SUM = C1_SUM1 + C2_SUM2 + C3_SUM3
        IF a ==0 AND j == 1 THEN
           CUT1 = 0
           CUT2 = 1
           temporary_var= TOTAL_SUM
        ENDIF
        IF TOTAL_SUM < temporary_var
           CUT1 = a
           CUT2 = j
           temporary_var = TOTAL_SUM
        ENDIF

     ENDFOR
  ENDFOR
  PRINT(CUT1)
  PRINT(array_d.[CUT1])
  PRINT(CUT2)
  PRINT(array_d.[CUT2])
ENDFUNCTION
```

4.4 Method

Let D be a collection of n data points $D = (\langle x_i, y_i \rangle)_{1 \leq i \leq n}$, where $\langle x_i, y_i \rangle$ are data points of real values. As per (1), the proposed measure *LSQM* computes appropriate cut-points. The number of cut-points is $k - 1$, where k is the number of partitions suggested by the user. The measure first calculates the squared difference of the y-value of each data point of the current partition. In (1), the condition

$(i_{j-1} < r \leq i_j)$ requires that the index of data points of the current partition should be greater than the highest index of the previous partition and less than or equal to the highest index of the current partition. After summing up the squared differences of the several partitions, it selects the minimum values, which correspond to the appropriate cut-points. For the second measure *LADM*, we just take the sum of the absolute differences of the several partitions. Next, we first provide the pseudo-code for the case of three partitions ($k = 3$); and then, we compute the cut-points for two partitions and three partitions. In Listing 1, we provide pseudo-code for the three-partitioning approach according to Definition 2.

5 Evaluation

In this section, we experimentally validate the proposed measure in terms of the quality of the resulting discretization and its ability to find the independent variable's impact on the target variable. In Fig. 2, a two-step function graph is shown. In the graph, manual selection of cut-points can be performed for two-partitions and three-partitions easily. However, we demonstrate the cut-points after implementing the proposed measure on the same step-data sample and then compare its cut-points with the human perceived manual methods.

5.1 Data Sets

We have conducted the experiment using twelve synthetic data sets and one real-world data set. The real-world data set, New Jersey (NJ) school teacher salaries (2016) [16] is sourced from the (NJ) Department of Education. It contains 138715 records and 15 attributes. However, we have reduced the number of rows from the data set to analyze the cut-points visually. We have taken only initial 350 rows from the data set. The Data set NJ Teacher Salaries (2016) consists of salary, job and experience data for the teachers and employees in New Jersey schools. We are interested in the column {experience_total} and {salary}. The column {experience_total} is a numeric and independent attribute, whereas {salary} is a numeric target attribute. The twelve synthetic data sets are DS1 to DS12[1]. These twelve synthetic data sets have only two attributes named *Age* and *Salary* which are numeric. These data sets have a different number of rows and different values of attributes. In Table 1, we describe the data sets. As the limit of pages, all the graphs for all data sets are not included in this article. Repository of data sets has been given on the GitHub (See footnote 1).

5.2 Results and Discussion

Two-Partitioning. For two-partitioning, $k = 2$, we need one cut-point. We provide a graph for two-step data points in Fig. 2. The data set DS5 is from the

[1] https://github.com/minakshikaushik/Least-square-measure.git.

Table 1. Data sets used in evaluation.

Dataset	Number of records	Number of attributes
NJ Teacher Salaries(2016)	347	15
DS1	31	2
DS2	31	2
DS3	35	2
DS4	24	2
DS5	30	2
DS6	100	2
DS7	40	2
DS8	31	2
DS9	30	2
DS10	30	2
DS11	45	2
DS12	30	2

list of synthetic data sets. The data set DS5 is a sample of two-step function data points. We use this data set for the manual selection method and later implement the *LSQM* and *LADM* on the same data set. In the given data set, the human would identify 44 as the natural cut-point of the two partitions 0–44 and 45–72, see Fig. 3. Next, we implement the proposed measures on the same data set and see the cut-points.

Table 2. Comparison of the two-step and the three-step function using manual selection of cut-points and using proposed measures.

Dataset	DS5 (Two-step function)	DS12 (Three-step function)
Two-partitioning		
Manual cut point	44	35
LSQM cut point	44	35
LADM cut point	44	35
Three-partitioning		
Manual cut point	(20,44)	(35,52)
LSQM cut point	(20,44)	(35,52)
LADM cut point	(20,44)	(35,52)

Three-Partitioning. For three-partitioning, $k = 3$, we need two cut-points. We provide a graph for three-step data points in Fig. 4. We use data set DS12 as an extreme case of a three-step function. Earlier described in two-partitioning, we

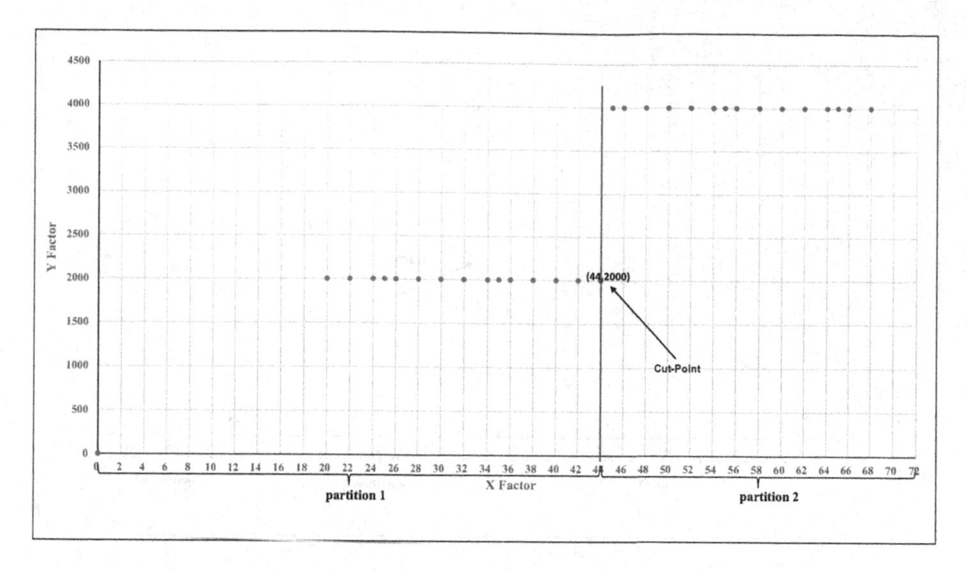

Fig. 3. Graph for showing the cut-point and two-partitions using manual method.

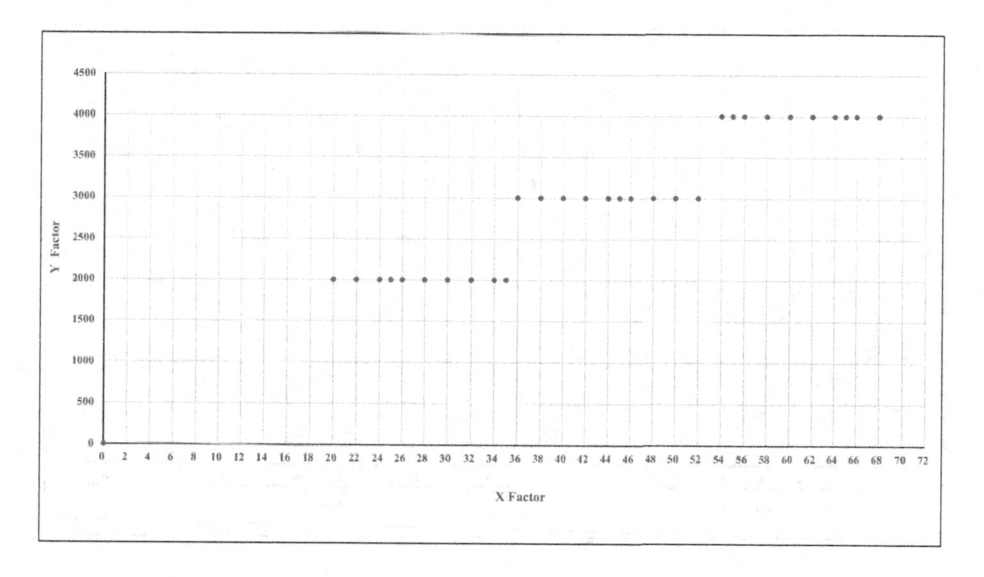

Fig. 4. Graph for three-step data-points (DS12 data set).

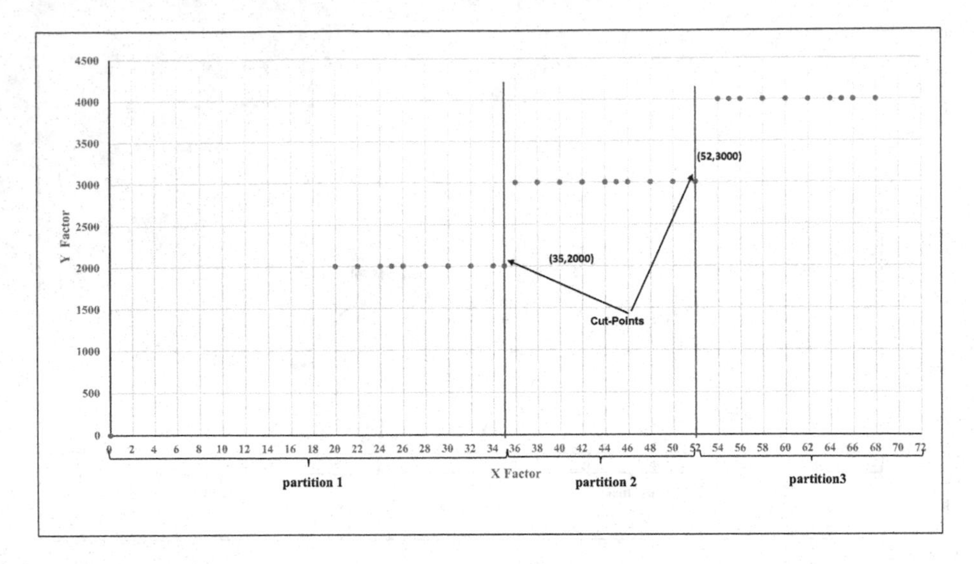

Fig. 5. Graph for showing the cut-points and three-partitions using manual method.

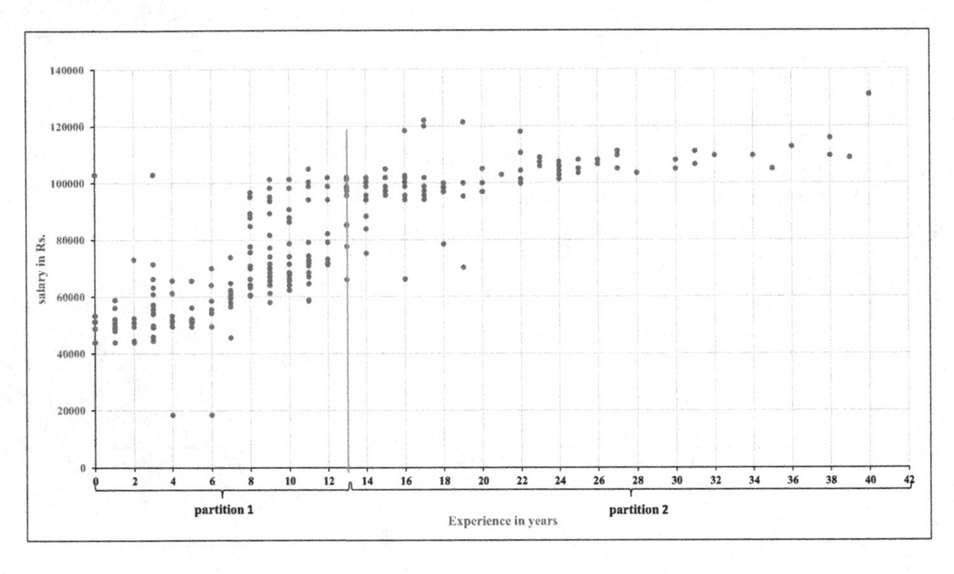

Fig. 6. Graph for showing the one cut-point and two-partitions using *Least Squared Ordinate-Directed Impact Measure* on real-world data set.

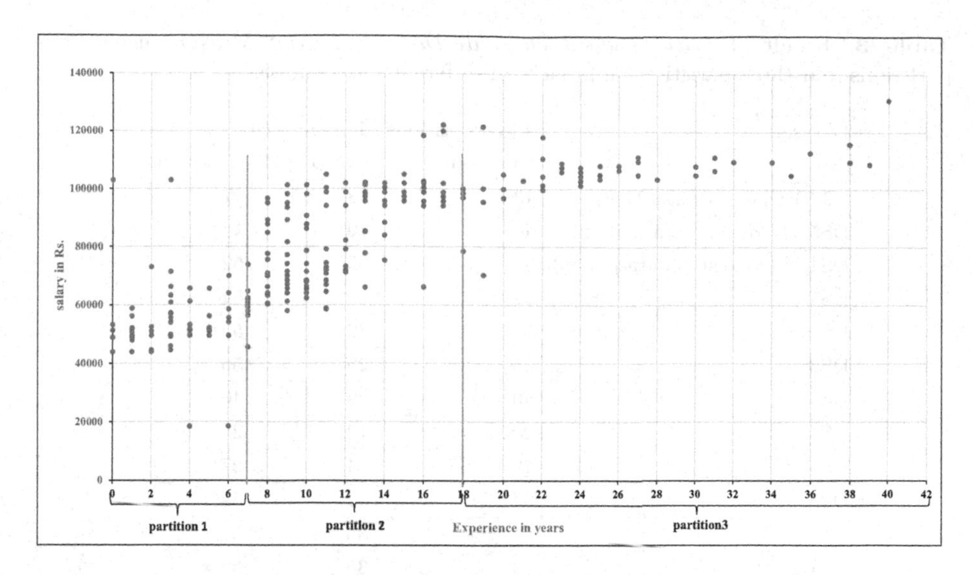

Fig. 7. Graph for showing the two cut-points and three-partitions using *Least Squared Ordinate-Directed Impact Measure* on real-world data set.

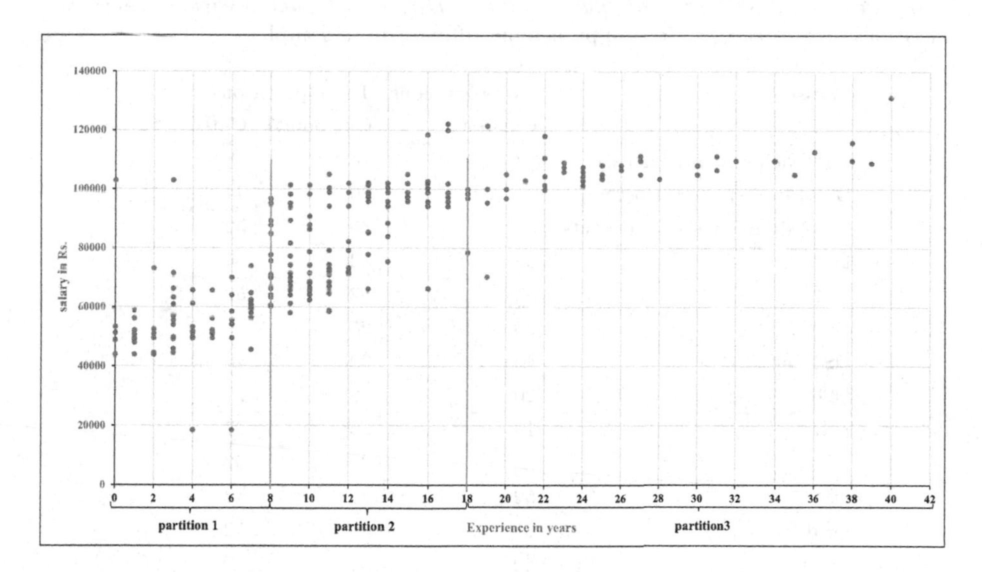

Fig. 8. Graph for showing the two cut-points and three-partitions using *Least Absolute-Difference Ordinate-Directed Impact Measure* on real-world data set.

Table 3. Result of *Least Squared Ordinate-Directed Impact Measure* using two-partitions and three-partitions approach on different data samples.

Dataset	Two-partitions	Three-partitions	
	Cut-point	Cut-point1	Cut-point2
NJ Teacher Salaries(2016)	13	18	7
DS5 (Two-step data-points)	44	20	44
DS12 (Three-step data-points)	35	35	52
DS1	52	52	54
DS2	52	32	52
DS3	25	25	56
DS4	40	29	40
DS6	20	12	24
DS7	19	14	27
DS8	32	32	52
DS9	52	35	52
DS10	35	35	52
DS11	42	32	42

Table 4. Result of *Least Absolute Ordinate-Directed Impact Measure* using two-partitions and three-partitions approach on different data samples.

Dataset	Two-partitions	Three-partitions	
	Cut-point	Cut-point1	Cut-point2
NJ Teacher Salaries(2016)	13	18	8
DS5 (Two-step data-points)	44	20	44
DS12 (Three-step data-points)	35	35	52
DS1	52	52	54
DS2	52	32	52
DS3	25	25	56
DS4	40	29	40
DS6	20	12	25
DS7	19	15	27
DS8	32	32	52
DS9	52	35	52
DS10	35	35	52
DS11	42	32	42

use this data set for the manual selection method and implement the proposed measures. By using the manual method human would identify 35 and 52 as two cut-points of the three-partitions 0–35, 36–52 and 53–72 in Fig. 5. After implementing the proposed measures on the same data set, we can verify the cut-points.

We implement both measures on data sets DS5 and DS12. We compare manually selected cut-points with cut-points provided by *LSQM* and *LADM* for two-partitioning and three-partitioning. As shown in Table 2, the cut-points for data set DS5 are the same for the manual selection method and *LSQM* and *LADM* measures. In the same way, cut-points for data set DS12 are also the same for manual method and proposed measures.

Next, we implement both measures on real-world data set NJ Teacher Salaries (2016) and the rest ten synthetic data sets (DS1, DS2, DS3, DS4, DS6, DS7, DS8, DS9, DS10, DS11). Figure 6 shows one cut-point for two-partitioning using *LSQM* measure on data set NJ Teacher Salaries(2016). Figure 7 shows two cut-points for three-partitioning using *LSQM* measure on data set NJ Teacher Salaries(2016). We received one cut-point 13 using *LSQM* for two-partitioning and received two cut-points 18 and 7 for three-partitioning for the same data set. Figure 8 is showing the two cut-points 18 and 8 after applying *LADM* measure. Both the measures *LSQM* and *LADM* provide the same cut-point for two-partitioning as given in Fig. 6. However, their cut-point for three partitioning is different and it is given in Figs. 7 and 8.

Table 5. Comparison of cut-points provided for measures *LSQM* and *LADM* for two-partitioning and three-partitioning.

Dataset	k=2		k=3				Deviation
	LSQM	LADM	LSQM		LADM		
	I	I	I	II	I	II	
NJ Teacher Salaries(2016)	13	13	18	7	18	8	Yes (In cut point2)
DS5 (Two-step data-points)	44	44	20	44	20	44	No
DS12 (Three-step data-points)	35	35	35	52	35	52	No
DS1	52	52	52	54	52	54	No
DS2	52	52	32	52	32	52	No
DS3	25	25	25	56	25	56	No
DS4	40	40	29	40	29	40	No
DS6	20	20	12	24	12	25	Yes (In cut point2)
DS7	19	19	14	27	15	27	Yes(In cut point1)
DS8	32	32	32	52	32	52	No
DS9	52	52	35	52	35	52	No
DS10	35	35	35	52	35	52	No
DS11	42	42	32	42	32	42	No

The results of *LSQM* and *LADM* measures for two-partitioning and three-partitioning on all the data sets are given in Tables 3 and 4, respectively. In Table 5, We have compared the results of both measures for $k = 2$ and $k = 3$. As we can see in the Table 5, NJ Teacher Salaries(2016) data set has one point deviation in cut-point2 for $k = 3$. The measure *LSQM* cut-point2 has a value of 7, whereas *LADM* cut-point2 has a value of 8. The data sets DS6 also have only one point difference in cut-point2 that is 24 and 25 whereas DS7 has one point difference in cut-point1 14 and 15. We observed a deviation in the result when $k = 3$. Except for these data sets, all the data sets have the same cut-points for both measures. After analyzing and comparing the results of both measures, we find out that the outcomes of both proposed measures are approximately similar. We analyzed that the proposed measures provide the cut-points which reflect best the impact of one independent numerical factor on a dependent numerical target factor.

6 Conclusion

This paper aimed to find the partitions that best reflect the impact of a numerical independent variable on a dependent numerical target variable. We proposed two *Least Squared Ordinate-Directed Impact Measure* and *Least Absolute-Difference Ordinate-Directed Impact Measure*. In the case of step functions, there is an immediate, intuitive understanding of best cut-points regarding human judgment. Therefore, we evaluated the performance of these measures for two-step staircase data sets (step functions), three-step staircase data set and arbitrary data set (non-step function). We examined that the proposed measures provide the same human perceived cut-points for two-step staircase and three-step staircase data sets. Furthermore, the results of both proposed measures on twelve synthetic data sets and one real-world data set are approximately similar. As future work, we plan to evaluate the proposed measures in long series of data repositories against respective human judgments (by data experts and domain experts). We also plan to implement the measure for arbitrary numbers of k-partitions beyond two- and three-partitions. A particular challenge will be to come up with *inter*-measures for comparing partitions of different numbers of k-partitions.

Acknowledgements. This work has been conducted in the project "ICT programme" which was supported by the European Union through the European Social Fund.

References

1. Abachi, H.M., Hosseini, S., Maskouni, M.A., Kangavari, M., Cheung, N.M.: Statistical discretization of continuous attributes using Kolmogorov-Smirnov test. In: Wang, J., Cong, G., Chen, J., Qi, J. (eds.) Databases Theory and Applications, pp. 309–315. Springer International Publishing, Cham (2018)

2. Bergerhoff, L., Weickert, J., Dar, Y.: Algorithms for piecewise constant signal approximations. In: 27th European Signal Processing Conference (EUSIPCO), pp. 1–5. IEEE (2019)
3. Catlett, J.: On changing continuous attributes into ordered discrete attributes. In: Kodratoff, Y. (ed.) EWSL 1991. LNCS, vol. 482, pp. 164–178. Springer, Heidelberg (1991). https://doi.org/10.1007/BFb0017012
4. Dougherty, J., Kohavi, R., Sahami, M.: Supervised and unsupervised discretization of continuous features. In: Machine learning proceedings 1995, pp. 194–202. Elsevier (1995)
5. Draheim, D.: Generalized Jeffrey Conditionalization: A Frequentist Semantics of Partial Conditionalization. Springer, Cham (2017). https://doi.org/10.1007/978-3-319-69868-7
6. Eubank, R.: Optimal grouping, spacing, stratification, and piecewise constant approximation. Siam Rev. **30**(3), 404–420 (1988)
7. Garcia, S., Luengo, J., Sáez, J.A., Lopez, V., Herrera, F.: A survey of discretization techniques: taxonomy and empirical analysis in supervised learning. IEEE Trans. Knowl. Data Eng. **25**(4), 734–750 (2012)
8. Gelman, A., et al.: Analysis of variance - why it is more important than ever. Ann. Stat. **33**(1), 1–53 (2005)
9. Kaushik, M., Sharma, R., Peious, S.A., Shahin, M., Ben Yahia, S., Draheim, D.: On the potential of numerical association rule mining. In: Dang, T.K., Küng, J., Takizawa, M., Chung, T.M. (eds.) FDSE 2020. CCIS, vol. 1306, pp. 3–20. Springer, Singapore (2020). https://doi.org/10.1007/978-981-33-4370-2_1
10. Konno, H., Kuno, T.: Best piecewise constant approximation of a function of single variable. Oper. Res. Lett. **7**(4), 205–210 (1988)
11. Kotsiantis, S., Kanellopoulos, D.: Discretization techniques: a recent survey. GESTS Int. Trans. Comput. Sci. Eng. **32**(1), 47–58 (2006)
12. Liu, H., Hussain, F., Tan, C.L., Dash, M.: Discretization: an enabling technique. Data Min. Knowl. Discov. **6**(4), 393–423 (2002)
13. Lud, M.C., Widmer, G.: Relative unsupervised discretization for association rule mining. In: Zighed, D.A., Komorowski, J., Żytkow, J. (eds.) PKDD 2000. LNCS (LNAI), vol. 1910, pp. 148–158. Springer, Heidelberg (2000). https://doi.org/10.1007/3-540-45372-5_15
14. Mehta, S., Parthasarathy, S., Yang, H.: Toward unsupervised correlation preserving discretization. IEEE Trans. Knowl. Data Eng. **17**(9), 1174–1185 (2005)
15. Montgomery, D.C., Peck, E.A., Vining, G.G.: Introduction to Linear Regression Analysis. John Wiley & Sons, Hoboken (2021)
16. Naik, S.: Nj teacher salaries (2016). https://data.world/sheilnaik/nj-teacher-salaries-2016
17. Pearson, K.: VII. Note on regression and inheritance in the case of two parents. Proc. Roy. Soc. London **58**(347–352), 240–242 (1895)
18. Arakkal Peious, S., Sharma, R., Kaushik, M., Shah, S.A., Yahia, S.B.: Grand reports: a tool for generalizing association rule mining to numeric target values. In: Song, M., Song, IY., Kotsis, G., Tjoa, A.M., Khalil, I. (eds.) DaWaK 2020. LNCS, vol. 12393, pp. 28–37. Springer, Cham (2020). https://doi.org/10.1007/978-3-030-59065-9_3
19. Quinlan, J.R.: Induction of decision trees. Mach. Learn. **1**(1), 81–106 (1986)
20. Shahin, M., et al.: Big data analytics in association rule mining: a systematic literature review. In: International Conference on Big Data Engineering and Technology (BDET), pp. 40–49. Association for Computing Machinery (2021)

21. Sharma, R., Kaushik, M., Peious, S.A., Yahia, S.B., Draheim, D.: Expected vs. unexpected: selecting right measures of interestingness. In: Song, M., Song, I.Y., Kotsis, G., Tjoa, A.M., Khalil, I. (eds.) DaWaK 2020. LNCS, vol. 12393, pp. 38–47. Springer, Cham (2020). https://doi.org/10.1007/978-3-030-59065-9_4
22. Srikant, R., Agrawal, R.: Mining quantitative association rules in large relational tables. In: Proceedings of the 1996 ACM SIGMOD International Conference on Management of Data, pp. 1–12 (1996)
23. Stigler, S.M.: Francis galton's account of the invention of correlation. Stat. Sci. **4**, 73–79 (1989)
24. Xun, Y., Yin, Q., Zhang, J., Yang, H., Cui, X.: A novel discretization algorithm based on multi-scale and information entropy. Appl. Intell. **51**(2), 991–1009 (2021)

Towards Machine Learning to Machine Wisdom: A Potential Quest

P. Nagabhushan, Sanjay Kumar Sonbhadra(✉), Narinder Singh Punn, and Sonali Agarwal

Indian Institute of Information Technology Allahabad, Prayagraj, U.P., India
{pnagabhushan,rsi2017502,pse2017002,sonali}@iiita.ac.in

Abstract. In the present era of artificial intelligence (AI) enabled solutions, the world is observing a tremendous influx in machine learning (ML) approaches across various application domains like healthcare, industry, document analysis, audio-video processing, etc. All existing machine learning approaches claim for intelligent solutions, but till date the learning is guided by the human wisdom i.e. all the proposed machine intelligence algorithms are data centric and infer knowledge without understanding the scenarios. The wisdom is an ability to take wise decisions based on the inferred knowledge to satisfy W5HH principle which outlines the series of answers to the questions such as why, what, who, when, where, how and how much, in a given context. This paper discusses the scope of machine wisdom (artificial wisdom)over conventional machine learning strategies along with its significance and how it can be achieved.

Keywords: Artificial intelligence · Machine learning · Machine wisdom · Knowledge. · W5HH principle

1 Introduction

In the present era, artificial intelligence (AI) based solutions are offered across various application domains like healthcare, industry, document analysis, etc. All proposed machine learning algorithms try to imitate human intelligence and claim to understand the data. But these algorithms are data centeric and the phenomenon of human wisdom is absent. To understand the data and scenario completely, the existing machine learning algorithms must be adaptive to simulate the human wisdom. During past decades, human intelligence and other implicit characteristics like morality and rationality are discussed by the research community in the context of AI [27]. It is evident from the exhaustive literature survey that scientists and philosophers continuously talked about machine intelligence, but the phenomenon of human wisdom in machine learning is not well explored by the AI research communities.

If we talk about machine learning approaches to make the system intelligent, the existing approaches majorly try to learn the pattern of the information obtained from the raw data, thus infer the context based decisions based on

S. N. Srirama et al. (Eds.): BDA 2021, LNCS 13147, pp. 261–275, 2021.
https://doi.org/10.1007/978-3-030-93620-4_19

the acquired domain dependent knowledge. To develop an intelligent system, the primary objective must be the context adaptation and must not be limited to learn only the context dependent scenarios. Thus, philosophically the process starts with data acquisition, then the useful informative patterns must be extracted. Furthermore, the model must learn the patterns of the acquired knowledge to make context specific decisions and finally, the magic of wisdom must be integrated to make the context sensitive decisions. In the present era of big data, the whole computation world is swimming in the ocean of data, where running out of it is not a problem, but drowning in it is. The data is being generated at phenomenal and extraordinary rate, where it acts like one of the natural resources, if unrefined it cannot really be used. Several implicit features of data such as volume, speed, etc. encouraged the computing world to devise new powerful, robust and scalable technologies for efficient storage, convenient management and efficient computation. Due to the technological advancements, presently we are equipped with powerful and efficient data processing/storage solutions. However, people not only require the data, but the insights and values are equally desired.

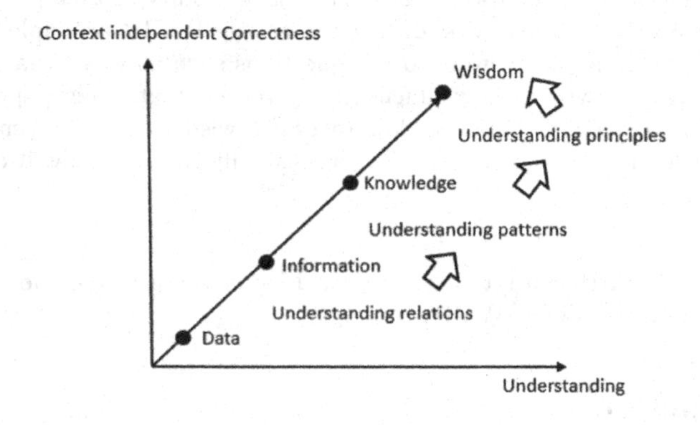

Fig. 1. From data to wisdom (wisdom/insight hierarchy)

Concerning to artificial intelligence equipped solutions to any real time problem, the primary objective must be to devise a model to take us from data to insight and value, via wisdom hierarchy which is conceptually a framework to store raw data and it is later transformed into information followed by the knowledge, thereby finally converges to wisdom. These steps as levels of insight are shown in Fig. 1 which depict all stages for transition of data to wisdom. To understand this, we can consider the example of software product development where it starts with specific error codes and outputs a robust software system, where each step can be described as follows:

- **Data:** Usually the system error logs.
- **Information:** An organized and structured report of associated errors.
- **Knowledge:** The ability to deal with errors for possible solution.

- **Wisdom:** A better understanding of different and mutually exclusive errors may cause a more complex problem.

From above example, it is clear that wisdom can be treated as deep subject matter expertise that outputs better decision-making, business value, and wise decisions.

From a human stance, the above discussed insight levels are not absolute, and the boundary between any two level is fuzzy. The growing insight for an optimal solution to a given problem usually does not follow a direct path and may refer to a circular path. For example, the acquired knowledge changes as data or nature of data changes. The obtained insights from data analysis are data themselves. Additionally, insight does not follow the up-direction in wisdom hierarchy, but arrives from all possible dimensions. For any AI based smart solution, the insight hierarchy passes through complex and continuous recursive operations. In the insight hierarchy, the humans usually climbs, whereas for any AI system it would be a complex process. All computing devices deal only with data that is always treated as input to any AI system. The present machine learning algorithms are data centric; and produces some more data as outputs at each level of insight hierarchy and need human interpretation (partially or fully). Machine wisdom enables a system to understand the whole environment to become adoptable for a wise decision or prediction. Data science and AI technocrats have an opportunity to develop artificial wisdom systems to imitate the nature of human flourishing and human excellence. While making a decision, the conventional machine learning approaches helps to design AI solutions that avoids causing harm while taking decision, but unable to answer the question that "What is human excellence?"

In present scenario, almost all businesses are directly/indirectly using AI based solution to make important decisions. Big giants in the field of information technology such as Google, Amazon and Facebook smartly choose what user wants, whereas several modern vehicles use driver-assist technology that aids drivers for breaking, steering, etc. Ola and Uber do matching of commuters with drivers and set prices. All above use cases have respective challenges and complexities, but share a common core: data driven machine learning algorithms; that helps to make a decision with minimum human inter-mediation.

Figure 2 depicts the flow of remaining paper, where context dependent learning strategies (intelligence) is covered first and late, the context independent learning strategy (wisdom) is discussed. And finally the to achieve artificial wisdom is discussed followed by the concluding remarks. Thus, the present paper is organized as follows: Sect. 2 discusses the intelligence (human intelligence followed by artificial intelligence) in detail, whereas Sect. 3 talks about wisdom concerning human and machines. Section 4 discusses the possibilities and principles of artificial wisdom, whereas the last section concludes the overall contributions of present paper.

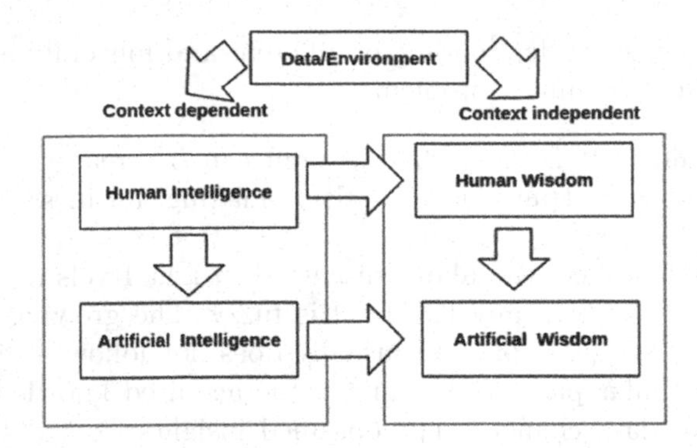

Fig. 2. Flow of the discussed concern (human intelligence to artificial intelligence, human wisdom to artificial wisdom, and intelligence to Wisdom).

2 Intelligence

Section 1 talks about the natural pattern of achieving wisdom. The intelligence comes from the understanding of the patterns exist in the observations and uses the domain knowledge (Fig. 1) to learn the context dependent challenges. The intelligence acts as the basic building block to achieve wisdom. This section discusses the human intelligence followed by the artificial intelligence in detail.

2.1 Human Intelligence

The following interesting question is very common since all of our childhood: "What is intelligence?". It is answered in different ways by different people; but till date there is no common answer available. However, conventionally it is defined as: "general mental capability than others; involves the ability to reason, plan, solve problems, think abstractly, comprehend complex ideas, learn quickly, and learn from experience" [5]. Basically, intelligence refers to the ability of knowledge acquisition and apply the obtained skills. Several theories of intelligence have been proposed by several scientists: 'g' (general intelligence) factor by Spearman, Gf-Gc (fluid-crystallized) theory by Cattell and theory of cognitive abilities by Cattell-Horn-Carroll's, etc. [20]. All these theory conclude that intelligence is a "multifactorial construct comprised of several components, and integrates a number of cognitive domains and abilities, particularly sensory processing (including attention and working memory), language, acquired knowledge, memory consolidation and retrieval, processing and psychomotor speed, and reasoning".

2.2 Artificial Intelligence

AI tries to imitate the human behaviour and tries to include human intelligence components such as reasoning, planning, etc. [4]. It is evident that there are several AI systems exist that exceed human intelligence, but only in context of processing speed and capacity to store information such as: google map provides a complete navigation information of any location; the IBM Watson defeated chess world champions, etc. Due to technological development the artificial intelligence passed through massive evaluation during past few decades (shown in Fig. 3).

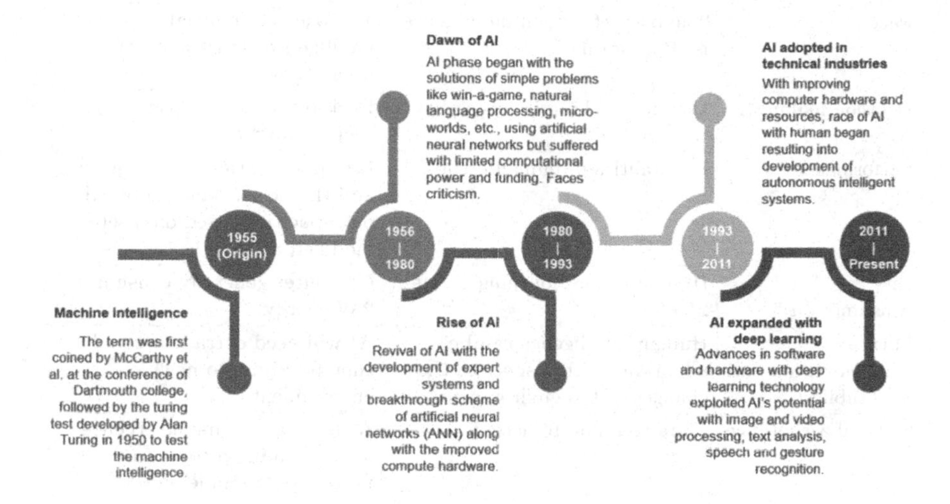

Fig. 3. Evolution of AI technologies over time.

Benefits and Limitations. Ultra-high speed analysis, smart storage and retrieval of big data are the major strength of today's world AI solution. The AI uses machine learning algorithms to identify high-level/hidden enunciable patterns and relationships that helps in decision making process [17]. Knowledge of hidden patterns is necessary for accurate predictions in all application areas including healthcare [6] and efforts are underway to predict advancement of disease, such as the transformation of mild cognitive impairment (MCI) to Alzheimer's, using objective clinical data [8]. However, till date AI is not able model human emotions, intentions, beliefs, or desires.

The majority of present AI solutions takes logical decisions and predict based on ontology (rules). However, for every decision wise thinking not intelligence whereas must incorporates human values. Therefore, AI is still way behind from human wisdom in terms of challenges, risks and limitations [7]. The recent technological and computational advancements uplifted the capability of AI solutions to a great extent, but at the same time opened a new area for development of societal guidelines about the ethics and morality of AI to ensure safety. The differences between human intelligence and artificial intelligence is shown in Table 1.

Table 1. Human intelligence vs artificial intelligence.

Feature	Human intelligence	Artificial intelligence
Rise	Human intelligence is based on the ability of thinking, reasoning and reviewing etc. It is perishable in nature	AI is technological implementation of human insights; initiated by Norbert Weiner. It is permanent in nature
Pace	The pace of human intelligence is slow and natural	The pace of artificial intelligence is fast and programmed
Decision making	Decision making is subjective	Decision making is based on accumulated data
Performance	Possibilities of human error	Less possibilities of human error and the performance is based on capacities based on a set of modified rules
Energy consumption	Human brains consume about 25 W	Computer generally consumes 2 W energy
Future transformation and adaptability	Human intelligence can be adaptable with respect to the changes in the environment	AI will need extra effort and time to adjust to new environment
Societal acceptance	More common to accept	AI has challenging emergence and becoming a tool of new normal post pandemic

3 Wisdom

Section 1 talks about the natural pattern to achieve wisdom. Wisdom comes through intelligence and increases with time and experience. This natural phenomenon is applicable for both human and machine wisdom. The wisdom comes from the understanding the principles of learning via context adaptation (shown in Fig. 1). This section discusses the human wisdom followed by the artificial wisdom in detail.

3.1 Natural Wisdom: Human Wisdom

"How one can define human intelligence?" This question seems very simple in AI, but it is not true. To understand this, consider a recent argument in machine learning initiated by Judea Pearl about causation [19]. Pearl argued that neural-nets based learning models are the present form of AI solutions that is not a favorable practice of artificial intelligence, because of inability to deal with causal/counterfactual reasoning that is the most important feature of human intelligence. Unlike intelligence, wisdom is adaptive that increases over the time and personal experience [24]. It is natural phenomenon that compared to the young people the older persons exhibit more components of wisdom that comes from experience [28]. Additionally, wisdom is more adaptive compared to intelligence that improves emotional regulation and pro-social properties [26].

After such a long successful journey of AI based solutions, still something is missing i.e. true replica of humans' strategic, linguistic, and intelligence. This fact raises a question that "would an AI solution imitate a paradigmatic human being?" Oracle of Delphi [3] said that, Socrates was the wisest in the world, and if Socratic ignorance is the best form of human wisdom, then AI solution to copy the Socratic wisdom is the best kind of AI [14]. Technically speaking, wisdom is next level of intelligence, where intelligence is associated with deviousness to find right means, whereas wisdom identifies the right ends. In the context of value alignment, the AI solutions must imitate a broader notion of intelligence that contains wisdom rather than an instrumental view of intelligence.

3.2 Artificial Wisdom: Beyond Artificial Intelligence

The objective to develop AI enabled technologies is to devise solutions to serve humanity in better and efficient way. Though, several AI systems have been designed to satisfy the need but still great advancements are required to go beyond the basic components of general intelligence. The "intelligence" is not sufficient enough to meet the present global requirements of society, because only "wisdom" can fulfill the desired objectives; therefore the term "artificial wisdom" (AW) came into existence.

Though, there were some futurists dreamed of artificial consciousness [2], earlier it was believed that only humans can be wise because of some implicit important properties like: consciousness, autonomy, will, and theory of mind [16] that are keys to cultivate/develop wisdom. Moreover, the culture and society helps to develop several important characteristics of wisdom in humans. The most important factor to develop the human wisdom is self-actualization that helps to develop above discussed features. However, the computers and AI solutions are developed by humans to perform wise actions [23]. The most important question is whether the computers can be advanced to adapt and become wiser in their algorithms. Differences between human wisdom and artificial wisdom is discussed in Table 2.

From Human Wisdom to Artificial Wisdom (Machine Wisdom). Asimov's three laws are the heart of whole robotics world [1]. The primary law that supersedes the other laws says that a robot never injure humans (directly or indirectly). The second principle says that a robot must follow human's orders and finally, the last law says that a robot should seek to protect its own existence. Later, these laws are further imposed to complex environment adaptation to achieve machine wisdom [12,22]. Table 2, provides a comparison of all components of human wisdom along with corresponding future artificial wisdom envisioned. The transition from human wisdom to machine wisdom is shown in Fig. 4.

Due to the technological advancements and milestone research contributions, AI emulated human intelligence significantly, especially computing and

Table 2. Human wisdom vs artificial wisdom.

Feature	Human wisdom	Artificial wisdom
Emergence	It is a composite characteristic of human including emotional/social appearance, self-strength to make decision as per context and circumstances	It is technology enabled characteristics to exhibit human like moral, ethical, and logical decision-making through facilitating instantaneous feedback from trusted advisers, or gathering input from, and disseminating data to, large numbers of people at once
Risk	Based on experience and less challenging in terms of privacy, security and ethical use	Challenging in terms of privacy, security and ethical use
Decision making	Decision making is subjective	Decision making is perception based
Assessment	Assessment is easy	Assessment is difficult
Future transformation and adaptability	Natural and more adaptive	Forced and difficult to evolve

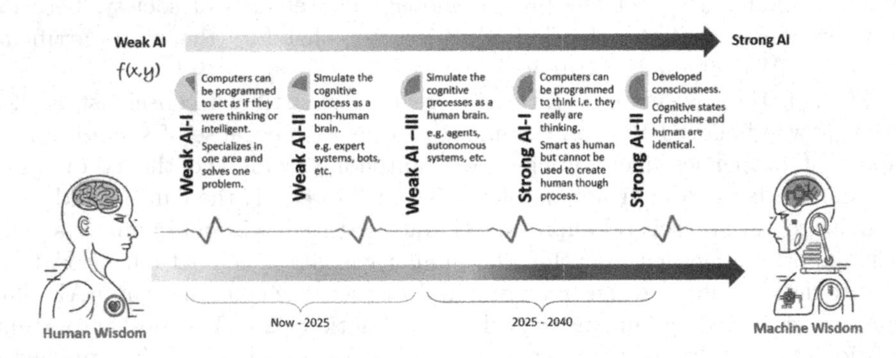

Fig. 4. Transition from human wisdom to machine wisdom.

planning intelligence. AI robots have became very common now a days in all application domains including defence, healthcare, industry, etc. Deep Blue and AlphaGo have defeted human world champions in chess. In 2019 the Google-owned AlphaStar won StarCraft II at a grand-master level, capable of beating 99.8% of all human opponents. Apple's Siri and Google Translate have shown the capabilities of AI solutions to imitate human linguistic intelligence. Boston Dynamics's humanoid robots have proven that AI can imitate kinetic intelligence too.

4 Transition Scope from Artificial Intelligence to Artificial Wisdom Systems

The artificial wisdom or machine wisdom is today's need to provide better human assistance for robust and error free operations. The most crucial issue is environment adaptation to achieve artificial wisdom for accurate predictions and actions. In this context the present section discusses the following: Artificial intelligence to artificial wisdom, human intelligence to human wisdom, and principles of wisdom systems:

Human Intelligence to Wisdom. The wisdom has been a common term in several philosophical and religious texts since ancient days; however, scientific study started in the early 1970s [9]. The amount of reported research on wisdom has been increased rapidly during last few decades [13]. The differences between human intelligence and wisdom are discussed in Table 3.

Table 3. Human intelligence vs human wisdom.

Description	Human intelligence	Human wisdom
Description	Mental capability then others, reasoning, planing, problem solving capability, abstraction, comprehend complex plans, quick learner, and self-learning	Characteristics like: emotions, social behaviors, self-reflection, etc.
Confirmation	Common tests, e.g. Stanford-Binet IQ test and WAIS	Rating scales, e.g. 3D-WS and SD-WISE
Biological	Strong hereditary	relatively inherited
Neuro analysis	Frontoparietal network for facets like perception, short-term memory storage, and language	Prefrontal cortex and limbic striatum for facets like decision-making, problem-solving, self control, and emotional regulation
Change in intelligence	Fluid intelligence decreases with age; crystallized intelligence remains stable until the 70s. Intelligence does not increase with aging	Wisdom is adaptive and may increase with age and personal experience until the 80s, when it begins to decline due to cognitive impairment
Health significance	IQ predicts health, educational achievement, job performance, and income, but alone is insufficient for the well-being of the individual or the society	IQ predicts health, educational achievement, job performance, and income, but alone is insufficient for the well-being of the individual or the society

Artificial Intelligence to Wisdom. Artificial intelligence must be promoted to artificial wisdom via means of environment adaptation. Robots can be considered as examples of offering artificial wisdom up-to some extent with interfacing with humans. For consumer-level wiser applications, a companion robot is required in the future that can be offered by various other forms (e.g. eye glasses, hearing aids, etc.) to support human well-being. In this context, social robots must be designed to interact with people and proactively communicate [18].

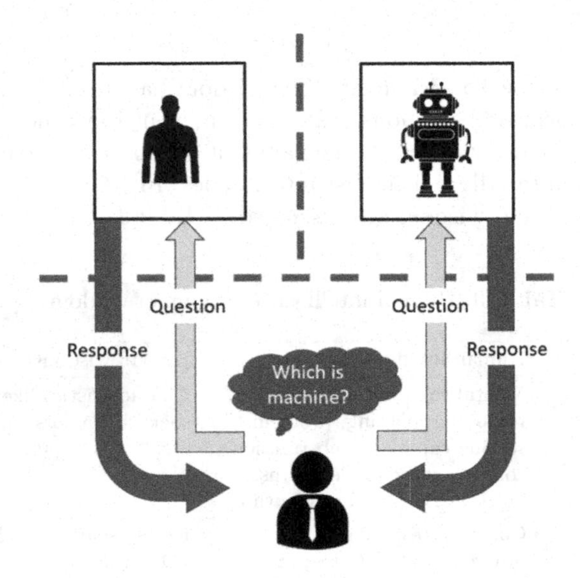

Fig. 5. Schematic representation of Turing test.

AI techniques for several real time applications are quickly being adopted for smart decisions. This fact raises the question of compatibility of AI and human values. Following are the major societal concerns: a) How to ensure the trustworthiness of AI based solutions so that it must not harm us?, b) The operations must be always controlled?, c) It must always serve as the best assistant to the human being and follow the values? With all these worries it can be concluded that the "value alignment" must be given highest priority in AI equipped solutions. Before seven decades, the famous mathematician Alan Turing wrote about adaptation of human values by the machines: "The machine must be allowed to have contact with human beings in order that it may adapt itself to their standards (as shown in Fig. 5") [11]. The overall comparison is between artificial intelligence and wisdom is shown in Table 4.

4.1 Principles of Artificial Wisdom Systems

To develop artificial wisdom following characteristics are needed: self-reflection (ability to learn from experience), divergent/diversity tolerant (integrating mul-

Table 4. Artificial intelligence vs artificial wisdom.

Study criteria	Artificial intelligence	Artificial wisdom
Goal	Perform complex human tasks faster	Act as personal assistants to humans
Components	Comprehension knowledge	Pro-social behaviors (empathy, compassion)
	Processing speed	Self-reflection
	Quantitative reasoning	Emotional regulation
	Memory	Accepting diversity of perspectives
	Visuospatial ability	Decisiveness
	Auditory processing	Social advising
Assessment	Turing test	Turing-like test for AW

tiple perspectives) and learning from historical experience. Therefore, the artificial wisdom system need an fast, organized and persistent memory system like human being. Like human being the an AW system must learn from their experience and mistakes to ensure more correct future operations (i.e. reinforcement learning) [25]. Recently, Qureshi et al. [21] proposed intrinsically motivated reinforcement learning for social robots that do not need any external reward system to learn from their mistakes, which makes the implementation of reinforcement learning more suitable for the "reward-sparse" real world.

A wise system must exhibit pro-social behaviors (empathy, compassion) and social decision-making (social advising). Artificial wisdom must also facilitate social skill development. Artificial wisdom must also recognize human emotions to help people for emotion regulation to make wise decisions. Enhancement of emotion regulation has been proven through several existing randomized controlled trials (RCTs) [15]. An artificial wisdom system must act as a "wisdom coach" to provide a cognitive valuation of a situation to decide a controlled action. In this context, Hao et al. [10] proposed a AW robot to manage emotions between positive and negative, during waiting scenario (waiting in queue, seating, etc.).

It is evident from above discussion that there is enough possibility to develop artificial wisdom to imitate the actual human wisdom. The environment adaptability is the major concern that should be focused using more intelligent reinforcement learning mechanism. This can be easily understood using the data, information, knowledge and wisdom framework with W5HH principles (DIKW-W5HH) shown in Fig. 6.

5 Challenges

The great potential of machine wisdom brings great challenges associated with it. The challenges that need to be addressed are shown in Fig. 7. The under-

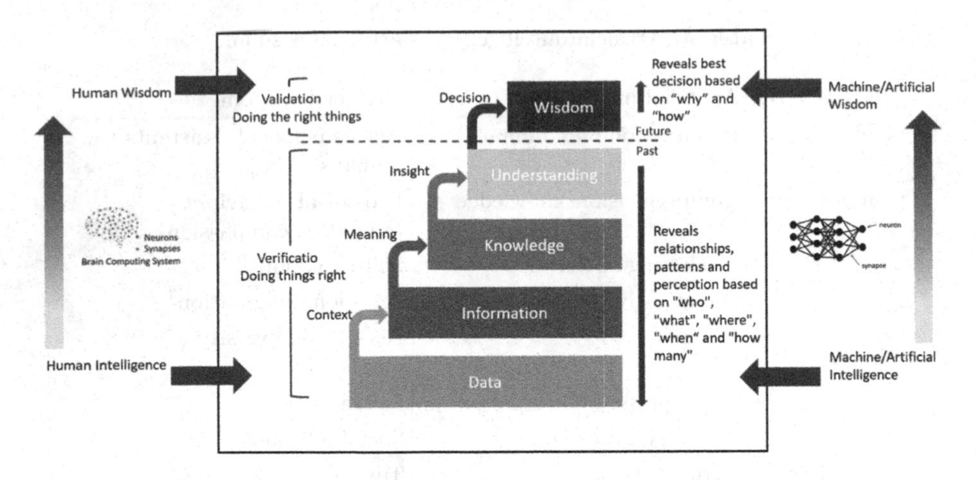

Fig. 6. DIKW-W5HH framework for machine/human intelligence/wisdom.

standing of the working environment is the most crucial aspect of developing machine wisdom. This is the most critical aspect where the system needs to adapt dynamically to surrounding changes and modify the process of its decision making accordingly. For instance, the slightest alteration in the path of a robot may require it to re-learn and adapt to the new environment. Moreover, developing a reliable technological solution with artificial intelligence is still a challenging task, where technology can be trusted blindly. In this regard, a huge amount of data or information with vivid diversity is required to learn every possibility of the task at hand and to annotate such data is again a challenging task.

At the present stage, AI is no match for human reasoning and hence is not completely dependable or trustworthy. Following this, a socially aware system is another major challenge that needs to be addressed to achieve human wisdom. For instance, if certain decision making procedures require interaction with certain individuals and making the unbiased verdict based on the collected viewpoints while also acknowledging the emotional states. Similarly with such development ethical issues also need to be addressed such as the unemployment rate with job displacement, preserving privacy, protection from unintended consequences, machine laws and regulations, cybersecurity, etc. Moreover, developing such advanced technological solutions would require a rich source of power supply and other resources to keep their motor running. However, at present, these systems are highly inefficient in utilizing energy. In this context, the field of soft mechanics is an emerging area that can help design AI enabled systems that are highly flexible and sustain well with the available resources.

Overall, we can say that to achieve machine wisdom the above discussed challenges are the most crucial to be addressed while developing such advanced systems for real world applications. Indeed, with regular advancements in technology, the gap to address such challenges is getting narrowed over time.

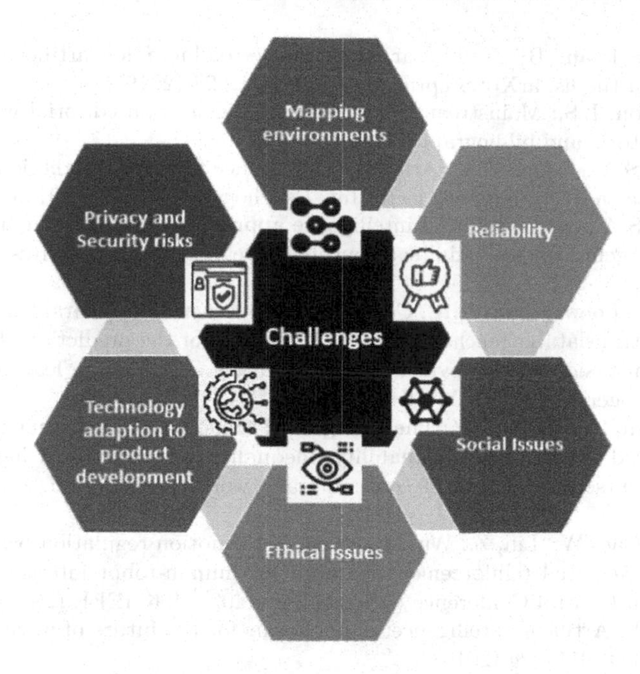

Fig. 7. Challenges of machine wisdom.

6 Conclusions

The present paper focuses on future transition of artificial intelligence to artificial wisdom. To achieve a machine wisdom the technologies must be designed to emulate human wisdom to serve the humanity. In every application domain, data and computational scientists are trying continuously to develop machine wisdom models. Intelligence is needed and necessary but not sufficient to achieve wisdom. Wisdom is critical for flourishing in the modern society. Human wisdom increases with time and experience and more modifiable than IQ. AI technologies of the future need different conceptual and computational models to emulate human wisdom to produce artificial wisdom. An artificial wisdom agent must act as peer, assistant or coach and can surely make humans wiser. A long-term trans-disciplinary collaboration is essential for the development of machine wisdom, that will have far greater positive impact on human.

References

1. Asimov, I.: Three laws of robotics. Asimov, I. Runaround (1941)
2. Buttazzo, G.: Artificial consciousness: utopia or real possibility? Computer **34**(7), 24–30 (2001)
3. Corey, D.D.: Socratic citizenship: Delphic oracle and divine sign. Rev. Politics **67**(2), 201–228 (2005)

4. Gil, Y., Selman, B.: A 20-year community roadmap for artificial intelligence research in the us. arXiv preprint arXiv:1908.02624 (2019)
5. Gottfredson, L.S.: Mainstream science on intelligence: an editorial with 52 signatories, history, and bibliography (1997)
6. Graham, S.A., Depp, C.A.: Artificial intelligence and risk prediction in geriatric mental health: what happens next? Int. Psychogeriatr. **31**(7), 921–923 (2019)
7. Graham, S.A., et al.: Artificial intelligence approaches to predicting and detecting cognitive decline in older adults: a conceptual review. Psychiatry Res. **284**, 112732 (2020)
8. Grassi, M., Loewenstein, D.A., Caldirola, D., Schruers, K., Duara, R., Perna, G.: A clinically-translatable machine learning algorithm for the prediction of Alzheimer's disease conversion: further evidence of its accuracy via a transfer learning approach. Int. Psychogeriatr. **31**(7), 937–945 (2019)
9. Grossmann, I., Brienza, J.P.: The strengths of wisdom provide unique contributions to improved leadership, sustainability, inequality, gross national happiness, and civic discourse in the face of contemporary world problems. J. Intell. **6**(2), 22 (2018)
10. Hao, M., Cao, W., Liu, Z., Wu, M., Yuan, Y.: Emotion regulation based on multi-objective weighted reinforcement learning for human-robot interaction. In: 2019 12th Asian Control Conference (ASCC), pp. 1402–1406. IEEE (2019)
11. Howard, J.: Artificial intelligence: implications for the future of work. Am. J. Ind. Med. **62**(11), 917–926 (2019)
12. Jeste, D.V., Graham, S.A., Nguyen, T.T., Depp, C.A., Lee, E.E., Kim, H.C.: Beyond artificial intelligence: exploring artificial wisdom. Int. Psychogeriatr. **32**(8), 993–1001 (2020)
13. Jeste, D.V., et al.: The new science of practical wisdom. Perspect. Biol. Med. **62**(2), 216 (2019)
14. Kim, T.W., Mejia, S.: From artificial intelligence to artificial wisdom: what socrates teaches us. Computer **52**(10), 70–74 (2019)
15. Lee, E.E., et al.: Outcomes of randomized clinical trials of interventions to enhance social, emotional, and spiritual components of wisdom: a systematic review and meta-analysis. JAMA Psychiat. **77**(9), 925–935 (2020)
16. Leslie, A.M.: Pretense and representation: the origins of "theory of mind". Psychol. Rev. **94**(4), 412 (1987)
17. Minsky, M.L.: Logical versus analogical or symbolic versus connectionist or neat versus scruffy. AI Mag. **12**(2), 34 (1991)
18. Nocentini, O., Fiorini, L., Acerbi, G., Sorrentino, A., Mancioppi, G., Cavallo, F.: A survey of behavioral models for social robots. Robotics **8**(3), 54 (2019)
19. Pearl, J.: Causality. Cambridge University Press, Cambridge (2009)
20. Plucker, J., Esping, A., Kaufman, J., Avitia, M.: Handbook of Intelligence: Evolutionary Theory, Historical Perspective, and Current Concepts (2015)
21. Qureshi, A.H., Nakamura, Y., Yoshikawa, Y., Ishiguro, H.: Intrinsically motivated reinforcement learning for human-robot interaction in the real-world. Neural Netw. **107**, 23–33 (2018)
22. Salge, C., Polani, D.: Empowerment as replacement for the three laws of robotics. Front. Robot. AI **4**, 25 (2017)
23. Sevilla, D.C.: The quest for artificial wisdom. AI Soc. **28**(2), 199–207 (2013)
24. Staudinger, U.M.: Older and wiser? Integrating results on the relationship between age and wisdom-related performance. Int. J. Behav. Dev. **23**(3), 641–664 (1999)
25. Thrun, S., Littman, M.L.: Reinforcement learning: an introduction. AI Mag. **21**(1), 103 (2000)

26. Treichler, E.B., et al.: A pragmatic trial of a group intervention in senior housing communities to increase resilience. Int. Psychogeriatr. **32**(2), 173–182 (2020)
27. Tsai, C.H.: Artificial wisdom: a philosophical framework. AI Soc. **35**(4), 937–944 (2020)
28. Worthy, D.A., Gorlick, M.A., Pacheco, J.L., Schnyer, D.M., Maddox, W.T.: With age comes wisdom: decision making in younger and older adults. Psychol. Sci. **22**(11), 1375–1380 (2011)

Pattern Mining and data Analytics

Big Data over Cloud: Enabling Drug Design Under Cellular Environment

B. S. Sanjeev(✉) and Dheeraj Chitara

Department of Applied Sciences, Indian Institute of Information Technology - Allahabad, Prayagraj 211012, UP, India
sanjeev@iiita.ac.in

Abstract. Molecular Dynamics (MD) simulations mimic the motion of atoms. While simulations often require weeks to complete, the terabytes of data generated in the process also present a challenge for analysis. Recent studies discovered drug-receptor interactions with the cellular environment. Comprising of hundreds of small and large molecules seen in cells, such complex studies are necessary for improving drug design. Current resources and techniques do not provide, in reasonable time, necessary insights into such systems. In this study, we developed an algorithm to identify molecules interacting with drug-receptor complexes. Using first ever application of big data framework to MD studies, we demonstrate that our approach enables rapid analysis of MD data. Finally, we propose a cloud-based, Spark-enabled, self-tuning, scalable and responsive framework to accomplish optimal MD studies for drug-development within available resources.

Keywords: Big data · Cloud infrastructure · Molecular dynamics simulations · Drug design · Spark

1 Introduction

Molecular Dynamics (MD) simulation is a powerful technique to study molecules in atomistic detail. It models the movement of atoms, ions and molecules subject to forces acting on them. Similar to a video capturing the motion with a series of images, *evolution* of a molecular system too is captured as a sequence of *frames*. A frame holds the necessary details of the system corresponding to the particular time it represents. Analysis of the frames provides insights into the dynamics of the system itself.

Standard software (e.g., AMBER [18]) are available to carry out MD simulations on biomolecules. Such studies have been used to examine a variety of phenomena such as protein folding [21], stability [14] and intermolecular interactions [3]. In particular, they are of great value in drug design and drug discovery, providing vital inputs on ligand docking, virtual screening, dynamics of drug-bound protein/receptor complexes, allosteric modulation and role of water molecules in drug binding [4,17]. However, the *length* of a simulation (in terms of time) needs to be adequate for meaningful interpretation of data. Smaller systems generally

S. N. Srirama et al. (Eds.): BDA 2021, LNCS 13147, pp. 279–295, 2021.
https://doi.org/10.1007/978-3-030-93620-4_20

attain equilibrium faster and larger systems require longer simulations. This places additional hurdles on simulating large systems, and often results in run-times ranging from days to weeks. Unsurprisingly, longer simulations of larger systems produce data running into terabytes.

Development of a new drug costs about a billion dollars [22]. While designing a drug *in silico* in itself is not an expensive task, the final hurdle that besets the acceptance of any drug involves clinical trials. The environment around the drug-receptor complex inside human body is very different from the simple theoretical models used for the drug design. Large number of detailed studies have consistently highlighted the fact that cellular environment could drastically alter the properties of biomolecules [9,15], and argued for the development of more realistic models in drug design [6,7]. A recent study simulated drug-bound enzyme in presence of hundreds of molecules seen in cells [19]. The analysis of the data required newer approach and is discussed here. Given the size of computations and the volume of data, applying big data framework (through Apache Spark) was a natural choice as it enables simultaneous loading of data from multiple frames along with concurrent computations. However, none of the current tools, including CPPTRAJ [16] and MDTraj [11], use big data frameworks (e.g., Spark or MapReduce) for the analysis of data from MD simulations.

Given the motivation to study complex MD simulations, the primary goal was to provide a scalable framework that could process voluminous data while concurrently performing computationally intensive tasks. The other goal was to design a more capable frame-work for drug design over Cloud infrastructure that could mix scalability with given priorities and resources.

In this paper, we demonstrate the necessity for a new framework in the study of MD simulations. Section 2 introduces the chosen simulated systems and the parameters for study. Section 3 demonstrates that big data approach facilitates rapid analysis of MD data. Based on this advance, we present a Spark-based approach to Cloud that provides a self-tuning, flexible and scalable framework to deliver speedy insights of immense value in drug design. Section 4 concludes the paper.

2 Materials and Methods

Proteins are polymers made of 20 types of units called amino acids. To perform biologically meaningful activities, they take different shapes in 3-D space. Certain calculations on proteins use only C_α atoms (see Fig. 1) to represent the corresponding amino acids. Often two or more polymers of amino acids combine to form a single functional protein.

Two systems based on a protein called Main protease (M^{pro}) of SARS-CoV-2 virus were covered for the current study. The first system (called M^{pro}-free hereafter) consisted of M^{pro} along with 10 different types metabolites present in the cell, potassium ions and water molecules. The second system (called M^{pro}-drug hereafter), had 3 different drugs (viz., Elbasvir, Glecaprevir and Ritonavir [19]) bound to M^{pro} and *crowder* proteins to emulate large molecules seen in

```
        R1                 R2                 R3
        ||                 ||                 ||
N  -  Cα  -  C  -  N  -  Cα  -  C  -  N  -  Cα  -  C ...
             ||                 ||                 ||
             O                  O                  O
```

Fig. 1. Protein is a polymer made of 20 types of amino acids. Chemically, they differ at C_α atoms shown as R_is. Hydrogen atoms are not shown.

cellular environment, in addition to the above molecules. For reasons beyond the scope of this work, 5 units of these drugs were *docked* (i.e., bound) to M^{pro}. A schematic diagram of M^{pro}-drug is shown in Fig. 2.

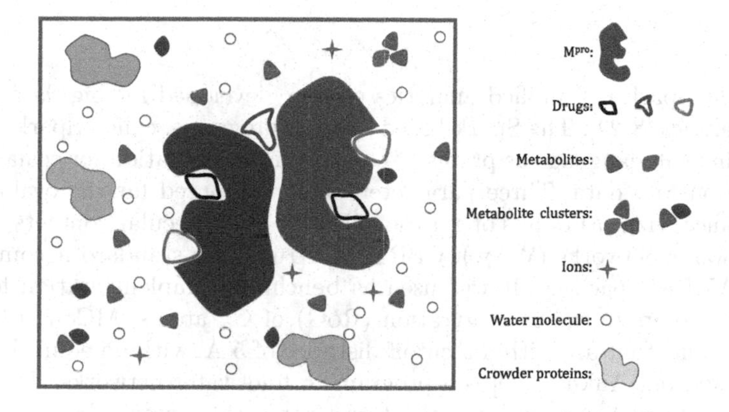

Fig. 2. A schematic model of M^{pro} system with 185,255 atoms. The more realistic model for simulation involved not only system of interest (M^{pro} protein bound to 5 drugs of 3 types) but also the *crowded* environment. This environment consisted of crowder proteins, metabolites seen in cells, ions and water. The second system, M^{pro}-free, did not have crowder proteins and drugs.

While a docking study identifies *where* drugs may bind on a protein, MD simulations help to identify *if* such a complex is viable. Analyzing the dynamics of simulated system involves processing numerous frames of data generated by MD simulation. Details of the system at a given time (such as coordinates and sizes of the simulated box) are captured through a frame associated with it. Sequence of frames thus reproduce evolution of system modelled by simulation. While simulations run over hundreds of nanoseconds (ns), simulated data is typically stored (as frames) at intervals of 1 picosecond (ps). In M^{pro}-drug, 8 protein GB1s (each with 56 amino acids) were used as protein crowders along with metabolites of 10 different types. Details of the two chosen systems is given in Table 1.

Table 1. Composition of MD simulations of M^{pro}-free and M^{pro}-drug systems is given below. Size of M^{pro} is given terms of amino acids (aa.). Crowders comprising of crowder proteins (Cr. proteins), metabolites and ions were incorporated in the simulations.

Item	M^{pro}-free	M^{pro}-drug
Protein	608 a.a.	608 a.a.
Drugs	0	5
Cr. proteins	0	8
Metabolites	95	170
K^+ ion	90	190
Total atoms	100,093	185,255
Frames	1,000,000	200,000
Data size of frames (GB)	1201	445

Apache Spark is a unified analytics engine, developed for big data applications analytics [8,24]. The Spark-based suite of programs, called SparkTraj, was written in Scala language as proofs-of-concept implementation for scalable computation on MD data. Three parameters were computed for the evaluation of performance, viz., (a) radius of gyration (RoG), (b) molecular contacts (MCon), and (c) water networks (WNw). CPPTRAJ [16] is the standard accompanying tool of AMBER package. It was used as benchmark implementation for serial version to compute radius of gyration (RoG) of C_α atoms. MCon is the total pairs of atoms that are within a cutoff distance of 5 Å, with no equivalent serial implementation. Ankush [20] is a program to find water networks. The parameters are defined below and are of interest for various reasons as discussed in Sect. 3.1 under **Parameter selection for benchmarking.**

Radius of Gyration: The compactness of a protein can be measured using radius of gyration, which would increase if the protein begins to unfold. In the current work, this was equivalent to finding the root-mean-squared deviation of C_α atoms. RoG of only the first chain of M^{pro} was computed.

$$\mathrm{RoG} = \sqrt{\frac{1}{N}\sum\nolimits_{i=1}^{N}(r\_\bar{r})^2}$$

Molecular Contacts: Typical MD simulations consider only biomolecule(s) of interest apart water molecules and ions. Cellular environment comprises of many other molecules that influence the dynamics of system of interest. Under such circumstances, it would be necessary to find molecules (such as metabolites) getting in proximity with protein of interest. As shown in Fig. 3, this information was captured using MCon algorithm that finds the number of pairs of atoms within a distance of 5 Å between two molecules.

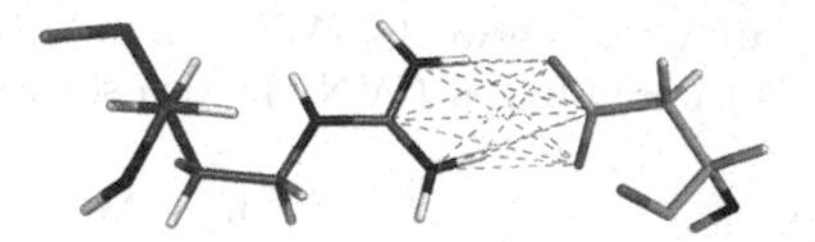

Fig. 3. Due to crowded environment within cells, molecules often bump into one another. The number of pairs of atoms that are within a distance of 5 Å is called as molecular contacts (MCon). MCon reflects proximity of molecules. The figure shows a few such contacts between two molecules represented with dotted lines.

Water Networks: A water network is a set of polar atoms (oxygen/nitrogen atoms of solute shown as P_i in Table 2) that are within 3.5 Å distance with a common water molecule. As shown in Fig. 4, some networks such as {P_1, P_2} are formed independently by different water atoms. Larger networks such as {P_a, P_b, P_c, P_d} give rise to smaller networks such as {P_b, P_d} and {P_a, P_c, P_d}. Occurrence (occ.) of a water network is defined as the number of frames in which a given network is seen. From all observed water networks (WNw^c in Table 2), we *selected* for benchmarking water networks with a minimum occurrence of 40% (WNw in Table 2), though lesser cutoffs were used for the study. *Trajectory* for each WNw, which shows its presence or the absence in each frame ordered by time, was computed. Using trajectory, Maximum Residence Time (MRT), defined as the highest number of sequential frames in which a particular WNw continues to exist, was also computed.

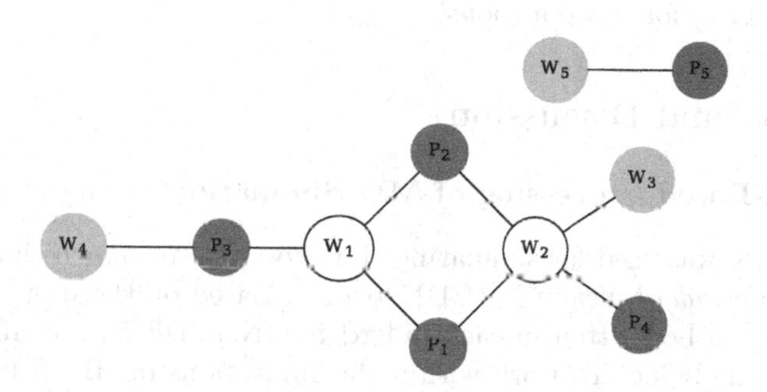

Fig. 4. Two nodes (i.e., atoms) are connected if, (a) their distance is within 3.5 Å, and (b) if least one node is oxygen of water. A water network is a set of 2 or more nitrogen/oxygen atoms of solute connected to a common water molecule. The networks seen here are {P_1, P_3}, {P_2, P_3}, {P_1, P_2, P_3} through W_1, {P_1, P_4}, {P_2, P_4}, {P_1, P_2, P_4} through W_2, and {P_1, P_2} independently through W_1 and W_2.

It is necessary to not just process but also to read data in parallel to obtain meaningful scaling of computation when size of data is large. To this end we used Apache Spark (referred also as 'Spark') with Scala. Spark is an open-source distributed general purpose framework that is designed for large-scale data analysis and offers implicit data parallelism with fault tolerance [8]. AMBER generates

Table 2. From all candidate water networks (WNw^c), only those networks with minimum occurrence (i.e., 40%) were filtered (WNw). Also see Fig. 4.

Item	M^{pro}-free	M^{pro}-drug
Polar atoms (P$_i$)	2,250	4,076
Water molecules (W$_i$)	29,593	54,968
WNw^c	5,566,453	3,261,310
WNw	655	1,317

data in NetCDF format [10]. As Spark does not have native support for NetCDF format, SciSpark was used to read the simulation data. SciSpark is a framework developed by NASA and UCLA for applications in earth and space sciences [13]. The basic approach for computation we used was to read the MD data into SciSpark's Resilient Distributed Datasets (RDDs) [23] first, followed by computation of necessary parameters through Spark Datasets and DataFrames. SciSpark supports HDFS [1] as well as native Linux file systems. In the next section, Spark's application to MD data analysis is discussed in detail.

NVIDIA Tesla V100 based GPU nodes were used for simulations. For benchmarking, a cluster with 20 data nodes was used. GRIDScaler SFA7700X appliance with 15 GB/s of throughput was the storage server for the two systems. GPU nodes and data nodes of the cluster have two Xeon Gold 6148 CPUs with 384 GB of RAM. The big data cluster supported Apache Spark (ver. 2.3.0) and Scala (ver. 2.11.8) for computations.

3 Results and Discussion

3.1 Spark-Based Processing of MD Simulation Data

AMBER18 [18] was used for simulating the chosen systems, which saves output in one or more *mdcrd* files in NetCDF format. Based on the user input, one or more frames can be written in each mdcrd file. NetCDF format allows parallel reading of a single file. To work within the limitations of HDFS block size [2], only a limited number of frames per file were stored. Each mdcrd file of M^{pro}-free and M^{pro}-drug systems contained data 50 frames. AMBER stores the detailed description of a simulated system (such as names of atoms, various molecules, types of bonds, etc.) in a separate file called *parmtop*.

Apache Spark is a framework designed for massive in-memory data processing with lazy evaluation. This framework was used to process the simulation data. To implement efficient solutions, both Spark operations as well as sequence of their application should be carefully chosen. Sub-optimal approach may lead to significant delays due to *data shuffling* (i.e., movement of massive data across nodes due to its redistribution). It may even lead to expensive recomputations of Spark's RDDs. Hence, we developed the solutions to circumvent these significant barriers. Another potential rate-limiting step avoided was the generation and use of intermediate data on storage.

Parameter Selection for Benchmarking: Three carefully chosen parameters shape the proof concept implementations. Radius of Gyration (RoG) is a trivial calculation akin to finding root-mean-square deviation of less than 0.5% of atoms. This computation allows to estimate the throughput of data available to Spark. Computation of Molecular Contacts (MCons) requires examination of hundreds of millions possible contacts in each frame. This is a straightforward computation can be done using either an efficient User-Defined Function (UDF) or a Dataset method. As each frame needs to be processed only once and results can be collected with relative ease. Given the resilient nature of Spark RDDs and the involvement of huge data across multiple phases, water networks demand careful computations to avoid performance hazards such as data shuffling.

Radius of Gyration: CPPTRAJ [16] was used to compute the radius of gyration for benchmarking of serial variant. It is the standard accompanying software for AMBER package and is used to compute a wide variety of parameters from simulation data.

Algorithm 1: Calculation of Radius of Gyration

read topology from parmtop
create *mdRDD* from mdcrds
obtain *time*, *RoG* through:
  flattening *mdRDD* into *frRDD*
  creating Dataset using additional parameters
  invoking Dataset's method for RoG
sort *RoG* w.r.t. *time*

Algorithm 1 shows the approach used for the Spark version. One should first note that if a system has N atoms, it has $3N$ coordinates (i.e., 3-D coordinates x_1, y_1, z_1, ..., x_N, y_N, z_N). Details of the system were read from the *parmtop* file (as discussed in the previous section). Then, the mdcrd files were read parallelly into a Resilient Distributed Dataset (RDD) called mdRDD loaded through SciSpark. mdRDD was then *flattened* into another RDD, called frRDD, so that every element was now a tuple that contained information of a particular frame in the form of ($time_i$, $box_{i,x}$, $box_{i,y}$, $box_{i,z}$, $x_{i,1}$, $x_{i,2}$, ... $x_{i,3N}$), where i is the frame number. Then, a Spark Dataset, with a method to find RoG for a frame, was created using frRDD. Invoking this method, RoGs of all frames were computed, data was collected into a Scala list, and then sorted w.r.t. the time of the frames. Sorting by time was not done using Spark itself in any of the implementations for reasons discussed later in this section. Complexity of the computing RoG within a frame in terms of the number of residues in the given chain (n_{res}) is $O(n_{n_{res}})$. Additional parameters include parmtop and cutoff data.

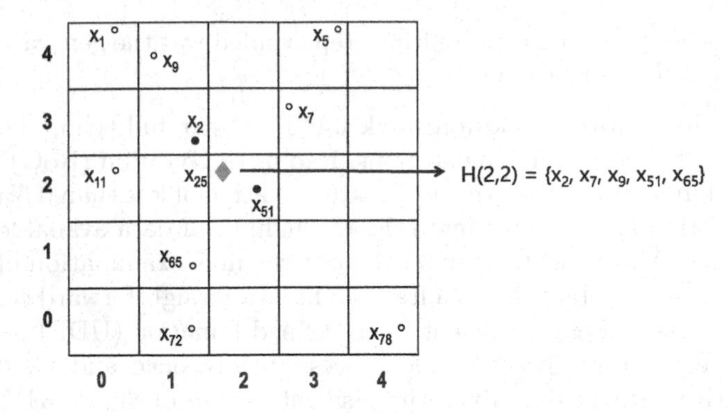

Fig. 5. The diamond shows the atom of interest, circles are other atoms, and only those represented with filled circles have relevant contacts. If we discretize space into squares with size cut-off distance (d) itself, examination of 9 squares is adequate to find potential contacts of an atom. A hash data structure, with tuples made of its discretized coordinates as key, can store all potential atoms that need to be examined for any atom in a given square. The same approach can be extended to 3-D for speedy computation of molecular contacts and water networks.

Molecular Contacts: Unlike the other two parameters, MCon could not be benchmarked with serial version due to absence of equivalent implementation. As shown in Algorithm 2, much of the procedure to compute molecular contacts was similar to that of the radius of gyration, but with two differences. The output from each frame was not a number (like RoG) but instead a string, which, when split, gave triplets (r_i, r_j, cnt) where r_ks were residue numbers and *cnt* was the number of contacts they shared. Also, unlike RoG, MCon requires finding contacts between two large sets of atoms. M^{pro}-drug system, for instance, has 9,796 atoms for the M^{pro}-drug complex and crowder proteins and metabolites constitute 10,364 atoms. This translates to over 101 million pairs of atoms. Clearly, a brute force way of counting is prohibitively expensive. As explained below, only potentially feasible pairs of atoms were scanned.

A 2-D illustration of the idea is shown in Fig. 5. First, we scanned every atom from one set and placed them in all possible (i.e., 9) squares where they may have contacts. Then, given an atom of interest from the other set, say x_{25}, we can immediately examine the list of atoms corresponding to the square. This approach allows us to linearize the computation time to find MCon as physically each square can only have a certain number of atoms at any time. The time taken for calculating MCon was comparable to that of RoG, which is not very surprising considering the overall complexity of finding MCon within a frame now being $O(L_xL_yL_z)+O(n_{systemAt})+O(n_{crowderAt})+O(n_{conacts}log(n_{conacts}))$ where L_is are lengths of simulation box along x, y and z directions.

Algorithm 2: Calculation of Molecular Contacts

read topology from parmtop
create $mdRDD$ from mdcrds
obtain $time$, $\{MCon\}$ through:
  flattening $mdRDD$ into $frRDD$
  creating Dataset using additional parameters
  invoking Dataset's method for $MCon$
sort $MCon$ w.r.t. $time$

Water Networks: Water networks were benchmarked against the serial version that works in two stages. In the first stage, it generates 2 files per frame; one that contains the contacts between polar atoms and water molecules, and the other between water molecules themselves. In the second stage, it processes these files to determine water networks. The first stage uses a C++ program and the second stage a Perl script. While this version worked well when it was originally developed, currently it is too inefficient for large systems with hundreds of thousands of frames. Hence, as shown in Algorithm 3, a Spark version was developed to update it.

Algorithm 3: Water network calculations

read topology from parmtop
create $mdRDD$ from mdcrds
create and cache $nwDF$ through:
  flattening $mdRDD$ into $frRDD$
  creating Dataset through additional parameters
  obtaining $time$, $\{WNw^c\}$ through Dataset's method
create $\{WNw\}$ using $nwDF$ and occ. cutoff
compute $WNwTraj$ using WNw and $nwDF$

Unlike the previous two algorithms, the study of water networks involves multiple phases. Phase 1 involves processing all frames to find all networks seen, called candidate water networks (WNw^c). The number of networks could be very high. For example, M^{pro}-free simulation generated over 5.5 million networks. In phase 2, the presence or absence of every network in each frame is used to compute occurrence of networks, and only networks with a minimum occurrence are selected (WNw). Finally, using the data available from phase 1, trajectory of the selected networks are be computed. RDDs have a tendency to recompute. Given the amount of data involved in every phase, a careful implementation is necessary to avoid data shuffling and expensive recomputations by Spark.

The approach we use is as follows. To prevent recomputations from slowing down the process, we first find and cache all (candidate) water networks seen in each frame {WNw^c}. The occurrence of each network is then computed, and only those networks of interest are selected based on occurrence {WNw}. The trajectories (WNwTraj), consequently maximum residence times (MRTs), are computed from the cached data. Network trajectories can be stored if desired. However, we found that in general, it was adequate to have the summary featuring occurrence, MRT, and the window in which MRT was observed. As discussed in the case of MCon, the hash-list approach described through Fig. 5 was applied to short-list the interacting polar atoms of each water. First, all polar atoms are scanned and placed in lists corresponding to cuboids where they may have contacts. Then, for each water, the corresponding list of its hash are scanned. Once all connected atoms for any particular water are known, all subsets (i.e., networks) with at least 2 elements can also be created. The complexity of finding water networks within a frame is $O(L_x L_y L_z) + O(n_{polarAt}) + O(n_{waterAt}) + O(n_{wnet} log(n_{wnet}))$.

A vital aspect of calculating distance between atoms involves taking into account the *periodic boundary condition* (PBC) of the simulation box. It means that during simulations a molecule does not face a wall at the edge of the box. Instead, it can move transparently from one side, say from right in Fig. 2, and appear from the left. (The rationale for PBC is beyond the scope of this work.) Hence, the two metabolite clusters seen here are not far away but are much closer. The size of the simulation box itself may change during simulation and hence frame-wise box lengths are stored in mdcrd files.

3.2 Benchmarks and Insights

The two chosen systems for the proof-of-concept implementations (M^{pro}-free and M^{pro}-drug) were benchmarked. Given the necessity for consistency and substantial resource requirements, three variants were tested on the GPFS-based DDN storage solution capable of 15 GBps throughput for both read and write operations. These were, (a) serial version (serial), (b) Spark-version with data on native GPFS filesystem (gpfs), and (c) Spark-version with data on HDFS (hdfs) filesystem. HDFS storage was available from the same GPFS server through a connector. Times of completion for the three parameters are given in Table 3.

Table 3. Time taken for completion of any job in serial mode (serial) is the longest, compared to Spark when input was read from native GPFS filesystem (gpfs) or HDFS filesystems (hdfs). All times are in seconds and speedups are shown in parenthesis when available.

Parameter	M^{pro}-free			M^{pro}-drug		
	serial	gpfs	hdfs	serial	gpfs	hdfs
RoG	8,691	1,889	273	1,931	584	131
MCon	–	1,928	735	–	584	176
WNw	667,000	1,953	619	251,010	614	218

It is clear from Table 3 that performance differs significantly when the input is read from HDFS, instead of the GPFS filesystem. In particular, *speedup*, as defined in Eq. (1), shows that close to 5 times speedier computation can be done if input is read from HDFS instead of GPFS filesystem.

$$speedup = \frac{time_{serial}}{time_{spark}} \tag{1}$$

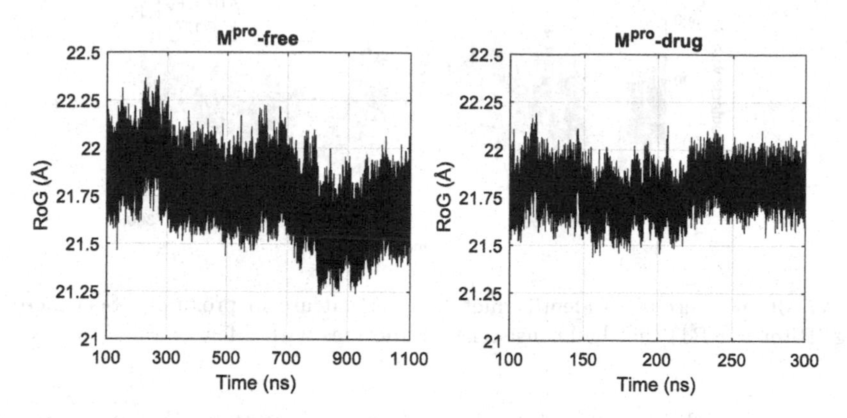

Fig. 6. The radii of gyration for M^{pro}s were stable, indicating that no unfolding of these proteins happened during course of simulations. Shown here is the RoG of one monomer.

The RoGs of the two systems are shown in Fig. 6. The values show that no unfolding of proteins occurred during the simulation. Speedups over 1000 were obtained for water network calculation. As the serial version could not be run over the entire trajectory, after confirming with shorter runs (with up to 20% of data), linear scaling was applied to estimate the time required for computations. MCon, another important parameter that requires only a single pass over each frame and is critical for the analysis of crowder-based simulations, performed faster than WNw. While it could be argued that serial version for water network could be improved, it is evident that the times for completion of MCon and WNw are comparable to that of RoG; all three of them requiring just a few minutes to complete their tasks. While CPPTRAJ required 8,691s to compute *RoG*, Spark (hdfs) for *WNw* was completed within 619s. In other words, compared to RoG, SparkTraj's WNw was 14 times faster over 1.2 TB!

Interesting events were captured using MCon. As Fig. 7 shows, multiple metabolites were seen interacting with drug R1. The most significant interactions were seen with only two drugs, a Ritonavir and a Glecaprevir. With further analysis based on this information, a cluster of 5 metabolites were identified that were seen blocking the movement of this drug, perhaps for the first time in any simulation. Ordinarily, such long-duration associations occur due to hydrogen bonds, salt bridges, π-π interactions, etc. Here, no long-lasting hydrogen bonds were seen as these metabolites were mobile, though still in the vicinity of the drug during the entire course of simulation. Such interactions need to be computed, and a parameter such as MCon is very useful for the same. Detailed insights and their significance obtained using SparkTraj is beyond the scope of this work and is discussed elsewhere [19].

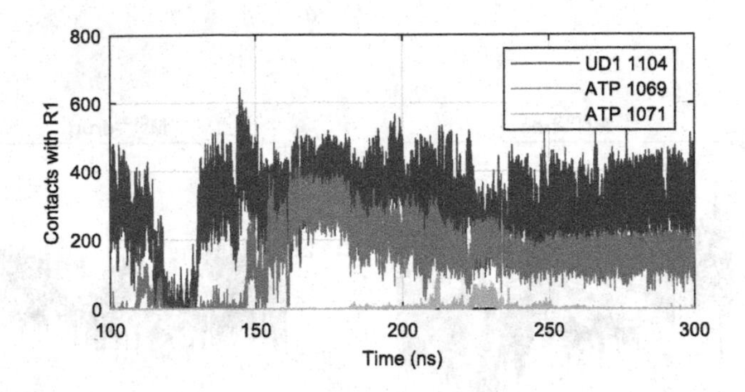

Fig. 7. MCon is useful to identify metabolites that are in proximity. Seen here is a drug (Ritonavir R1) that had consistent interactions with a few metabolites.

As already discussed in Introduction, water networks (WNw), especially those that stabilize drug binding to proteins and other receptors, are of interest. With the current study, we have been able to quickly identify such water networks. As shown in Fig. 8, they were also seen between drugs and M^{pro}.

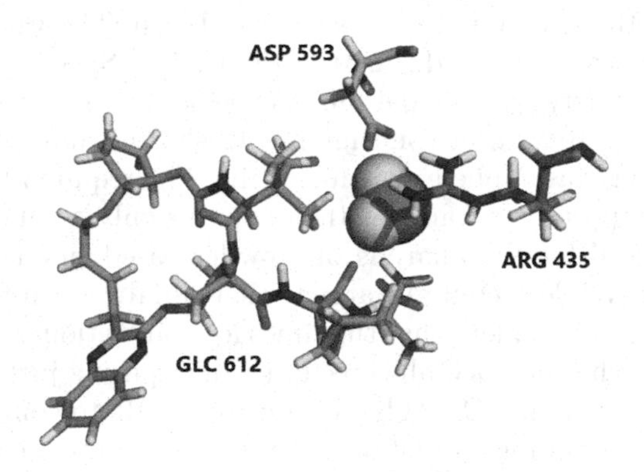

Fig. 8. Water Network (WNw) present between a drug (GLC 612) and two amino acids of M^{pro}.

Another crucial aspect of benchmarking software would be to find how the time for completion *scales* with size of the data. Normally, processing time being same for each frame, increasing number of frames should lead to an increased time for completion by the same factor. For a large system of terabyte size such as M^{pro}-free, increasing input size by 10-fold matched the expected performance where GPFS filesystem was used (see Table 4). In case of M^{pro}-drug system (20K vs 200K frames), the escalations were smaller. It is very interesting that the escalation of times were close to half in case of HDFS environment w.r.t. GPFS filesystem.

Table 4. With a 10-fold increase in data size, escalation in times for completion was almost by the same factor for the larger M^{pro}-free system when data was read from GPFS filesystem, and by much less in case of HDFS. M^{pro}-drug had lesser increment.

Parameter	M^{pro}-free		M^{pro}-drug	
	gpfs	hdfs	gpfs	hdfs
RoG	8.7	3.5	6.0	2.3
MCon	7.8	7.7	6.5	2.6
WNw	8.5	4.8	5.9	2.9

To improve the performance of Spark implementation, many variations were considered and examined. For instance, sorting (by time) using Spark was less efficient due to data shuffling. So the data was first collected and was sorted directly using Scala. While we contemplated different ways to improve the performance further, it was not pursued to avoid possible over-engineering; reckon that time required to read mdcrds was at least 60% of the total runtime.

To appreciate the immense value and contribution of such studies, it would be pertinent to reckon a few insights from the simulations of M^{pro} complex. Apart from metabolites and metabolite clusters, identified initially through MCon, the simulations in crowded environment yielded other fascinating insights as well. They include, possible preference of certain metabolites to particular sites, movement of a free amino acid in a probable (drug) binding site, and *crawling* of a drug over the surface from one pocket to another in presence of crowders (both proteins and metabolites).

3.3 Framework for Cloud-Based MD Simulation Service

The unique advantage of Cloud infrastructure over typical high-performance computing servers is the ability to scale resources on demand. However, the operating costs of Cloud infrastructure vary based on type, quantity and duration [5,12]. We present a framework to expedite the best MD simulations studies possible within the available resources.

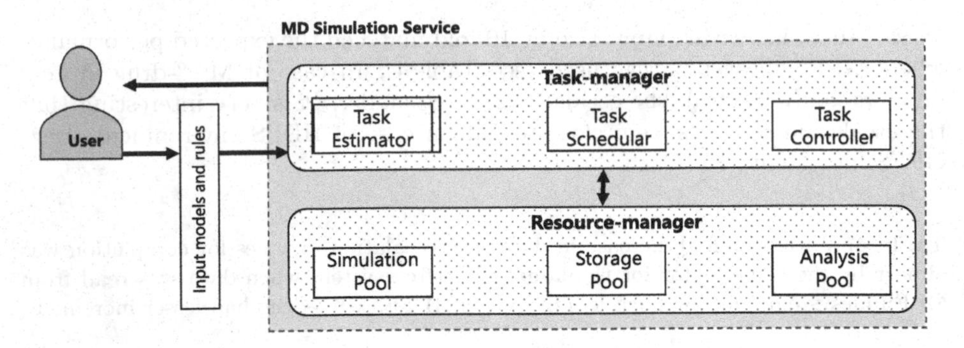

Fig. 9. Block diagram of the proposed framework for MD studies. Task-manager estimates task size, generates an execution plan, and in concert with Resource-manager, dynamically orchestrates tasks and resources. Periodic updates and final results are sent to user for information and feedback.

As shown in Fig. 9, given the models to simulate and the accompanying rules (such as given priorities and available funds), Task Estimator first runs sample jobs to assess the time required on various platforms, determines the number, types of servers, infrastructure, and services that could be acquired within given constraints. Task Schedular then prepares a schedule of jobs to be submitted. In certain cases there could more specific rules. For instance, consider *screening* of drug-candidates. A user may require only those drug-candidates that retained interactions with the target protein throughout the simulation(s). This could be monitored by Task Controller that can preemptively terminate simulations in which drugs lose interactions. Other reasons to pre-emptively discard a model from simulation studies include extreme changes in RoG, insufficient binding energy between drug and protein, etc. This ability to dynamically preempt simulations of poor drug-candidates allows users to channel the resulting savings into examination of additional drug-candidates or to enhance infrastructure for the computations. Task-manager updates Resource-manager on inputs, schedules, and resources to be deployed. Resource-manager then acquires, manages and frees appropriate software, services or (virtual) hardware from Cloud. The necessary software and hardware resources for simulation, analysis, and storage together form corresponding pools. As computations progress, Task-manager updates user on the latest status of simulations/analysis (scheduled, underway, completed, or terminated), available results, deployed resources, and the projected time for completion of the work. If required, the user may opt to intervene and update inputs to alter or improvise the study undertaken, or expand its scope and allocated funding.

This above framework can be implemented by the Cloud provider (First-party API) or could be developed by third parties (Third-party API). Competent clients may develop solutions through their own efforts and deploy them either on public or private Clouds. In principle, it is of practical utility, and with relative ease, the solution can be deployed by providers of Cloud infrastructure to identify even those drug-receptor candidates that may work well in aqueous conditions but fail in physiological environment. This would definitely help in containing both time and phenomenal costs involved in the development of new drugs.

3.4 Limitations

Installing and using software designed for Apache Spark requires more expertise compared to their serial counterparts. Further, we have studied different systems on the only state-of-the-art commercial storage that was available. It was selected as the server provides access to both local as well as HDFS filesystems on the same hardware, and was connected a dedicated big data server with 20 nodes. The performance gains may wary between different storage solutions. Like for any distributed computing, a few issues are critical for performance. One is the latency and bandwidth available to load the input files. The second is the amount of time required for computation itself. For instance, latency and (data) bandwidth could be significantly improved using more expensive solid-state storage devices such as NVMe SSDs even on less expensive desktop systems. This could improve the performance of parameters that have little computation, such as RoG. However when a frame requires significant amount of computation, then having higher core count, RAM and networking bandwidth would be more useful. Nevertheless, the above proposed cloud framework could be designed to select optimal storage and computational resources based on available funds.

SparkTraj is freely available for the interested researchers.

4 Conclusions

In this paper, using Apache Spark, we demonstrated that big data approach provides substantial speedups in MD studies. This is especially needed in the more cell-like environment involving a plethora of biomolecules. A responsive, scalable, and self-tuning framework that pairs Spark with the flexibility of Cloud infrastructure for the MD studies is presented. This framework enables users to optimally utilize the available resources. Insights obtained using the proposed approach were discussed. Complex MD studies, accompanied by such newer and more versatile tools, would bridge the gap between theoretical modelling and experimental observations. Such studies may soon become an imminent requirement in the capital intensive yet time-constrained pharmaceutical industry. The relevance of such advances, especially in these pandemic times, cannot be overstated.

Acknowledgment. We thank the Central Computing Facility of IIIT Allahabad (IIIT-A) for all the computations carried out. We thank Dr. Muneendra Ojha (IIIT-A) and Dr. Jagpreet Singh (IIIT-A) for proof-reading the manuscript. BSS thanks Mr. Asari Ramprasad (IIIT-A) and Ms. Shivani Maheshwari (Postman Inc.) for useful discussions on SciSpark and Cloud infrastructure. Dr. Arumugam Madhumalar and Mr. Zahoor Ahmad Bhat (JMI, New Delhi) are also gratefully acknowledged for the help provided regarding M^{pro} simulations. SparkTraj was developed by one of the authors (BSS), and analysis was done jointly by the two.

References

1. Borthakur, D.: The hadoop distributed file system: architecture and design. Hadoop Project Website **11**(2007), 21 (2007)
2. Borthakur, D., et al.: HDFS architecture guide. Hadoop Apache Project **53**(1–13), 2 (2008)
3. Bux, K., Moin, S.T.: Solvation of cholesterol in different solvents: a molecular dynamics simulation study. Phys. Chem. Chem. Phys. **22**(3), 1154–1167 (2020)
4. De Vivo, M., Masetti, M., Bottegoni, G., Cavalli, A.: Role of molecular dynamics and related methods in drug discovery. J. Med. Chem. **59**(9), 4035–4061 (2016)
5. Amazon EC2: Amazon Web Services (2006). http://aws.amazon.com/
6. Feig, M., Yu, I., Wang, P.h., Nawrocki, G., Sugita, Y.: Crowding in cellular environments at an atomistic level from computer simulations. J. Phys. Chem. B **121**(34), 8009–8025 (2017). https://doi.org/10.1021/acs.jpcb.7b03570. PMID: 28666087
7. Gao, M., et al.: Modulation of human IAPP fibrillation: cosolutes, crowders and chaperones. Phys. Chem. Chem. Phys. **17**(13), 8338–8348 (2015)
8. Karau, H., Warren, R.: High Performance Spark: Best Practices for Scaling and Optimizing Apache Spark. O'Reilly Media Inc., Sebastopol (2017)
9. Kuznetsova, I.M., Turoverov, K.K., Uversky, V.N.: What macromolecular crowding can do to a protein. Int. J. Mol. Sci. **15**(12), 23090–23140 (2014)
10. Li, J., et al.: Parallel netCDF: a high-performance scientific I/O interface. In: SC 2003: Proceedings of the 2003 ACM/IEEE Conference on Supercomputing, p. 39. IEEE (2003)
11. McGibbon, R.T., et al.: MDTraj: a modern open library for the analysis of molecular dynamics trajectories. Biophys. J . **109**(8), 1528–1532 (2015)
12. Microsoft: Microsoft Azure (2014). https://azure.microsoft.com/
13. Palamuttam, R., et al.: Scispark: applying in-memory distributed computing to weather event detection and tracking. In: 2015 IEEE International Conference on Big Data (Big Data), pp. 2020–2026. IEEE (2015)
14. Pikkemaat, M.G., Linssen, A.B., Berendsen, H.J., Janssen, D.B.: Molecular dynamics simulations as a tool for improving protein stability. Protein Eng. **15**(3), 185–192 (2002)
15. Rincón, V., Bocanegra, R., Rodríguez-Huete, A., Rivas, G., Mateu, M.G.: Effects of macromolecular crowding on the inhibition of virus assembly and virus-cell receptor recognition. Biophys. J . **100**(3), 738–746 (2011)
16. Roe, D.R., Cheatham, T.E., III.: PTRAJ and CPPTRAJ: software for processing and analysis of molecular dynamics trajectory data. J. Chem. Theory Comput. **9**(7), 3084–3095 (2013)
17. Salo-Ahen, O.M., et al.: Molecular dynamics simulations in drug discovery and pharmaceutical development. Processes **9**(1), 71 (2021)
18. Salomon-Ferrer, R., Case, D.A., Walker, R.C.: An overview of the amber biomolecular simulation package. Wiley Interdiscip. Rev. Comput. Mol. Sci. **3**(2), 198–210 (2013)
19. Sanjeev, B.S., Chitara, D., Arumugam, M.: Physiological models to study the effect of molecular crowding on multi-drug bound proteins: insights from SARS-CoV-2 main protease. J. Biomol. Struct. Dyn. (2021). https://doi.org/10.1080/07391102.2021.1993342

20. Sanjeev, B.: Ankush. Indian Institute of Science (2004)
21. Shaw, D.E., et al.: Atomic-level characterization of the structural dynamics of proteins. Science **330**(6002), 341–346 (2010)
22. Wouters, O.J., McKee, M., Luyten, J.: Estimated research and development investment needed to bring a new medicine to market, 2009–2018. JAMA **323**(9), 844–853 (2020)
23. Zaharia, M., et al.: Resilient distributed datasets: a fault-tolerant abstraction for in-memory cluster computing. In: 9th {USENIX} Symposium on Networked Systems Design and Implementation ({NSDI} 2012), pp. 15–28 (2012)
24. Zaharia, M., Chowdhury, M., Franklin, M.J., Shenker, S., Stoica, I., et al.: Spark: cluster computing with working sets. In: HotCloud 2010, no. 10, p. 95 (2010)

Predictive Analytics for Recognizing Human Activities Using Residual Network and Fine-Tuning

Alok Negi[1], Krishan Kumar[1(✉)], Narendra S. Chaudhari[2], Navjot Singh[3], and Prachi Chauhan[4]

[1] National Institute of Technology, Uttarakhand, Srinagar (Garhwal) 246174, India
kkberwal@nituk.ac.in

[2] Indian Institute of Technology Indore, Indore 453552, India
nsc@iiti.ac.in

[3] Indian Institute of Information Technology, Allahabad, Allahabad 211015, UP, India
navjot@iiita.ac.in

[4] G. B. Pant University of Agriculture and Technology, Pantnagar 263145, Uttarakhand, India

Abstract. Human Action Recognition (HAR) is a rapidly growing study area in computer vision due to its wide applicability. Because of their varied appearance and the broad range of stances that they can assume, detecting individuals in images is a difficult undertaking. Due to its superior performance over existing machine learning methods and high universality over raw inputs, deep learning is now widely used in a range of study fields. For many visual recognition tasks, the depth of representations is critical. For better model robustness and performance, more complex features can represent using deep neural networks but the training of these model are hard due to vanishing gradients problem. The use of skip connections in residual networks (ResNet) helps to address this problem and easy to learn identity function by residual block. So, ResNet overcomes the performance degradation issue with deep networks. This paper proposes an intelligent human action recognition system using residual learning-based framework "ResNet-50" with transfer learning which can automatically recognize daily human activities. The proposed work presents extensive empirical evidence demonstrating that residual networks are simpler to optimize and can gain accuracy from significantly higher depth. The experiments are performed using the UTKinect Action-3D public dataset of human daily activities. According to the experimental results, the proposed system outperforms other state-of-the-art methods and recorded high recognition accuracy of 98.25% with a 0.11 loss score in 200 epochs.

1 Introduction

Action Recognition (HAR) from a set of video sequences is a difficult problem in computer vision technology that is essential to a wide range of applications in

S. N. Srirama et al. (Eds.): BDA 2021, LNCS 13147, pp. 296–310, 2021.
https://doi.org/10.1007/978-3-030-93620-4_21

industries, health-care monitoring systems [9,10], consumer electronics and consumer behaviour analysis, video summarization [4,5], sports videos, academia, robot-human interaction, video surveillance [2,12] and security, and elderly patient monitoring. The major issue of HAR is to accurately distinguish action in the presence of clutter, opacity, and viewpoint shifts. Also, due to vanishing gradient effect, curse of dimensionality, covariate shift and degradation problem, adding extra layers to the network abruptly decreased or saturated the accuracy value while training deep neural networks. To address these issues, a number of researchers presented alternative techniques.

The machine learning field has seen a paradigm change towards deep learning approaches to produce better outcomes, notably in the computer vision sector. Deep learning techniques have improved HAR performance on various benchmark datasets in recent years. In addition, as research into these applications progresses, the potential of deep learning has become clear. The infamous vanishing gradient problem makes it difficult to train deep neural networks. During back propagation of gradients to prior layers, repetitive multiplication can result in exceedingly small gradients. As the network becomes more complex, its performance becomes saturated or even degrades quickly.

To address this issue, residual networks developed the skip connection, which allows the gradient to flow along a short cut path while allowing the model to learn an identity function that assures the top layer performs at least as well as the lower layer. So, adding original input to the output of the convolution block are the skip connections as shown in Fig. 1. The normal network learns from the Y but in residual network learn from the F(x) so the main idea behind the residual network is to make Y = X by making the F(X) = 0. When input to the network and output of the network is same then identity block can be used else convolution layer can be added on the shortcut path.

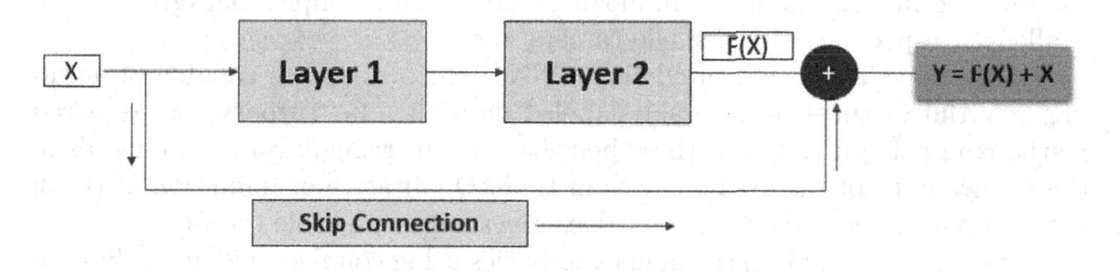

Fig. 1. Residual block

The residual deep neural networks are easy to optimize and solve the vanishing gradient problem by using the skip connection and identity connection that skip one or more layers. Due to identity mapping, shortcut connection does not increase additional parameters and computational complexity ensuring that deeper networks perform just as well as their shallower counterparts. The proposed work utilizes all the benefits from residual network and implement the novel approach for ten type of human activity recognition on UT Kinect Action-3D dataset using transfer learning approach by unfreezing the selective layers from ResNet-50 for training and weight updation. The work described in this paper translates the following contributions to the field of HAR.

- To design a residual learning-based deep neural networks with transfer learning approach for training that is easier to optimise and can gain accuracy from increased depth.
- To show the importance of residual learning against covariate shift and vanishing gradient problem using skip connection.
- To compare the results with existing approach and accuracy, categorical crossentropy loss, precision and recall, f1 score based evaluation.

The rest of the paper is structured as follows: a brief overview of previous surveys is provided in Sect. 2 and includes the related work on activity recognition in RGB images, depth data, and skeleton data. Section 3 describes the proposed approach followed by experimental outcome and discussion in Sect. 4. Section 5 describes the performance comparision with the esixting approaches. Finally Conclusion with future work are described in Sect. 6.

2 Related Work

Popoola et al. reviewed the [12] highlights of recent trends in the research on video-based human abnormal behavior detection. The definition anomaly can have some degree of ambiguity within a domain of application. Visual behavior [8] are complex and have much variety in an unconstrained environment. The influence of noisy data, the choice, and representation of low-level features, significantly influences the discriminative power of the classifier, and video quality, shadows, occlusion, illumination, moving camera, and complex backgrounds are challenges, especially with a single-camera view.

Karpathy et al. [3] indicated that CNN architectures are capable of learning powerful features from weakly-labeled data that far surpass feature-based methods in performance and these benefits are surprisingly robust to details of the connectivity of the architectures in time. Qualitative examination of [1,13] network outputs and confusion matrices reveals interpretable errors.

Liu et al. [6] recognized the human activities using coupled hidden conditional random field approach using RGB and depth information on DHA, UTKinect action, and TJU dataset. The author designed the graph structure and potential function and recorded 95.9%, 92.0%, and 92.5% accuracy on DHA, UTKinect action, and TJU dataset respectively.

Phyo et al. [11] introduced skeleton information-based daily human activities recognition using image processing and deep learning. Firstly the skeleton joints of the human body are extracted. Then Color Skeleton Motion History Image and Relative Joint Image is created for obtaining the motion history features and relative distance features. Finally, output fusion is performed for human activity classification using deep learning. The experiments are performed on UTKinect

Action-3D Dataset and CAD-60 Daily Activity Dataset. This approach recorded 97% accuracy on the UTKinect Action-3D dataset while 96.15% on the CAD-60 dataset.

Vemulapalli et al. [14] recognized the human activities by representing 3D Skeletons as Points in a Lie Group. Fourier temporal pyramid representation, dynamic time warping, and linear SVMare used for classification purposes. The experiments are performed on the UTKinect-Action dataset and Florence3D-Action dataset with 97.08% and 90.88% accuracy respectively.

Verma et al. [15] implemented the CNN model for recognizing the human activities from the UTKinect-Action dataset. In the first step, the author separated the single limb and multi limb activity. In the second step, individual activities are separated using sequence classification. The author recorded an overall accuracy of 97.88% using this two-stage framework.

Xia et al. [16] introduced the histograms of the 3D Joints-based approach for human activity recognition. The proposed work recorded 90.92% accuracy on the UTKinect-Action dataset. For the same dataset, Zhao et al. [17] proposed a 3D space-time convolutional neural network and recorded 97.29% accuracy. Our proposed method is trained and validated on the UTKinect-Action 3D dataset for RGB information, yielding promising results and advancements over earlier research.

3 Proposed Approach

The proposed work recognize the ten type of human activity wave hands, pull, walk, sit down, push, stand up, throw, pick up, carry, and clap hands using ResNet-50 with tranfer learning approach and this makes it multiclass classification problem in machine learning terms. Sit down, standup, pull, push and throw are categorized into human single limb activities and walk, pickup, carry, wave hands, clap hands into human multi-limb activities category. Stage III and IV of ResNet-40 model are unfrozen and rest of the stages are frozen during the training process as shown in Fig. 2. Then fully connected layer from the orginal architecture is replaced with five new fully connected layers.

3.1 Data Preprocessing

The original network expected the input in (B, G, R) format but python expect the (R, G, B) order by default so images are converted into RGB format using the preprocess input function which is imported from the Application Module Keras. Afterward, images are resized into $224 \times 224 \times 3$ to reduce the computation time. For the experiment UTKinect-Action Dataset consist of total 1896 images out of which training set have 1610 while 286 in validation set. The training set have 178, 158, 160, 158, 163, 168, 160, 160, 153 and 152 images while test set have 32, 25, 29, 29, 29, 30, 29, 29, 28 and 26 images for carry, claphands, pickup, pull, push, sitdown, standup, throw, walk and wavehands respectively. The distribution of dataset among different classes are shown in Fig. 3.

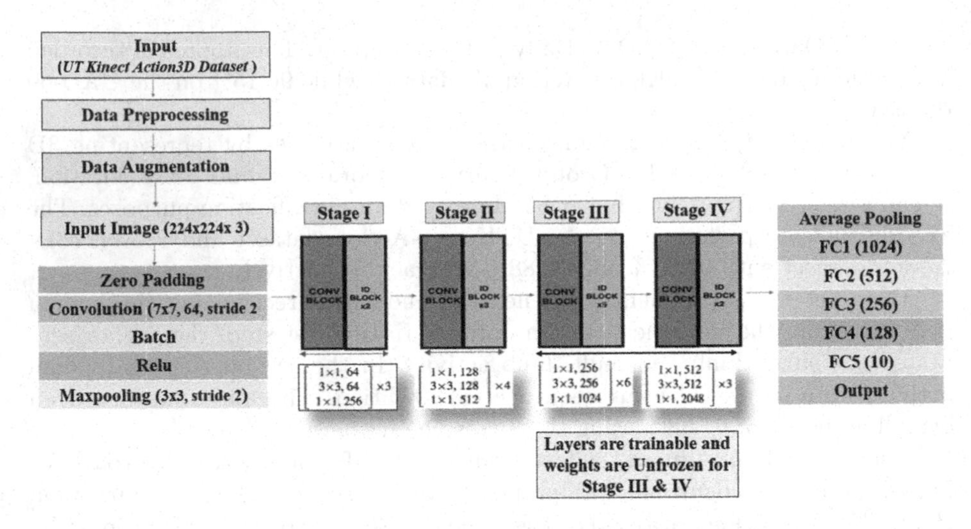

Fig. 2. Proposed approach (Improved ResNet-50)

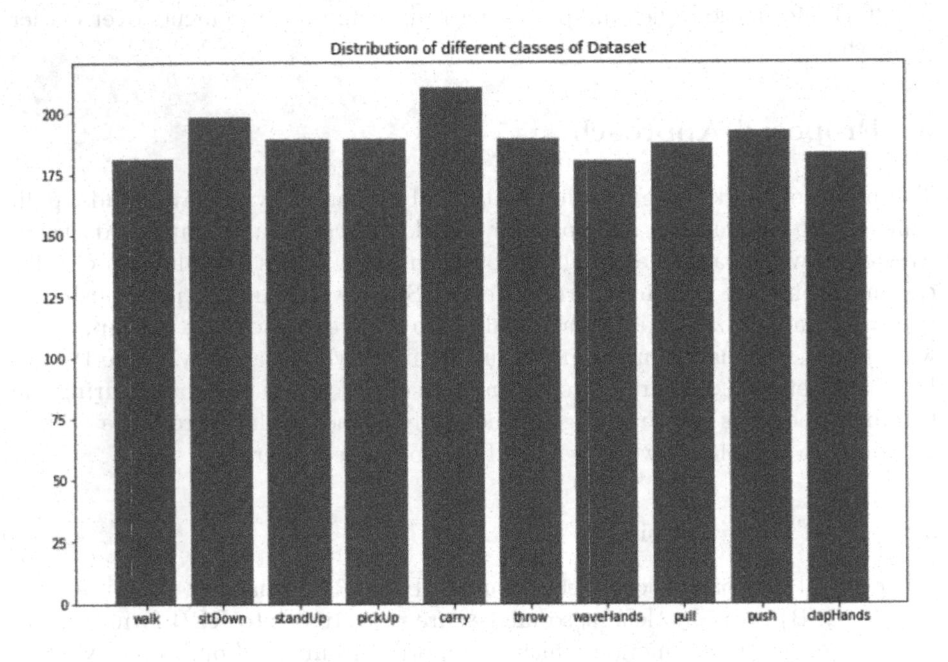

Fig. 3. Dataset distribution

3.2 Data Augmentation

for the diversity and quality of the training data, data augmentation technique is used which apply different transformation on the existing data to synthesize new data. Data augmentation also overcome the problem of data imbalancing and the overfitting. The proposed work use the rotation with 30°, horizontal flip and Zoom range with 0.4 as augmentation parameter using the ImageDataGenerator Keras framework. The sample images after the augmentation are shown in Fig. 4 and 5.

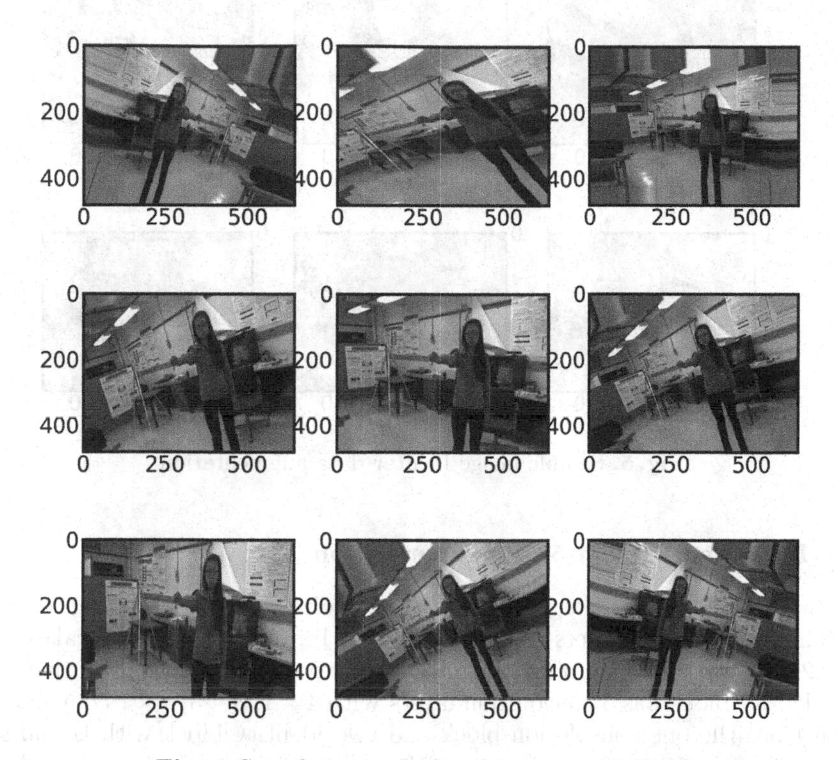

Fig. 4. Sample image I after data augmentation

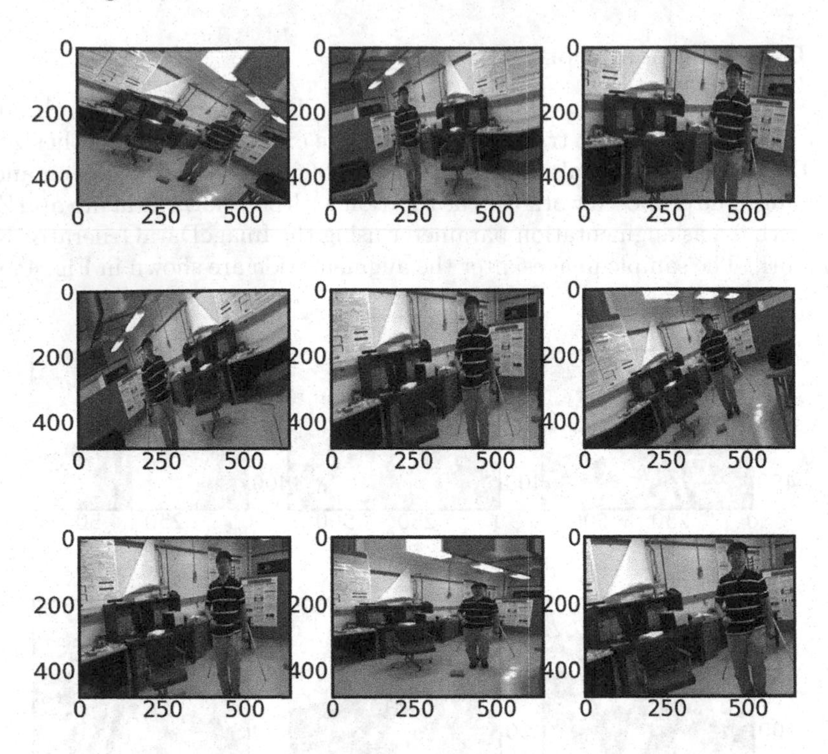

Fig. 5. Sample image II after data augmentation

3.3 Improved ResNet-50 Implementation

The ResNet-50 has the 4 stages each with a convolution and identity block and 23 million trainable parameters as shown in Fig. 2. Initially the network takes input $224\times224\times3$ with 7×7 convolution and 3×3 maxpooling. Each convolution block and identity block has 3 convolution layers with 1×1, 3×3, 1×1 convolutions. Stage I has the one convolotion block and two identity block with kernal sizes of 64, 64, and 256 respectively. Stage II consist of one convolution block and three identity block with kernal size of 128, 128 and 512 respectively. Stage III consist of one convolution block and five identity block with kernal size of 256, 256 and 1024 respectively. Afterward, Stage IV consist of one convolution block and two identity block with kernal size of 512, 512 and 2048 respectively. Then base ResNet50 has the average pooling layer and the fully connected layer with 1000 hidden nodes. In the proposed work, last fully connected layer is replaced with 5 new dense layers. The output size are calculated based on the Eq. 1.

$$OutputSize = \frac{n+2p-f}{s} + 1 \times \frac{n+2p-f}{s} + 1 \tag{1}$$

Where, n, p, f and s denote the input size, padding size, filter size and stride respectively.

3.4 Fine Tuning

For fine tuning, Last fully connected layer of the pre-trained model is replaced with 5 new dense layers FC1, FC2, FC3, FC4 and FC5. Each new layer has the 1024, 512, 256, 128 and 10 hideen nodes respectively. Generic features are learned by the initial layers so weights of the initial layers are frozen which are not updated during of the training. Task specific features are learned by the higher layer so the Stage III and IV layers are unfrozen and weights are updated during the training which improves the model performance. The original ResNet-50 architecture has the total 175 layers and Stage III is started from 82 to 143 while Stage IV is started from 144 to 175. The Stage III consist of total six block with kernal size of 256, 256 and 1024 respectively while Stage IV have total three blocks with kernal size of 512, 512 and 2048 respectively. The status of the Stage III and IV layers during training are true as shown in Figs. 6, 7 and 8.

4 Experiment and Results

We performed experiments on the proposed HAR under various conditions using the UTKinect Action-3D dataset. The datasets included daily activities with high intra-class and viewpoint variations. In this dataset same activity is performed by different people and each performing the activity in a different way. For training, the proposed work used the 16 batch size, 200 epochs and SGD optimizer with learning rate 0.001. The traditional gradient descent approach has a long training time, so the SGD approach is utilised for training.

82	<keras.layers.convolutional.Conv2D object at 0x7f99aee4f250>	conv4_block1_1_conv	TRUE
83	<keras.layers.normalization.batch_normalization.BatchNormalization object at 0x7f99aee4ba10>	conv4_block1_1_bn	TRUE
84	<keras.layers.core.Activation object at 0x7f99aee5e410>	conv4_block1_1_relu	TRUE
85	<keras.layers.convolutional.Conv2D object at 0x7f99aee66b10>	conv4_block1_2_conv	TRUE
86	<keras.layers.normalization.batch_normalization.BatchNormalization object at 0x7f99aee5e710>	conv4_block1_2_bn	TRUE
87	<keras.layers.core.Activation object at 0x7f99aee77950>	conv4_block1_2_relu	TRUE
88	<keras.layers.convolutional.Conv2D object at 0x7f99aeeacc90>	conv4_block1_0_conv	TRUE
89	<keras.layers.convolutional.Conv2D object at 0x7f99aee77dd0>	conv4_block1_3_conv	TRUE
90	<keras.layers.normalization.batch_normalization.BatchNormalization object at 0x7f99aeec5ed0>	conv4_block1_0_bn	TRUE
91	<keras.layers.normalization.batch_normalization.BatchNormalization object at 0x7f99aeec2e50>	conv4_block1_3_bn	TRUE
92	<keras.layers.merge.Add object at 0x7f99aeef0850>	conv4_block1_add	TRUE
93	<keras.layers.core.Activation object at 0x7f99aeee4210>	conv4_block1_out	TRUE
94	<keras.layers.convolutional.Conv2D object at 0x7f99aeefb6d0>	conv4_block2_1_conv	TRUE
95	<keras.layers.normalization.batch_normalization.BatchNormalization object at 0x7f99aeeaa510>	conv4_block2_1_bn	TRUE
96	<keras.layers.core.Activation object at 0x7f99aeeed6d0>	conv4_block2_1_relu	TRUE
97	<keras.layers.convolutional.Conv2D object at 0x7f99aeed3d50>	conv4_block2_2_conv	TRUE
98	<keras.layers.normalization.batch_normalization.BatchNormalization object at 0x7f99aeeb6cd0>	conv4_block2_2_bn	TRUE
99	<keras.layers.core.Activation object at 0x7f99b0d5df50>	conv4_block2_2_relu	TRUE
100	<keras.layers.convolutional.Conv2D object at 0x7f99aef2bf10>	conv4_block2_3_conv	TRUE
101	<keras.layers.normalization.batch_normalization.BatchNormalization object at 0x7f99aeefba90>	conv4_block2_3_bn	TRUE
102	<keras.layers.merge.Add object at 0x7f99aee88b50>	conv4_block2_add	TRUE
103	<keras.layers.core.Activation object at 0x7f99aeee81d0>	conv4_block2_out	TRUE
104	<keras.layers.convolutional.Conv2D object at 0x7f99aeed9450>	conv4_block3_1_conv	TRUE
105	<keras.layers.normalization.batch_normalization.BatchNormalization object at 0x7f99aee0ca50>	conv4_block3_1_bn	TRUE
106	<keras.layers.core.Activation object at 0x7f99aee12250>	conv4_block3_1_relu	TRUE
107	<keras.layers.convolutional.Conv2D object at 0x7f99aee838d0>	conv4_block3_2_conv	TRUE

Fig. 6. Stage III unfrozen layers (82–107)

108	<keras.layers.normalization.batch_normalization.BatchNormalization object at 0x7f99aee16e90>	conv4_block3_2_bn	TRUE
109	<keras.layers.core.Activation object at 0x7f99aee24110>	conv4_block3_2_relu	TRUE
110	<keras.layers.convolutional.Conv2D object at 0x7f99af2ba210>	conv4_block3_3_conv	TRUE
111	<keras.layers.normalization.batch_normalization.BatchNormalization object at 0x7f99aee2bb10>	conv4_block3_3_bn	TRUE
112	<keras.layers.merge.Add object at 0x7f99afb92490>	conv4_block3_add	TRUE
113	<keras.layers.core.Activation object at 0x7f99aee37fd0>	conv4_block3_out	TRUE
114	<keras.layers.convolutional.Conv2D object at 0x7f99aee24850>	conv4_block4_1_conv	TRUE
115	<keras.layers.normalization.batch_normalization.BatchNormalization object at 0x7f99aee42910>	conv4_block4_1_bn	TRUE
116	<keras.layers.core.Activation object at 0x7f99aedcc090>	conv4_block4_1_relu	TRUE
117	<keras.layers.convolutional.Conv2D object at 0x7f99aee37e90>	conv4_block4_2_conv	TRUE
118	<keras.layers.normalization.batch_normalization.BatchNormalization object at 0x7f99aedd0d10>	conv4_block4_2_bn	TRUE
119	<keras.layers.core.Activation object at 0x7f99aeddd1d0>	conv4_block4_2_relu	TRUE
120	<keras.layers.convolutional.Conv2D object at 0x7f99aeddd0d0>	conv4_block4_3_conv	TRUE
121	<keras.layers.normalization.batch_normalization.BatchNormalization object at 0x7f99aede7990>	conv4_block4_3_bn	TRUE
122	<keras.layers.merge.Add object at 0x7f99aedef910>	conv4_block4_add	TRUE
123	<keras.layers.core.Activation object at 0x7f99aedefe10>	conv4_block4_out	TRUE
124	<keras.layers.convolutional.Conv2D object at 0x7f99aede7390>	conv4_block5_1_conv	TRUE
125	<keras.layers.normalization.batch_normalization.BatchNormalization object at 0x7f99aee007d0>	conv4_block5_1_bn	TRUE
126	<keras.layers.core.Activation object at 0x7f99aee09110>	conv4_block5_1_relu	TRUE
127	<keras.layers.convolutional.Conv2D object at 0x7f99aedf15d0>	conv4_block5_2_conv	TRUE
128	<keras.layers.normalization.batch_normalization.BatchNormalization object at 0x7f99aed8dc50>	conv4_block5_2_bn	TRUE
129	<keras.layers.core.Activation object at 0x7f99aed91290>	conv4_block5_2_relu	TRUE
130	<keras.layers.convolutional.Conv2D object at 0x7f99aed9bf90>	conv4_block5_3_conv	TRUE
131	<keras.layers.normalization.batch_normalization.BatchNormalization object at 0x7f99aee09850>	conv4_block5_3_bn	TRUE
132	<keras.layers.merge.Add object at 0x7f99aede7ed0>	conv4_block5_add	TRUE
133	<keras.layers.core.Activation object at 0x7f99aeddd850>	conv4_block5_out	TRUE
134	<keras.layers.convolutional.Conv2D object at 0x7f99aee02650>	conv4_block6_1_conv	TRUE
135	<keras.layers.normalization.batch_normalization.BatchNormalization object at 0x7f99aee3c7d0>	conv4_block6_1_bn	TRUE
136	<keras.layers.core.Activation object at 0x7f99aeef2cd0>	conv4_block6_1_relu	TRUE
137	<keras.layers.convolutional.Conv2D object at 0x7f99aee1ff90>	conv4_block6_2_conv	TRUE
138	<keras.layers.normalization.batch_normalization.BatchNormalization object at 0x7f99aee0b190>	conv4_block6_2_bn	TRUE
139	<keras.layers.core.Activation object at 0x7f99aee83ad0>	conv4_block6_2_relu	TRUE
140	<keras.layers.convolutional.Conv2D object at 0x7f99aee548d0>	conv4_block6_3_conv	TRUE
141	<keras.layers.normalization.batch_normalization.BatchNormalization object at 0x7f99aef47b50>	conv4_block6_3_bn	TRUE
142	<keras.layers.merge.Add object at 0x7f99aeda2050>	conv4_block6_add	TRUE
143	<keras.layers.core.Activation object at 0x7f99aeda86d0>	conv4_block6_out	TRUE

Fig. 7. Stage III unfrozen layers (108–143)

144	<keras.layers.convolutional.Conv2D object at 0x7f99aedb50d0>	conv5_block1_1_conv	TRUE
145	<keras.layers.normalization.batch_normalization.BatchNormalization object at 0x7f99aedaea50>	conv5_block1_1_bn	TRUE
146	<keras.layers.core.Activation object at 0x7f99aedc4290>	conv5_block1_1_relu	TRUE
147	<keras.layers.convolutional.Conv2D object at 0x7f99aedc4ad0>	conv5_block1_2_conv	TRUE
148	<keras.layers.normalization.batch_normalization.BatchNormalization object at 0x7f99aedbed10>	conv5_block1_2_bn	TRUE
149	<keras.layers.core.Activation object at 0x7f99aed541d0>	conv5_block1_2_relu	TRUE
150	<keras.layers.convolutional.Conv2D object at 0x7f99aee1bc10>	conv5_block1_0_conv	TRUE
151	<keras.layers.convolutional.Conv2D object at 0x7f99aed54ad0>	conv5_block1_3_conv	TRUE
152	<keras.layers.normalization.batch_normalization.BatchNormalization object at 0x7f99aedaad90>	conv5_block1_0_bn	TRUE
153	<keras.layers.normalization.batch_normalization.BatchNormalization object at 0x7f99aed5c750>	conv5_block1_3_bn	TRUE
154	<keras.layers.merge.Add object at 0x7f99aedc6fd0>	conv5_block1_add	TRUE
155	<keras.layers.core.Activation object at 0x7f99aed4b1d0>	conv5_block1_out	TRUE
156	<keras.layers.convolutional.Conv2D object at 0x7f99aed62a50>	conv5_block2_1_conv	TRUE
157	<keras.layers.normalization.batch_normalization.BatchNormalization object at 0x7f99aed6b050>	conv5_block2_1_bn	TRUE
158	<keras.layers.core.Activation object at 0x7f99aed76150>	conv5_block2_1_relu	TRUE
159	<keras.layers.convolutional.Conv2D object at 0x7f99aed76a50>	conv5_block2_2_conv	TRUE
160	<keras.layers.normalization.batch_normalization.BatchNormalization object at 0x7f99aed7f6d0>	conv5_block2_2_bn	TRUE
161	<keras.layers.core.Activation object at 0x7f99aed65e90>	conv5_block2_2_relu	TRUE
162	<keras.layers.convolutional.Conv2D object at 0x7f99aed6ec90>	conv5_block2_3_conv	TRUE
163	<keras.layers.normalization.batch_normalization.BatchNormalization object at 0x7f99aed0dd90>	conv5_block2_3_bn	TRUE
164	<keras.layers.merge.Add object at 0x7f99aed65d10>	conv5_block2_add	TRUE
165	<keras.layers.core.Activation object at 0x7f99aed0f810>	conv5_block2_out	TRUE
166	<keras.layers.convolutional.Conv2D object at 0x7f99aed22350>	conv5_block3_1_conv	TRUE
167	<keras.layers.normalization.batch_normalization.BatchNormalization object at 0x7f99aed22c10>	conv5_block3_1_bn	TRUE
168	<keras.layers.core.Activation object at 0x7f99aed198d0>	conv5_block3_1_relu	TRUE
169	<keras.layers.convolutional.Conv2D object at 0x7f99aed27950>	conv5_block3_2_conv	TRUE
170	<keras.layers.normalization.batch_normalization.BatchNormalization object at 0x7f99aed364d0>	conv5_block3_2_bn	TRUE
171	<keras.layers.core.Activation object at 0x7f99aed2bf10>	conv5_block3_2_relu	TRUE
172	<keras.layers.convolutional.Conv2D object at 0x7f99aed19d50>	conv5_block3_3_conv	TRUE
173	<keras.layers.normalization.batch_normalization.BatchNormalization object at 0x7f99aed86890>	conv5_block3_3_bn	TRUE
174	<keras.layers.merge.Add object at 0x7f99aed76090>	conv5_block3_add	TRUE
175	<keras.layers.core.Activation object at 0x7f99aed88c10>	conv5_block3_out	TRUE

Fig. 8. Stage IV unfrozen layers (144–175) status

During the training, Stage III and IV layers are unfrozen. Base ResNet-50 has the total 175 layers and Stage III is started from 82 to 143 while Stage IV is started from 144 to 175 as shown in Fig. 6, 7 and 8. The total parameter during the training recorded 127,039,498 out of which 125,536,394 trainable and 1,503,104 non trainable. The trainable parameters obtained from the fully connected and unfrozen layers.

The accuracy, loss, precision, recall and f1 score are calculated based on the following Eq. 2, 3, 4, 5 and 6 respectively.

$$Accuracy = \frac{TP + TN}{FN + TP + TN + FP} \tag{2}$$

$$logloss = -\frac{1}{N}\sum_{i=1}^{N}\sum_{j=1}^{M} y_{ij}log(p_{ij}) \tag{3}$$

$$Precision = \frac{TP}{TP + FP} \tag{4}$$

$$Recall = \frac{TP}{FN + TP} \tag{5}$$

$$f_1Score = \frac{2 \times Precision \times Recall}{Precision + Recall} \tag{6}$$

The overall accuracy and loss based on training epoch is shown in Fig. 9 and 10. The proposed work recored the overall accuracy 99.01% with 0.02 loss for training data and 98.25% with 0.11 loss for validating data with 200 epochs. The confusion matrix of all activity of validation data using 200 epochs was described in Fig. 11 and 12. The classification report for each class are shown in Table 1.

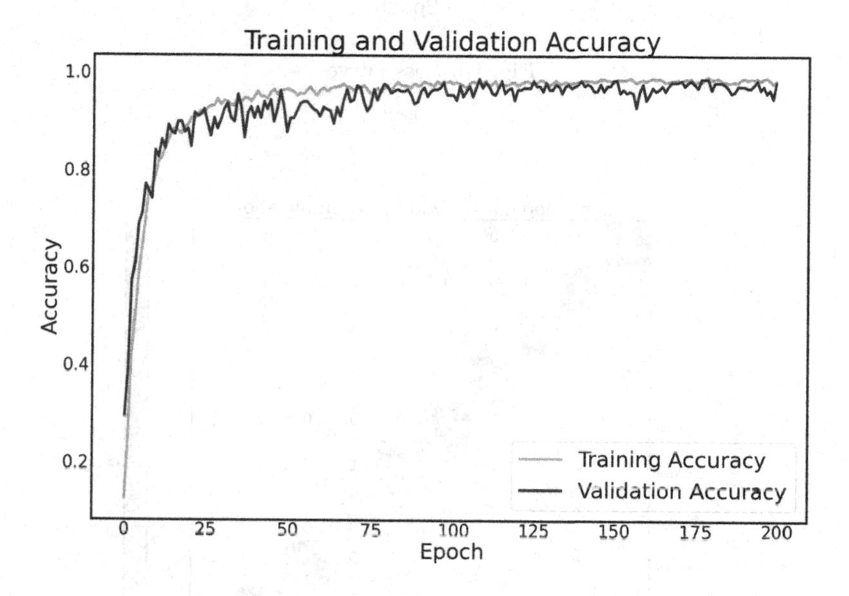

Fig. 9. Accuracy curve

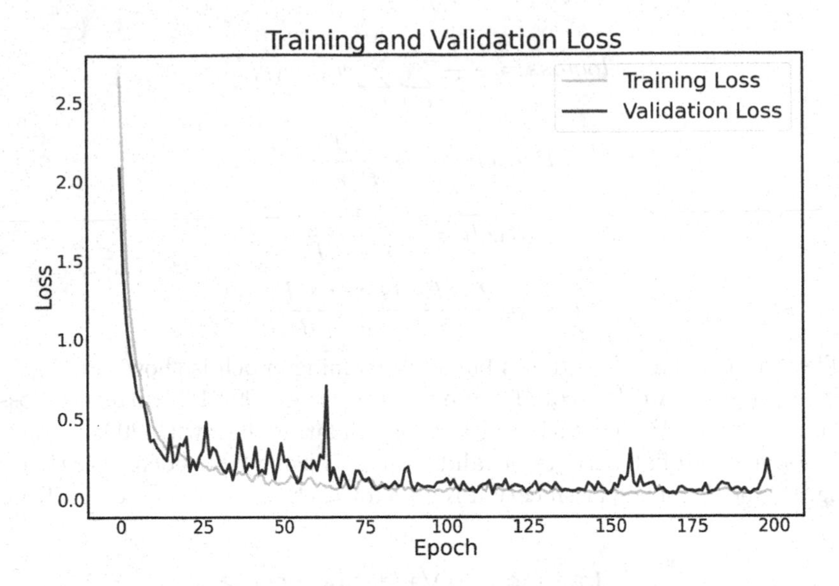

Fig. 10. Loss curve

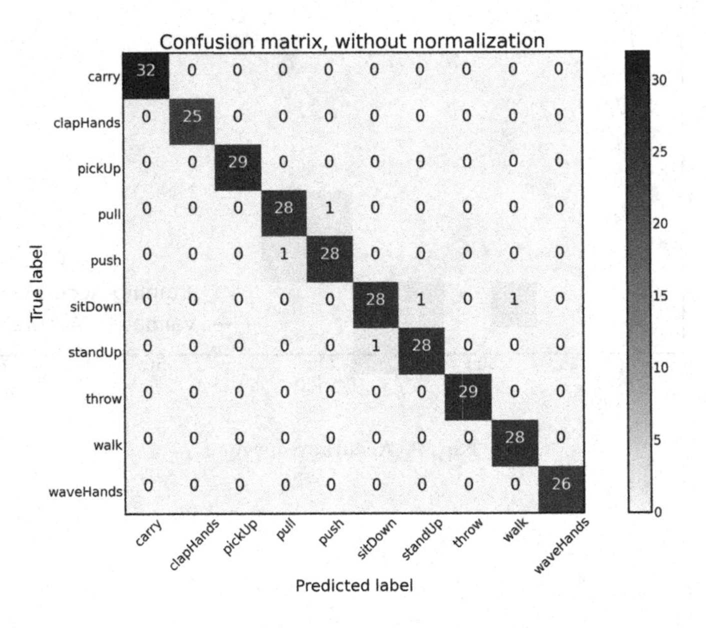

Fig. 11. Confusion matrix without normalization

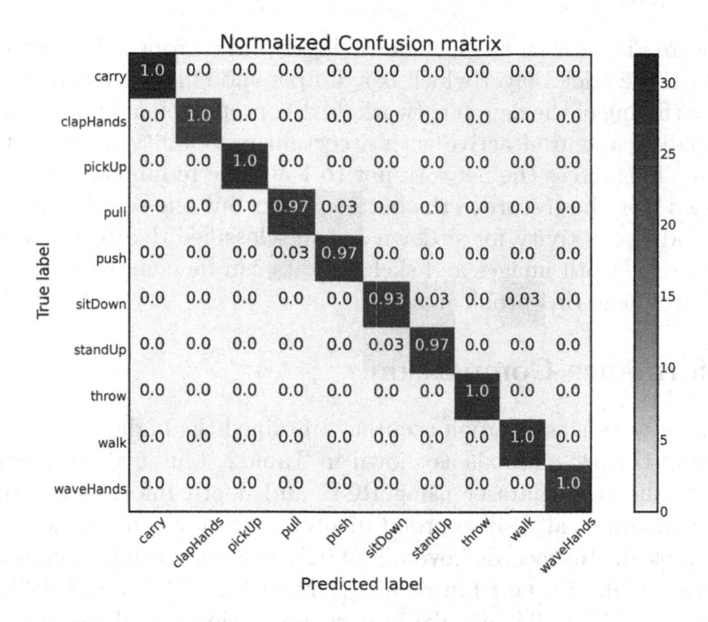

Fig. 12. Confusion matrix with normalization

Table 1. Precision, recall, f1 score (in percent) and support

Class and avg type	Precision	Recall	f1 score	Support
carry	100	100	100	32
clapHands	100	100	100	25
pickUp	100	100	100	29
pull	97	97	97	29
push	97	97	97	29
sitDown	97	93	95	30
standUp	97	97	97	29
throw	100	100	100	29
walk	97	100	98	28
waveHands	100	100	100	26
macro avg	98	98	98	286
weighted avg	98	98	98	286

The proposed work used the residual based learning by unfreezing the Stage III and IV of the ResNet-50 architecture to identify human behavior in order to achieve this end-to-end solution. Data Augmentation is used to minimize the loss as well as to prevent overfitting. At last fine tuning applied by replacing the fully connected layer with five new dense layers. The proposed architecture uses drop-outwith the added layer which is a simple and efficient technique designed to avoid overfitting of the neural network during preparation. Dropout is enforced only by holding a neuron active with a certain probability p, and setting it to 0 otherwise. This forces the network not to learn the redundant data. Although our proposed work has recorded the better results but one activity for push, pull, stand up and two activity for sitdown are misclassified due to the similarity of the activity. So Depth images and skelton joits can be combined with the RGB images to overcome this problem.

5 Performance Comparison

The proposed work classification accuracy obtained the highest accuracy among other state-of-the-art methods as shown in Table 2. Liu et al. [6] recorded 92% accuracy on the same dataset using RGB and depth information. In another work, Vemulapalli et al. [14] recorded 97.08% accuracy using skeletal representations. Xia et al. [16] recorded overall 90.92% accuracy using Histograms of 3D Joints. Zhao et al. [17] and Liu et al. [7] recorded 97.29% and 95% accuracy respectively. In [15], a 4% mistake in carrying action was discovered, with 1% misclassified as pick up and 2% misclassified as walk. Similarly, there was a 7% misclassification error in pickup activity, with 5% predicted as carrying and 2% as walking. Simultaneously, a 2% mistake was discovered in the walking activity, and the model projected it to be a carry activity because both activities contain motion information. Similarly, there was a 2% misclassification in wave hands activity, which was projected to be 1% wave hands activity and 1% walk activity due to their similarities. This work also overcome the activity misclassification in [11,15] as shown in Figs. 11 and 12.

Table 2. Comparison with the state-of-the-art models on UTKinect Action Dataset

Study	Accuracy (Percent)
Liu et al. [6]	92.00
Vemulapalli et al. [14]	97.08
Verma et al. [15]	97.88
Xia et al. [16]	90.92
Zhao et al. [17]	97.29
Liu et al. [7]	95.00
Phyo et al. [11]	97.00
Proposed work	**98.25**

6 Conclusion

Residual learning-based deep neural networks are a powerful backbone model for computer vision applications that use the skip connection to overcome the vanishing gradient problem. The proposed work implemented a ResNet-50 framework with transfer learning for the recognition of ten human activity. Stage III and IV of the model unfrozen during the training and the experimental results for the proposed work show that it is outperforming other state-of-the-art models on UTKinect Action-3D dataset. The proposed work recorded the highest accuracy 98.25% with 0.11 loss score for the 200 epochs. In future, more experments will be performed with more complex activity under the view variant environment using variety of deep neural networks.

References

1. Aggarwal, J.K., Cai, Q., Liao, W., Sabata, B.: Nonrigid motion analysis: articulated and elastic motion. Comput. Vis. Image Underst. **70**(2), 142–156 (1998)
2. Haering, N., Venetianer, P.L., Lipton, A.: The evolution of video surveillance: an overview. Mach. Vis. Appl. **19**(5–6), 279–290 (2008)
3. Karpathy, A., Toderici, G., Shetty, S., Leung, T., Sukthankar, R., Fei-Fei, L.: Large-scale video classification with convolutional neural networks. In: Proceedings of the IEEE conference on Computer Vision and Pattern Recognition, pp. 1725–1732 (2014)
4. Kumar, K.: EVS-DK: event video skimming using deep keyframe. J. Vis. Commun. Image Represent. **58**, 345–352 (2019)
5. Kumar, K., Shrimankar, D.D.: F-DES: fast and deep event summarization. IEEE Trans. Multimedia **20**(2), 323–334 (2017)
6. Liu, A.A., Nie, W.Z., Su, Y.T., Ma, L., Hao, T., Yang, Z.X.: Coupled hidden conditional random fields for RGB-D human action recognition. Signal Process. **112**, 74–82 (2015)
7. Liu, Z., Feng, X., Tian, Y.: An effective view and time-invariant action recognition method based on depth videos. In: 2015 Visual Communications and Image Processing (VCIP), pp. 1–4. IEEE (2015)
8. McNally, W., Wong, A., McPhee, J.: Star-net: action recognition using spatio-temporal activation reprojection. In: 2019 16th Conference on Computer and Robot Vision (CRV), pp. 49–56. IEEE (2019)
9. Negi, A., Chauhan, P., Kumar, K., Rajput, R.: Face mask detection classifier and model pruning with keras-surgeon. In: 2020 5th IEEE International Conference on Recent Advances and Innovations in Engineering (ICRAIE), pp. 1–6. IEEE (2020)
10. Negi, A., Kumar, K., Chauhan, P., Rajput, R.: Deep neural architecture for face mask detection on simulated masked face dataset against COVID-19 pandemic. In: 2021 International Conference on Computing, Communication, and Intelligent Systems (ICCCIS), pp. 595–600. IEEE (2021)
11. Phyo, C.N., Zin, T.T., Tin, P.: Deep learning for recognizing human activities using motions of skeletal joints. IEEE Trans. Consum. Electron. **65**(2), 243–252 (2019)
12. Popoola, O.P., Wang, K.: Video-based abnormal human behavior recognition-a review. IEEE Trans. Syst. Man Cybern. Part C (Appl. Rev.) **42**(6), 865–878 (2012)
13. Taylor, G.W., Fergus, R., LeCun, Y., Bregler, C.: Convolutional learning of spatio-temporal features. In: Daniilidis, K., Maragos, P., Paragios, N. (eds.) ECCV 2010. LNCS, vol. 6316, pp. 140–153. Springer, Heidelberg (2010). https://doi.org/10.1007/978-3-642-15567-3_11

14. Vemulapalli, R., Arrate, F., Chellappa, R.: Human action recognition by representing 3D skeletons as points in a lie group. In: Proceedings of the IEEE Conference on Computer Vision and Pattern Recognition, pp. 588–595 (2014)
15. Verma, K.K., Singh, B.M., Mandoria, H.L., Chauhan, P.: Two-stage human activity recognition using 2D-convnet. Int. J. Interact. Multimedia Artif. Intell. **6**(2) (2020)
16. Xia, L., Chen, C.C., Aggarwal, J.K.: View invariant human action recognition using histograms of 3D joints. In: 2012 IEEE Computer Society Conference on Computer Vision and Pattern Recognition Workshops, pp. 20–27. IEEE (2012)
17. Zhao, C., Chen, M., Zhao, J., Wang, Q., Shen, Y.: 3D behavior recognition based on multi-modal deep space-time learning. Appl. Sci. **9**(4), 716 (2019)

DXML: Distributed Extreme Multilabel Classification

Pawan Kumar(✉)

International Institute of Information Technology,
Hyderabad, Hyderabad 500032, India
pawan.kumar@iiit.ac.in

Abstract. As a big data application, extreme multilabel classification has emerged as an important research topic with applications in ranking and recommendation of products and items. A scalable hybrid distributed and shared memory implementation of extreme classification for large scale ranking and recommendation is proposed. In particular, the implementation is a mix of message passing using MPI across nodes and using multithreading on the nodes using OpenMP. The expression for communication latency and communication volume is derived. Parallelism using work-span model is derived for shared memory architecture. This throws light on the expected scalability of similar extreme classification methods. Experiments show that the implementation is relatively faster to train and test on some large datasets. In some cases, model size is relatively small.

Code: https://github.com/misterpawan/DXML

Keywords: Extreme multilabel classification · Distributed memory · Multithreading

1 Introduction

Extreme multi-label learning is an active research problem in big data applications with several applications in recommendation, tagging, and ranking. Consider a features matrix of n examples $X \in \mathbb{R}^{n \times d}$ with d-dimensional features with their corresponding labels matrix $Y \in \mathbb{R}^{n \times \ell}$, the aim of multilabel classification problem is to assign some relevant labels out of a total of ℓ-labels to new data sample. The extreme multilabel classification problem refers to the setting when n, d, and ℓ quickly scale to large numbers often upto several millions. Moreover, number of data samples also are in several millions. This is one of the classic challenge problems of big data analytics.

There are several challenges in designing algorithms for extreme multilabel classification. It is found that the average number of labels per data points is usually very small, moreover, it is know that label frequencies follow the so-called Zipf's law, which means that there are only small number of labels which are found in large number of samples, such labels are called head labels. On the other hand, there are large number of lables that occur less frequently, and such

S. N. Srirama et al. (Eds.): BDA 2021, LNCS 13147, pp. 311–321, 2021.
https://doi.org/10.1007/978-3-030-93620-4_22

labels are called tail labels. Such a distribution creates a bias in the classifier, because since head labels occur more frequently, the classifier may learn more robustly about predicting head labels compared to tail labels, thereby, leading to a classifier that is biased towards predicting head labels better. Despite all these challenges, one of the major concerns is designing scalable algorithms for modern day hardwares.

Looking at the increasing trend of number of labels going upto millions, a hybrid parallel design and implementation of extreme classification algorithms is essential that can exploit shared as well as distributed memory architectures [3,11–14]. In this paper, we show a parallel design and implementation for a hybrid distributed memory and shared memory implementation using MPI and OpenMP. In our knowledge, this is the first time a hybrid parallel implementation has been shown for extreme classification. We derive the communication bounds for distributed memory, and parallelism bound for shared memory implementation. Our preliminary numerical experiments suggest that we have fastest training and test times when compared to some of the existing parallel implementations in C/C++.

Following paper is organized as follows. In Sect. 2, we discuss previous related work. In Sect. 3, we discuss distributed memory implementation. We derive expressions for communication volume and latency. Finally, in Sect. 4, we discuss numerical experiments on multilabel datasets.

2 Previous Work

Some papers for multilabel classification were proposed during 2006–2014 [18,25]. The later is a review on multilabel learning algorithms. Some of these papers explored k nearest neighbours [26]. The idea of using random forest were also proposed [10]. With the recent demand for scaling the algorithms to millions of labels, new class of methods have been proposed, where scalability is achieved by either parallelism or exploiting the fact that label matrix is sparse with proposed methods PD-Sparse and PPDSparse [22,23], some form of dimension reduction using label clustering, nearest neighbor search, etc. [4,17,20,24], or by hierarchical embedding [6–9,15,21].

3 Distributed Memory Implementation

We show a hybrid parallel (MPI+OpenMP) implementation of CRAFTML [16], and call it DXML. During training, DXML computes a forest F of m_F k-ary instance trees, which are constructed by recursive partitioning. The training algorithm is shown in Algorithm 1. In line 1, the input is a feature matrix X and a label matrix Y. We wish to build a label-tree with nodes denoted by v. We then apply the termination condition. The termination condition of the recursive partitioning are the following

1. if the number of elements in the instance subset of the node is less than a given threshold n_{leaf}

2. all the given instances have the same features
3. all the given instances have the same labels

If the stop condition is false, and the current node v is not a leaf as in line 4, then a multi-class classifier is built using Algorithm 2.

The sequential node training stage in DXML will be decomposed into following three consecutive steps:

- a random projection into lower dimensional spaces of the features and label vectors corresponding to the node's instances. That is, in Algorithm 2, in line 2, we first sample X_s and Y_s from X_v and Y_v with a sample size n_s. Then in lines 3 and 4, we do the random projection using projection feature matrix P_x, and a random label projection matrix P_y.
- from projected labels, partitioning of the corresponding instances into k temporary subsets using $k-$means.
- A multiclass classifier is trained to assign each instance to the relevant temporary subset (that is, the cluster index found at step 2 above) from the corresponding feature vector. The classifier then partitiones the instances into k final subsets or child nodes. This is achieved in by first calling a splitting (line 6, Algorithm 1) of the instances into child nodes using output of k-means of Algorithm 2. Then trainTree function is called recursively on the child nodes.

Similar to FastXML, in DXML, we partition the nodes by regrouping instances with common labels in a same subset, but the computation is different. After a tree has been trained, the average label vector of corresponding instances are stored in leaves. The partition strategy is decided by considering two constraints: computation of partition should be based on randomly projected instances to make sure that there is diversity, and should involve low complexity operations to ensure scalability. The training algorithm is given below.

Algorithm 1. trainTree [16]

1: **Input:** Feature matrix X and label matrix Y.
2: Initialize v.
3: v.isLeaf $\leftarrow$ StopCondition(X, Y)
4: if v.isLeaf $=$ false then
5: $\quad v$.classif $\leftarrow$ trainNodeClassifier(X, Y)
6: $\quad (X_{child_i}, Y_{child_i})_{i=0,\cdots,k-1} \leftarrow$ split(v.classif,X, Y)
7: $\quad$ for i from 0 to $k-1$ do
8: $\quad\quad v.child_i \leftarrow$ trainTree$(X_{child_i}, Y_{child_i})$
9: $\quad$ end for
10: else
11: $\quad v.\hat{y} \leftarrow$ calculateMeanLabelVector(Y)
12: end if
13: **Output:** node v

Algorithm 2. trainNodeClassifier [16]

1: **Input:** feature matrix (X_v) and label matrix (Y_v) of the instance set corresponding to node v.
2: $X_s, Y_s \leftarrow$ sampleFromRows(X_v, Y_v, n_s) ▷ n_s is the sample size
3: $X_s^{'} \leftarrow X_s P_x$ ▷ random feature projection
4: $Y_s^{'} \leftarrow Y_s P_y$ ▷ random label projection
5: $c \leftarrow k$-means$(Y_s^{'}, k)$ ▷ $c \in \{0, \cdots, k-1\}^{\min(n_v, n_s)}$
6: ▷ c is a vector where the jth component c_j is the cluster idx of the jth instance corresponding to $(X_s^{'})_{j,.}$ and $(Y_s^{'})_{j,.}$.
7: for i from 0 to $k-1$ do
8: $(classif)_{i,.} \leftarrow$calculateCentroid$((X_s^{'})_{j,.} | c_j = i)$
9: end for
10: **Output:** Classifier $classif (\in \mathbb{R}^{k \times d_x^{'}})$

3.1 Some More Detail on Training

Step 1: Random Projections of the Instances of v**:** The feature vector x and the label vectors y of each instance in v are projected into a space with a lower dimensionality: $x^{'} = XP_x$ and $y^{'} = yP_y$ where P_x and P_y are random projection matrices of $\mathbb{R}^{d_x \times d_x^{'}}$ and $\mathbb{R}^{d_y \times d_y^{'}}$ respectively, and $d_x^{'}$ and $d_y^{'}$ are the dimensions of the reduced feature and label spaces resp. The projection matrices are kept different from each tree. The random projection considered is a sparse orthogonal projection matrix [19] with values of -1 or $+1$ on each row. The sparsity of the projections lead to faster computations, hence, faster projections. To retain the sparsity we use the hashing; we describe it next.

In this so-called hashing trick algorithm, high dimensional dataset is projected into a lower dimension. Projection is done row after row in this algorithm. In a row of original matrix each element index is considered as a key and each element index of corresponding row in the projected matrix is considered as the bucket. Each index of original row (key) is mapped to index of lower dimension projected row (bucket) using the hash function. Similarly each key is also mapped to one of the two signs (+ or −) using another hash function. Multiple keys may be mapped to the same bucket, in that case elements in the same bucket are multiplied by their respective signs (obtained from second hash function) and added. Below is the algorithm of Hashing Trick.

Algorithm 3. Hashing Trick

```
1: Input: X_s, projectedSpaceDimension, S_x, SS_x
2: X'_s ← 0                                  ▷ initialize projected data matrix
3: p ← projectedSpaceDimension
4: nR ← X_s.numberOfRows()
5: nC ← X_s.numberOfCols()
6: for i ← 0 to nR :
7:     for key ← 0 to nC :
8:         Index ← hash1(key, S_x)%p
9:         Sign ← 2 × (hash2(key, SS_x)%2) − 1
10:        X'_s[i, Index] ← X'_s[i, Index] + (Sign * X_s[i, key])
11: Output: X'_s
```

This Algorithm 3 takes X_s (or Y_s) and the projectedSpaceDimension, and S_x, SS_x (seeds for hash1, hash2 functions respectively), and returns projected samples X'_s (or Y'_s). In line 4, nR is the number of rows in X_s, and nC is the number of columns in X_s. The lines 6 and 7 are loops with i iterating over the row indices of X_s, and **key** iterating over the column indices in the row (with index i). In line 8, the Column Index (key) is hashed using hash1 function and if its more than the projectedSpaceDimension, then remainder when divided by p is considered to map it in the range$(0, p)$. In line 9, $2 \times hash2(key)\%2 - 1$ maps key to $+1$ or -1, and finally in line 10, the element $X_s[i, key]$ is multiplied by the sign, and then added to already present element at $X'_s[i, Index]$.

We consider the case where feature and label projections are the same in each node of T. We may have tried different projections per node, but we don't consider that in this paper.

Step 2: Partitioning of Instance into k Temporary Subsets: Let Y_s denote the label matrix of a sample drawn without replacement. Let the sample size be at most n_s. A spherical k-means is applied on $Y_s P_y$ to partition a sample space. The use of spherical k-means is motivated by the facts that it is well-adapted to sparse data, moreover, the cosine metric is fast to compute.

We initialize the cluster centroids using the k-means++ to ameliorate the cluster stability, and to improve the algorithm performance against a random initialization.

The $k$means++ initialization strategy is as follows:

1. Among the label centers, choose one center uniformly at random.
2. For each label vector y, compute the distance $D(y)$, defined to be the distance between y, and the closest center that was chosen already.
3. Choose a new data point randomly as a new center, using a weighted probability distribution where a point y will be chosen with probability proportional to $D(y)^2$.
4. Repeat steps 2 and 3 above until all centers have been chosen.
5. After all the centers are chosen, proceed with the spherical k-means clustering.

Step 3: From the Projected Features, Assign a Subset: In each temporary subset the centroid of the projected feature vectors is computed. During the prediction phase, if the centroid of the subset is nearest to the projected feature vector, then the classifier assigns this subset. For computing the closeness, the cosine measure is used.

Prediction: During prediction, for each tree, the input sample goes from root to leaf. The successive classifiers determine this passage of input from root to leaf. When the input sample reaches a leaf, then the average label vector saved in this leaf is reported as prediction. The forest of trees takes all such predictions from each tree and averages it for the final prediction.

Algorithm Analysis. Before we proceed, let the average number of non-zero elements in feature vector be denoted by s_x, and similarly let s_y denote the average number of non-zeros in label vectors for a given instance. As a result of the hashing trick, the feature and label vectors after projection have less than s_x and s_y non-zero elements in average. Let v denote node of a tree T, and let n_v denote the number of instances of the subset associated to v. Let i be the number of iterations required for spherical k-means algorithm.

Lemma 1. *Given a node v of a tree T, let C_v denote the time complexity, then C_v is bounded by $O(n_v \times C)$, where*

$$C = k \times (i \times s_y + s_x)$$

is the complexity per instance.

Proof. See [16].

Let T be a strictly k-ary tree and ℓ_T be its number of leaves. Let $m_T = \frac{\ell_T - 1}{k-1}$ be its number of nodes and $\bar{n}_T = \frac{\sum_{v \in T} n_v}{m_T}$ be the average number of instances in its nodes.

Proposition 1. *Assume balanced tree T, then its time complexity for training denoted by C_T is bounded by*

$$O\left(\log_k \left(\frac{n}{n_{leaf}}\right) \times n \times C\right).$$

Otherwise, C_T is equal to

$$O\left(\frac{\ell_T - 1}{k-1} \times \bar{n}_T \times C\right).$$

Proof. See [16].

It can be observed that the time complexities are independent of the projection dimensions $d_x^{'}$ and $d_y^{'}$. The clustering is done after sampling from the instances, thus the training and clustering complexity are further reduced.

Proposition 2. *For a tree T, the bound on memory complexity is given by*

$$O(n \times s_y + m_T \times k \times d_x^{'}).$$

Proof. See [16].

3.2 Hybrid MPI and OpenMP Parallel Implementation

We use message passing interface MPI [1] to train for each learner. Each learner or process reads the data, and calls trainTree in Algorithm 1. The classification for each node (child) at a given level is processed by multiple threads. Each learner stores their own mean label vector computed in line 11 of Algorithm 1. The model parameters for each tree is sent to master node for faster prediction. Let n_t be the model size for each tree, and there are m_F trees, then the communication volume from the worker nodes to the master node is $O(m_F n_t)$. The latency cost is the number of messages passed. In this case, each processor communicates once to master node. Hence the latency cost is $O(m_F)$. Let P denote the number of processors, then since there are as many processors as number of trees in the forest, $P = m_F$. We have the following results.

Proposition 3. *Let there be P processors and $m_F = P$, we have the following communication costs*

$$\begin{aligned} \text{communication\_volume} &= O(Pn_t) \\ \textit{latency} &= O(P) \end{aligned}$$

We also exploit shared memory parallelism using OpenMP [2] when training each tree. This follows a spawn-sync model. At root, a master thread launches k child processes, and each of the k child process calls trainNodeClassifier. We use work-span model [5] to do parallel complexity analysis of trainTree. The span denoted by T_∞, which corresponds to critical path, i.e., the longest path from root node to a leaf node. Now we calculate the work denoted by T_1, which is the total work done to train the classifier for all the nodes. We define the parallelism to be T_1/T_∞. We have the following proposition.

Proposition 4. *Let T be a strictly $k-$ary tree, ℓ_T be its number of leaves. Let $m_T, \bar{n}_T$, and C be defined as above, then the parallelism for a tree is bounded by*

$$T_1/T_\infty = O\left(\frac{1}{\log_k m_T} \log_k \left(\frac{n}{n_{leaf}}\right) \times n \times C\right).$$

Proof. From Proposition 1, the total work done, which is denoted by C is given by the following bound

$$T_1 = C = O\left(\log_k \left(\frac{n}{n_{leaf}}\right) \times n \times C\right).$$

We also have $T_\infty = \log_k m_T$. This gives the required parallelism.

4 Numerical Experiments

We did our MPI+OpenMP experiments on Intel Xeon architecture with 10 nodes with 120GB RAM. We used 10 MPI processes and 5 threads per processes. We choose $m_F = 50$ and $n_{leaf} = 10$. The feature and label projection dimensions are $d'_x = \min(d_y, 10000)$ and $d'_y = \min(d_y, 10000)$. We show the precision scores in Table 2. The best precision scores among the tree-based classifiers are indicated in bold. For example, the DXML has highest P@1 precision scores for Mediamill, EURLex-4K, Delicious-200K among the tree based classifiers. In other cases, it is close to the precisions of other classifiers. In Table 1, we show the train time, test time, and model size. The train times for DXML was best among all methods for all the datasets. The model size for DXML is same as the one for CRAFTML. The model size for DXML/CRAFTML was best for Amazon-670 and Delicious-200K. The model size for PPDSp is very large for Amazon-670. For DXML, the learned model parameters remain distributed, hence, there is an additional cost of reduction operation.

Table 1. Train time, test time, and model size. Here NA means not available.

Language		C/C++				C++
Machine		50 cores	1 core			100 cores
Algorithm		DXML	FastXML	PFastReXML	SLEEC	DISMEC
EURLex-4K	Train	**38.01**	315.9	324.4	4543.4	76.07
	Test (ms)	**1.29**	3.65	5.43	3.67	2.26
	Model (MB)	30	384	NA	121	**15**
Delicious-200K	Train	**2929.0**	8832.46	8807.51	4838.7	38814
	Test (ms)	10.40	1.28	7.4	2.685	311.4
	Model (GB)	**0.346**	1.3	20	2.1	18
Amazon-670K	Train	**752.65**	5624	6559	20904	174135
	Test (ms)	3.65	**1.41**	1.98	6.94	148
	Model (GB)	**0.494**	4.0	6.3	6.6	8.1
AmazonCat-13K	Train	**1164.13**	11535	13985	119840	11828
	Test (ms)	8.98	1.21	1.34	13.36	0.2
	Model (GB)	**0.659**	9.7	11	12	2.1

Table 2. Precision scores. Here NA stands for not available.

Method type		Tree based				Other	
Algorithm	Scores	DXML	PFastReXML	FastXML	LPSR	SLEEC	DISMEC
Mediamill	P@1	**87.20**	83.98	84.22	83.57	87.82	84.83
	P@3	**71.52**	67.37	67.33	65.78	73.45	67.17
	P@5	**57.56**	53.02	53.04	49.97	59.17	52.80
Delicious	P@1	67.28	67.13	**69.61**	65.01	67.59	NA
	P@3	61.64	62.33	64.12	58.96	61.38	NA
	P@5	57.17	58.62	59.27	53.49	56.56	NA
EURLex-4K	P@1	**78.20**	75.45	71.36	76.37	79.26	82.40
	P@3	64.2	62.70	59.90	63.36	64.30	68.50
	P@5	53.26	52.51	50.39	52.03	52.33	57.70
Wiki-10	P@1	**84.68**	83.57	83.03	72.72	85.88	85.20
	P@3	72.54	68.61	67.47	58.51	72.98	74.60
	P@5	62.47	59.10	57.76	49.50	62.70	65.90
Delicious-200K	P@1	**47.86**	41.72	43.07	18.59	47.85	45.50
	P@3	41.25	37.83	38.66	15.43	42.21	38.70
	P@5	38.00	35.58	36.19	14.07	39.43	35.50
Amazon-670K	P@1	37.31	**39.46**	36.99	28.65	35.05	44.70
	P@3	33.30	35.81	33.28	24.88	31.25	39.70
	P@5	30.50	33.05	30.53	22.37	28.56	36.10
AmazonCat-13K	P@1	92.69	91.75	**93.11**	NA	90.53	93.40
	P@3	78.43	77.97	78.20	NA	76.33	79.10
	P@5	63.56	63.68	63.41	NA	61.52	64.10

5 Conclusion

For large scale recommendation problem using extreme multilabel classification, a scalable recommender model is essential. We proposed a hybrid parallel implementation of extreme classification using MPI and OpenMP that can scale to arbitrary number of processors. The best part is that this does not involve any loss function or iterative gradient methods. Our preliminary results show that our training time is fastest for some datasets. We derived the communication complexity analysis bounds for both the shared and distributed memory implementations. With more cores, our parallel analysis suggests that the proposed implementation has the potential to scale further.

As a future work, we may explore federated learning approach as an extension of this work. In this case, the recommendation data corresponding to certain users remain separated from the other users, and learning happens with decentralized data. Note that in this work, we assumed centralized data. Also, since this recommendation method does not involve gradient method, it may have certain advantages compared to existing gradient based training of extreme multilabel classifier systems.

Acknowledgement. This work was done at IIIT, Hyderabad using IIIT seed grant. The author acknowledges all the support by institute. This project was partially supported by RIPPLE center of excellence at IIIT, Hyderabad.

References

1. Open MPI: Open source high performance computing. https://www.open-mpi.org/
2. Openmp. https://www.openmp.org/
3. Kumar, P., Markidis, S., Lapenta, G., Meerbergen, K., Roose, D.: High performance solvers for implicit particle in cell simulation (special issue). Procedia Comput. Sci. **18**, 2251–2258 (2013). https://doi.org/10.1016/j.procs.2013.05.396. https://www.sciencedirect.com/science/article/pii/S1877050913005395. 2013 International Conference on Computational Science
4. Bhatia, K., Jain, H., Kar, P., Varma, M., Jain, P.: Sparse local embeddings for extreme multi-label classification. In: Proceedings of the 28th International Conference on Neural Information Processing Systems, NIPS 2015, vol. 1, pp. 730–738. MIT Press, Cambridge (2015)
5. Blumofe, R.D., Joerg, C.F., Kuszmaul, B.C., Leiserson, C.E., Randall, K.H., Zhou, Y.: Cilk: an efficient multithreaded runtime system. SIGPLAN Not. **30**(8), 207–216 (1995). https://doi.org/10.1145/209937.209958
6. Jain, H., Prabhu, Y., Varma, M.: Extreme multi-label loss functions for recommendation, tagging, ranking and other missing label applications. In: Proceedings of the 22nd ACM SIGKDD International Conference on Knowledge Discovery and Data Mining, KDD 2016, pp. 935–944. Association for Computing Machinery, New York (2016). https://doi.org/10.1145/2939672.2939756
7. Jasinska, K., Dembczynski, K., Busa-Fekete, R., Pfannschmidt, K., Klerx, T., Hullermeier, E.: Extreme f-measure maximization using sparse probability estimates. In: Proceedings of the 33rd International Conference on International Conference on Machine Learning, ICML 2016, vol. 48, pp. 1435–1444. JMLR.org (2016)
8. Jayadev, N., Tanmay, S., Pawan, K.: A riemannian approach for constrained optimization problem in extreme classification problems. CoRR abs/2109.15021 (2021). https://arxiv.org/abs/2109.15021
9. Jayadev, N., Tanmay, S., Pawan, K.: A riemannian approach for extreme classification problems. In: CODS-COMAD 2021 (2021)
10. Kocev, D., Vens, C., Struyf, J., Džeroski, S.: Ensembles of multi-objective decision trees. In: Kok, J.N., Koronacki, J., Mantaras, R.L., Matwin, S., Mladenič, D., Skowron, A. (eds.) ECML 2007. LNCS (LNAI), vol. 4701, pp. 624–631. Springer, Heidelberg (2007). https://doi.org/10.1007/978-3-540-74958-5_61
11. Kumar, P.: Communication optimal least squares solver. In: 2014 IEEE International Conference on High Performance Computing and Communications, 2014 IEEE 6th Intl Symposium on Cyberspace Safety and Security, 2014 IEEE 11th International Conference on Embedded Software and Syst (HPCC, CSS, ICESS), pp. 316–319 (2014). https://doi.org/10.1109/HPCC.2014.55
12. Kumar, P.: Multithreaded direction preserving preconditioners. In: 2014 IEEE 13th International Symposium on Parallel and Distributed Computing, pp. 148–155 (2014). https://doi.org/10.1109/ISPDC.2014.23

13. Kumar, P.: Multilevel communication optimal least squares (special issue). Procedia Comput. Sci. **51**, 1838–1847 (2015). https://doi.org/10.1016/j.procs.2015.05.410. https://www.sciencedirect.com/science/article/pii/S1877050915012181. International Conference On Computational Science, ICCS 2015
14. Kumar, P., Meerbergen, K., Roose, D.: Multi-threaded nested filtering factorization preconditioner. In: Manninen, P., Öster, P. (eds.) PARA 2012. LNCS, vol. 7782, pp. 220–234. Springer, Heidelberg (2013). https://doi.org/10.1007/978-3-642-36803-5_16
15. Prabhu, Y., Varma, M.: Fastxml: a fast, accurate and stable tree-classifier for extreme multi-label learning, KDD 2014, pp. 263–272. Association for Computing Machinery, New York (2014). https://doi.org/10.1145/2623330.2623651
16. Siblini, W., Meyer, F., Kuntz, P.: Craftml, an efficient clustering-based random forest for extreme multi-label learning. In: Dy, J.G., Krause, A. (eds.) Proceedings of the 35th International Conference on Machine Learning, ICML 2018, Stockholmsmässan, Stockholm, Sweden, 10–15 July 2018. Proceedings of Machine Learning Research, vol. 80, pp. 4671–4680. PMLR (2018). http://proceedings.mlr.press/v80/siblini18a.html
17. Tagami, Y.: Annexml: approximate nearest neighbor search for extreme multi-label classification. In: Proceedings of the 23rd ACM SIGKDD International Conference on Knowledge Discovery and Data Mining, KDD 2017, pp. 455–464. Association for Computing Machinery, New York (2017). https://doi.org/10.1145/3097983.3097987
18. Tsoumakas, G., Katakis, I.: Multi-label classification: an overview. Int. J. Data Warehous. Min. **3**, 1–13 (2007)
19. Weinberger, K.Q., Dasgupta, A., Attenberg, J., Langford, J., Smola, A.J.: Feature hashing for large scale multitask learning. CoRR abs/0902.2206 (2009). http://arxiv.org/abs/0902.2206
20. Weston, J., Bengio, S., Usunier, N.: Wsabie: scaling up to large vocabulary image annotation, IJCAI 2011, pp. 2764–2770. AAAI Press (2011)
21. Weston, J., Makadia, A., Yee, H.: Label partitioning for sublinear ranking. In: Proceedings of the 30th International Conference on International Conference on Machine Learning, ICML 2013, vol. 28, pp. II-181–II-189. JMLR.org (2013)
22. Yen, I.E.H., Huang, X., Zhong, K., Ravikumar, P., Dhillon, I.S.: PD-sparse: a primal and dual sparse approach to extreme multiclass and multilabel classification. In: Proceedings of the 33rd International Conference on International Conference on Machine Learning, ICML 2016, vol. 48, pp. 3069–3077. JMLR.org (2016)
23. Yen, I.E., Huang, X., Dai, W., Ravikumar, P., Dhillon, I., Xing, E.: PPDSparse: a parallel primal-dual sparse method for extreme classification. In: Proceedings of the 23rd ACM SIGKDD International Conference on Knowledge Discovery and Data Mining, KDD 2017, pp. 545–553. Association for Computing Machinery, New York (2017). https://doi.org/10.1145/3097983.3098083
24. Yu, H.F., Jain, P., Kar, P., Dhillon, I.S.: Large-scale multi-label learning with missing labels. In: Proceedings of the 31st International Conference on International Conference on Machine Learning, ICML 2014, vol. 32, pp. I-593–I-601. JMLR.org (2014)
25. Zhang, M., Zhou, Z.: A review on multi-label learning algorithms. IEEE Trans. Knowl. Data Eng. **26**(8), 1819–1837 (2014). https://doi.org/10.1109/TKDE.2013.39
26. Zhang, M.L., Zhou, Z.H.: ML-KNN: a lazy learning approach to multi-label learning. Pattern Recogn. **40**(7), 2038–2048 (2007). https://doi.org/10.1016/j.patcog.2006.12.019

An Efficient Distributed Coverage Pattern Mining Algorithm

Preetham Sathineni[1(✉)], A. Srinivas Reddy[1(✉)], P. Krishna Reddy[1(✉)], and Anirban Mondal[2(✉)]

[1] IIIT, Hyderabad, Hyderabad, India
{preethamreddy.s,srinivas.annappalli}@research.iiit.ac.in, pkreddy@iiit.ac.in

[2] Ashoka University, Delhi, India
anirban.mondal@ashoka.edu.in

Abstract. Mining of coverage patterns from transactional databases is one of the data mining tasks. It has applications in banner advertising, search engine advertising and visibility computation. In general, most real-world transactional databases are typically large. Mining of coverage patterns from large transactional databases such as query log transactions on a single computer is challenging and time-consuming. In this paper, we propose Distributed Coverage Pattern Mining (*DCPM*) approach. In this approach, we employ a notion of the summarized form of Inverse Transactional Database (ITD) and replicate it at every node. We also employ an efficient clustering-based method to distribute the computational load of extracting coverage patterns among the Worker nodes. We performed extensive experiments using two real-world datasets and one synthetic dataset. The results show that the proposed approach significantly improves the performance over the state-of-the-art approaches in terms of execution time and data shuffled.

Keywords: Data mining · Coverage patterns · MapReduce · Clustered distribution

1 Introduction

Mining the knowledge of coverage patterns from transactional databases [24,25] is one of the important data mining tasks. It has applications in the areas of banner advertisement placement [13], search engine advertising [3–5] and visibility query processing [9].

Given a Transactional DataBase (TDB) over a set of items, a coverage pattern (CP) [24] is a set of items or a pattern, whose items cover a certain percentage of transactions in TDB. A coverage pattern is associated with Relative Frequency (RF), Coverage Support (CS) and Overlap Ratio (OR) measures (the details of these measures will be provided in Sect. 3). Given TDB and user-specified CS, OR, and RF constraints, the problem is to extract all CPs from TDB. In the literature, a level-wise candidate pruning approach, which

S. N. Srirama et al. (Eds.): BDA 2021, LNCS 13147, pp. 322–340, 2021.
https://doi.org/10.1007/978-3-030-93620-4_23

we designate as Coverage Pattern Mining (*CPM*) algorithm was proposed in [24]. Moreover, a Coverage Pattern Projected Growth (*CPPG*) approach was proposed in [25] by exploiting the pattern growth paradigm.

The existing *CPM* algorithm [24] was proposed by considering a single machine using a level-wise candidate generation approach to extract CPs from the given TDB. Typically, the number of candidate patterns at each level is very large. Given limited main memory and CPU power, computing and validating such a huge number of candidate patterns on a single machine is very time-consuming. To improve the performance, we propose a Distributed Coverage Patterns Mining (*DCPM*) algorithm. In this approach, we employ the notion of Inverse Transactional Database (ITD), a vertical form of TDB and replicate it at every node. Moreover, we propose a clustering-based method to distribute the task of extracting CPs among the worker nodes. Notably, the performance could be significantly improved by replicating TDB using the notion of ITD and employing a clustering-based approach to compute and validate candidate patterns in a distributed manner.

An effort has been made in [19] to employ a MapReduce paradigm to compute CPs. In that approach, the TDB is partitioned among the participant sites. As the size of the TDB and the number of items increase, this approach would encounter performance problems because a large number of candidate patterns are to be transmitted at each level. In the literature, the notion of ITD is employed for extracting frequent patterns by the ECLAT [16] approach, which extracts frequent itemsets in a depth-first search manner. An effort has been made to propose a distributed ECLAT [18,23] based on the Spark framework. The notion of ITD is employed in ECLAT to improve the performance of extracting frequent itemsets by computing and validating the frequent itemsets in a distributed manner. So far, the notion of ITD has not been extended to extract CPs.

The key contributions of this paper are two-fold:

1. We propose a Distributed Coverage Pattern Mining (*DCPM*) approach with the clustered distribution of patterns and Inverse Transactional Database (ITD) modelling of the database for efficiently extracting coverage patterns.
2. We performed extensive experiments using two real-world datasets and one synthetic dataset. The results show that the performance of the proposed approach significantly improves over the state-of-the-art approaches in terms of execution time and data shuffled across nodes.

In this paper, we propose a distributed *CPM* algorithm. A distributed pattern-growth approach to extract CPs will be investigated as a part of future work. The remainder of this paper is organized as follows. In Sect. 2, we discuss the related works. In Sect. 3, we discuss background information about coverage patterns. In Sect. 4, we discuss our proposed approach. In Sect. 5, we report the performance evaluation. Finally, we conclude in Sect. 6 with directions for future work.

2 Related Work

An effort [27] has been made to propose a Hadoop-based two-phase approach for extracting all frequent k-items from transactional datasets. The work in [12] presents an approach by projecting the transactional database into a tabular form with transaction identifiers as rows and items as columns. The work in [6] proposes the MapReduce implementation of the Eclat algorithm [16]. Another approach is also proposed in [6] for mining large-scale datasets using a combination of an Apriori variant and Eclat algorithm. An approach is proposed in [7], which uses the partitioning technique and bitmap representation of the data partitions to achieve better scaling. The work in [26] proposes a MapReduce based high fuzzy utility itemset mining algorithm. The work in [14] proposes a fuzzy mining algorithm (EFM), which uses a compressed data structure CFL and several effective pruning strategies to reduce the search space.

The work in [17] proposes YAFIM (Yet Another Frequent Itemset Mining), a parallel Apriori algorithm based on the Spark RDD framework [29]. YAFIM is specially designed for an in-memory parallel computing model, which supports iterative algorithms and interactive data mining. The work in [28] proposes an improved method to prune the candidate patterns by introducing a mapping model and combining Spark's features. The work in [11] proposes a novel pruning technique for matrix-based frequent itemset mining, which greatly reduces the memory usage by using the Spark distributed environment. The work in [21] presents Reduced-Apriori (R-Apriori), a parallel Apriori algorithm based on the Spark RDD framework, which prunes the candidate patterns and achieves higher speedups in comparison to YAFIM. The work Adaptive-Miner in [20] is built on R-Apriori, which uses the Bloom filter data structure instead of a tree structure. The Adaptive-Miner broadcasts the database to all the nodes and dynamically computes the execution plans for higher accuracy and efficiency. The work in [22] proposes a Hybrid Frequent Itemset Mining (HFIM), which is a mixture of both Eclat and apriori approaches. The work in [30] presents a distributed ECLAT (MREclat), which first converts frequent 1-itemsets into balanced groups.

A coverage pattern mining (*CPM*) algorithm [24] was proposed to extract coverage patterns from transactional databases. *CPM* uses the level-wise apriori method to generate candidate patterns and sorted closure property based candidate pruning technique. The work in [25] proposes CPPG, a pattern growth based approach for efficient extractions of coverage patterns. The work in [19] proposed a MapReduce-based approach for mining coverage patterns in a distributed environment.

In all preceding approaches, several efforts have been made to extract frequent patterns and utility patterns based on various distributions, compression and pruning techniques. Regarding the MapReduce-based *CPM* algorithm, it has been observed that as the size of the TDB and the number of items increase, the approach encounters performance problems due to the transmission of a large number of candidate patterns over the network at each level. The proposed approach employs notions of replicating ITD and compression for proposing an efficient distributed *CPM* approach.

3 About Coverage Patterns

Let $I=\{i_1, i_2, \ldots, i_m\}$ be a set of m items and let TDB be the transactional database, where each transaction $T_i \in TDB$ is a subset of I, i.e., $T_i \subseteq I$. The notation $|TDB|$ denotes the total number of transactions in the transactional database and TID represents the transaction identifier. The notation T^{i_p} represents the set of $TIDs$ that contain the item i_p and notation $|T^{i_p}|$ gives the number of transactions that contain item i_p. We shall now discuss Relative Frequency (RF), Coverage Set (CSet), Coverage Support (CS) and Overlap Ratio (OR).

The RF value of an item i_p ($RF(i_p)$) is the fraction of the transactional database containing the item i_p. Formally, $RF(i_p) = \frac{|T^{i_p}|}{|TDB|}$. Here, $0 \leq RF(i_p) \leq 1$. When $RF(i_p) \geq minRF$, item i_p is considered as a *frequent item*. Here, $minRF$ is a user-specified *minimum Relative Frequency* threshold. Let a pattern $X = \{i_p, \ldots, i_q, i_r\}$, where $(1 \leq p, q, r \leq m)$. The coverage set $CSet(X)$ of a pattern X is defined as the set of all transactions containing at least one item of the pattern X. Formally, $CSet(X) = T^{i_p} \cup T^{i_q} \cup \ldots \cup T^{i_r}$. The coverage support of a pattern X ($CS(X)$) is defined as the ratio of $|CSet(X)|$ to the size of database $|TDB|$. Formally, $CS(X) = \frac{|CSet(X)|}{|TDB|}$. Here, $0 \leq CS(X) \leq 1$. A pattern X is considered to be interesting if $CS(X) \geq minCS$, where $minCS$ is a user-specified *minimum Coverage Support* threshold.

A pattern with high CS is considered more valuable since it covers a significant number of transactions. We wish to expand the coverage of the pattern X by adding a new item i_k to the pattern X. In order to add a new item to the pattern X such that the coverage support increases significantly, the notion of Overlap Ratio (OR) has been introduced in [24].

Given a pattern X, suppose the items in X are sorted in decreasing order of their relative frequency values i.e., $|T^{i_p}| \geq \ldots \geq |T^{i_q}| \geq |T^{i_r}|$. The overlap ratio $OR(X)$ of a pattern X is defined as the ratio of the number of common transactions in $CSet(X - i_r)$ and T^{i_r} to the number of transactions having item i_r. Formally, $OR(X) = \frac{|CSet(X-i_r) \cap T^{i_r}|}{|T^{i_r}|}$. Here, $0 \leq OR(X) \leq 1$. A pattern X is considered to be interesting if $OR(X) \leq maxOR$, where $maxOR$ is a user-specified *maximum Overlap Ratio* threshold. A pattern X is called as a non-overlap pattern if $OR(X) \leq maxOR$ and $RF(i_k) \geq minRF$ $\forall i_k \in X$.

The coverage pattern is defined as follows. Given a TDB, let a pattern $X = \{i_p, \cdots, i_q, i_r\}$, where $(1 \leq p, q, r \leq m)$ and $|T^{i_p}| \geq \cdots \geq |T^{i_q}| \geq |T^{i_r}|$. The pattern X is called a coverage pattern when $OR(X) \leq maxOR$, $CS(X) \geq minCS$ and $RF(i_k) \geq minRF$ $\forall i_k \in X$.

Problem Statement: Given a transactional database TDB and user-specified $minRF$, $minCS$, and $maxOR$ constraints, the problem is to extract all coverage patterns from the transactional database TDB.

Notably, while extracting coverage patterns from a given TDB, we use overlap ratio as a pruning heuristic, which satisfies the sorted closure property [15,24]. We shall now define the sorted closure property.

Definition 1 (*Sorted closure property*). *Let us consider a pattern* $X = \{i_p, \ldots, i_q, i_r\}$, *where* $(1 \leq p, q, r \leq m)$ *such that* $|T^{i_p}| \geq \ldots \geq |T^{i_q}| \geq |T^{i_r}|$. *When* $OR(X) \leq maxOR$, *all the non-empty subsets of* X *containing* i_r *and with size* ≥ 2 *will also have an overlap ratio less than or equal to* $maxOR$.

For extracting CPs, the sorted closure property of overlap ratio has been explored in conjunction with a level-wise pruning approach in [24]. This approach is referred to as Coverage Pattern Mining (*CPM*) algorithm. Moreover, Coverage Pattern Projected Growth (CPPG) approach was proposed in [25] based on the pattern growth paradigm.

About *CPM* Approach: The *CPM* algorithm uses level-wise pattern generation method, where k-size patterns are generated from $(k-1)$-size patterns. Let C_k, NOP_k and CP_k denote k-size candidate patterns, k-size non-overlap patterns and k-size coverage patterns respectively. First, the relative frequency value of each item is computed by scanning TDB. For each item i_k, when $RF(i_k) \geq minRF$, i_k is added to NOP_1 and when $RF(i_k) \geq minCS$, i_k is added to CP_1. Starting with k=2 and at any k^{th} level, k-size candidate patterns C_k are generated by performing self-join operation on NOP_{k-1} i.e., $C_k = NOP_{k-1} \bowtie NOP_{k-1}$. For each pattern X in C_k, X is added to NOP_k if $OR(X) \leq maxOR$. Further, X is added to CP_k when $CS(X) \geq minCS$ and $OR(X) \leq maxOR$. The above step is repeated until no candidate patterns are generated i.e., $|C_k| = 0$.

4 Proposed Approach

In this section, we explain the basic idea and present the proposed approach.

4.1 Basic Idea

In the centralized *CPM* algorithm, candidate patterns are generated at each level. These candidate patterns are validated by accessing the entire TDB. To improve the performance, we incorporate the following ideas in the proposed Distributed *CPM* (*DCPM*) approach.

First, we efficiently distribute the TDB by computing Inverted Transactional Database (ITD), which is a vertical form of the database, where each item is associated with a set of transaction identifiers to which it belongs. The ITD constitutes the database of all items, which satisfies the $minRF$ threshold. As a result, a reduced ITD is replicated to each node. Second, at each iteration, we distribute the $NOPs$ in a balanced manner. While distributing, we group $NOPs$, which differ only in the last item and send the group to each node. With this, computation of coverage support of given candidate pattern can exploit the reuse at the local node. For example, if {a, b, c} and {a, b, d} are candidate patterns, coverage set of {a, b}, which is being calculated while computing the coverage of {a, b, c} can be re-used to calculate the coverage of {a, b, d}. Moreover, significant savings in the transmission cost is achieved due to sending of a group of candidate patterns in a clustered form instead of individual candidate patterns.

The overview of the proposed *DCPM* approach is as follows. We model the given transactional dataset into ITD and broadcast it to all Worker nodes. Each item in the ITD that satisfy the $minRF$ constraint forms the first level $NOPs$, and items that meet the $minCS$ constraint form the coverage patterns. For the second level iteration, we distribute the first level $NOPs$ and compute the second level $NOPs$ by joining the first level $NOPs$ with corresponding lesser frequent items in the ITD. From the next level onwards, the NOP clusters from the previous level are distributed, and next-level $NOPs$ and CPs are computed by utilizing the ITD.

4.2 The DCPM Approach

We first present the terminologies used and introduce the cluster data-structure to explain the proposed approach.

Definition 2 (*Inverted Transactional Database (ITD)*). *Given a TDB over a set I of items, ITD is a list of the form* $< x,\ L(x) >$, *where* $x \in I$ *and* $L(x)$ *is a set of TIDs in which the item* x *belongs.*

Definition 3 (*Penultimate Prefix (PP)*). *Consider a pattern* $X = \{i_p, i_q, \ldots, i_r\}$, *where* $|T^{i_p}| \geq |T^{i_q}| \geq \ldots \geq |T^{i_r}|$. *The penultimate prefix of a pattern* X, *which we refer to as* $PP(X)$, *is equal to the set of items excluding the last item* i_r, *i.e.,* $PP(X) = X - \{i_r\}$.

Definition 4 (*Penultimate Prefix Cluster (PPC)*). *Consider a set S of patterns of equal size and each pattern* $X \in S$ *is of the form* $X = \{i_p, i_q, \ldots, i_r\}$, *where* $|T^{i_p}| \geq |T^{i_q}| \geq \ldots \geq |T^{i_r}|$. *Consider that all patterns in S are having a common* PP. *Then, the Penultimate Prefix Cluster of S, which we refer to as PPC(X) is equal to* $< CPP,\ LLI >$, *where set* CPP *is the common penultimate prefix, which contains the set of items in* PP *and set* LLI *is the list of last items, which contains the last item of each pattern in* S.

Given a transactional dataset TDB, number of machines N, $minRF$, $minCS$ and $maxOR$, the proposed *DCPM* approach extracts all coverage patterns from TDB in distributed environment. We adapt a Master-Slave architecture for the distributed environment. We designate one of the nodes as the Master node and the remaining nodes as Worker nodes. The Master node receives the input, distributes it among the Workers and coordinates the computation. The computation is carried out in an iterative manner starting from first level ($k = 1$). The Master terminates the computation when there are no more $NOPs$ or candidate patterns at the given level.

We explain the procedures followed by both Master and Worker for the first iteration, second iteration and other iterations. The steps of *DCPM* at level $k = 1$, 2 and beyond 2 for Master and Worker are provided in Procedure 1 and Procedure 2 respectively.

Procedure 1: Master()

Variables: N: Number of Worker nodes, NOP_k: Set of k-size non-overlap patterns, CP_k: Set of k-size coverage patterns, PPC: Penultimate Prefix Cluster, ITD: Inverse Transactional Database, CPP: Common Penultimate Prefix, LLI: List of Last Items.

k=1 (First Level)

1: Partition the dataset into N equal parts
2: Distribute the N partitions among N Workers
3: Collect Partial ITD from all the Workers
4: Merge all the partial ITDs and sort it in decreasing order based on the item RF.
5: Prune items in the ITD based on $minRF$ constraint
6: Compute NOP_1, CP_1 from ITD
7: Broadcast ITD to all nodes

k=2 (Second Level)

1: Distribute NOP_1 among N Workers
2: Collect partial NOP_2 from each Worker ▷ Items in NOP_2 are modeled in the form of PPC
3: Merge all partial NOP_2 to form the complete NOP_2
4: Compute CP_2 by filtering patterns in NOP_2 with $minCS$ threshold.

k>2 (Beyond Second Level)

1: Distribute the $PPCs$ in NOP_{k-1} among N Workers
2: Recieve partial NOP_k from all the Workers ▷ Items in NOP_k are modeled in the form of set of PPC
3: Merge all the partial NOP_k to form the complete NOP_k
4: Compute CP_k by filtering the patterns in NOP_k that satisfy the $minCS$ threshold to obtain CP_k

Termination condition: The Master terminates the procedure at level k when either $NOP_k = \{\phi\}$ or the total number of $(k+1)$ size candidate patterns $= 0$.

(i) First level (k = 1): This level involves computation of inverse transactional database ITD, NOP_1 and CP_1 (1-sized $NOPs$ and CPs). (At $k = 1$, the algorithm for Master is given in Procedure 1 and the algorithm for Worker is given in Procedure 2).

Master: Master distributes the TDB among N Worker nodes equally, i.e., each Worker receives a total number of $|TDB|/N$ transactions to compute ITD.

Worker: Worker receives a TDB partition and computes the partial ITD of the TDB partition received and returns it to the Master.

Master: The Master receives all the partial ITDs from the Workers and merges them to compute the final ITD. Next, NOP_1 and CP_1 are computed by comparing the RF of the items in the ITD with $minRF$ and $minCS$ respectively. The items in the ITD, which have lesser frequency values than $minRF$ are pruned. The resultant ITD is then sorted in decreasing order of RF and broadcasted to all the Workers for further levels.

Worker: It receives the final ITD and saves it.

(ii) Second level (k = 2): In this level, we compute 2-size $NOPs$ and CPs from NOP_1 received from the first level. (At $k = 2$, the algorithm for Master is given in Procedure 1 and the algorithm for Worker is given in Procedure 2.).

Master: Master distributes NOP_1 among Workers such that each Worker receives only one item in NOP_1. The Master continues to assign the remaining items in the NOP_1 to Workers upon completion of the assigned task. The assignment process continues till all of the items in NOP_1 are exhausted.

Procedure 2: Worker()

Variables: N: Number of Worker nodes, NOP_k: Set of k-size non-overlap patterns, CP_k: Set of k-size coverage patterns, PPC: Penultimate Prefix Cluster, ITD: Inverse Transactional Database, CPP: Common Penultimate Prefix, LLI: List of Last Items.

k=1 (First Level)

1: Receive partial TDB from the Master
2: Compute partial ITD from the partial TDB
3: Send the partial ITD to the Master
4: Receive final merged ITD from Master
5: Store the ITD for further levels

k=2 (Second Level)

1: Receive partial NOP_1 from the Master
2: **for all** items i in partial NOP_1 **do**
3: Join item i with items in partial NOP_1 which have lesser frequencies to form new candidate patterns
4: Model the candidate patterns in the form of PPC: ($\{CPP, LLI\}$) ▷ Refer Level 2 in Illustrative Example
5: Compute CS and OR using ITD for each candidate pattern
6: Prunes all candidate patterns that do not satisfy $maxOR$ threshold and forms partial NOP_2
7: Send partial NOP_2 to Master

k>2 (Beyond Second Level)

1: Receive a PPC: $\{CPP, LLI\}$ in NOP_{k-1} from the Master
2: Compute the Coverage Set (CSet) of CPP.
3: **for all** items i in LLI **do**
4: Join CPP and i and calculate the new coverage set of $\{CPP, i\}$, $CSet$
5: **for all** item j in LLI where $RF(i) \geq RF(j)$ **do**
6: Compute CS and OR of the set $\{CPP, i, j\}$ using CSet
7: If the combination $\{CPP, i, j\}$ satisfies the $maxOR$ constraint, then add j into LLI of the PPC with new CPP :$\{CPP, i\}$.
8: Update and add all the newly formed $PPCs$ into partial NOP_k
9: Send partial NOP_k to Master

Worker: The Worker receives one item from NOP_1 say x. The Worker joins x with item in ITD say y such that $RF(x) \geq RF(y)$ and forms a 2-size candidate pattern $\{x, y\}$. The generated candidate pattern $\{x, y\}$ is said to be a 2-size NOP when it satisfies the $maxOR$ constraint. Note that, all the lesser frequent items w.r.t. item x will be after the position of x in ITD as the ITD is sorted in decreasing order of RF. The Worker utilizes the ITD to compute the OR and CS for each candidate patterns. The computed $NOPs$ forms a set of $PPCs$ (Refer to the definition of PPC from Definition 4). The computed CS value of a pattern is attached with each NOP. The final set of $NOPs$ along with their respective CS value is sent to the Master.

Master: It receives the partial NOP_2 along with their respective CS value and merges them to form NOP_2. The CP_2 (2-sized CPs) are computed by the Master by filtering out NOP_2 with $minCS$ constraint.

(iii) k^{th} level (k > 2): This level computes NOP_k and CP_k from NOP_{k-1}. (At k>2 (Beyond Second Level), the algorithm for Master is given in Procedure 1 and the algorithm for Worker is given in Procedure 2.).

Master: The Master distributes the $PPCs$ in NOP_{k-1} among the Workers such that each Worker receives one PPC. The Master continues to assign the remaining clusters to Workers upon completion of the assigned task. The assignment process continues till NOP_{k-1} is exhausted.

Worker: The Worker receives a PPC from the Master. The Worker generates candidate patterns by joining item x with item y such that $RF(x) \geq RF(y)$ and forms a candidate pattern $\{A, x, y\}$. Here, $x, y \in LLI$ and A is the CPP of the PPC. (Refer the definitions of LLI and CPP from the Definition 4). Note that, all the items after item x in LLI have lesser frequencies than x because LLI is already sorted in decreasing order of relative frequency values. Here, LLI in any PPC is already sorted in decreasing order because the candidate patterns generation is done in an ordered manner. Further, the Worker computes OR and CS of the pattern $\{A, x, y\}$. The set of candidate patterns that satisfy the $maxOR$ constraint will form the partial NOP_k and are sent to Master along with their individual CS values. The Worker computes the OR and CS in an efficient manner by avoiding the redundant set operations through reuse (Refer to lines 4, 5, 6 under Worker in Procedure 2).

Master: The Master receives all the partial NOP_k along with CS values from Workers and merges all the partial NOP_k to compute NOP_k. Further, the Master extracts CP_k from the NOP_k subject to $minCS$ constraint. Master terminates the process at a level, when there is no possibility of generating candidate patterns for the next level.

4.3 An Illustrative Example

Table 1. Transactional database

TID	1	2	3	4	5	6
Items	a, d	a, b, c	a, d	b, c	c	a, b, e

We demonstrate the working of *DCPM* through an example. The sample *TDB* is shown in Table 1. Consider one Master node and 3 Worker nodes $\{W_1, W_2, W_3\}$. Let the values of *minRF*, *minCS* and *maxOR* be 0.2, 0.8 and 0.8 respectively.

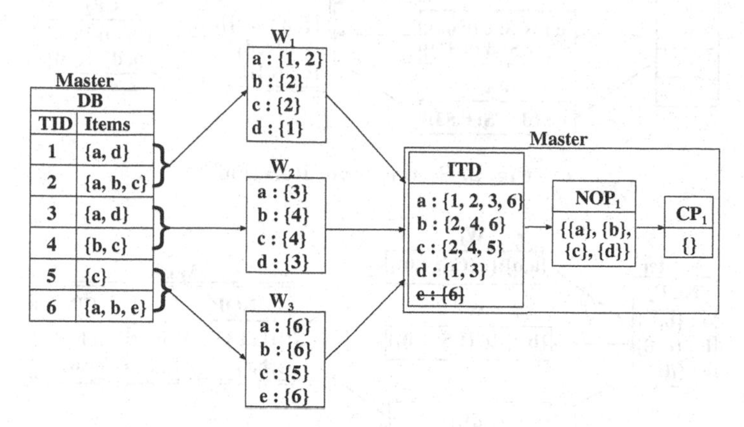

Fig. 1. First level iteration

Level 1. Figure 1 depicts the first level of *DCPM*. The Master distributes the transactions with TIDs {1, 2} to W_1, {3, 4} to W_2 and {5, 6} to W_3. Each Worker computes partial ITD for the transactions it has received and sends the partial ITD to the Master as shown in the Fig. 1. The Master merges partial ITDs that are received from the three Workers and filters out item e as $RF(e) = |ITD(e)|/|TDB| = 0.16$, which does not satisfy the *minRF* constraint. The items in the resultant ITD are sorted in a decreasing order of their RF values. The resultant ITD is $\{a{:}\{1, 2, 3, 6\}, b{:}\{2, 4, 6\}, c{:}\{2, 4, 5\}, d{:}\{1, 3\}\}$ and it is sent to all Worker nodes. All the items in the filtered ITD form NOP_1 i.e., $NOP_1 = \{\{a\}, \{b\}, \{c\}, \{d\}\}$. For 1-size patterns, the CS of any item, say a is the same as RF i.e., $CS(a) = RF(a) = |ITD(a)|/|TDB| = 0.66$. Since no item in NOP_1 satisfy the *minCS* constraint, $CP_1 = \{\phi\}$.

Level 2. Figure 2 depicts the second level iteration of *DCPM*. The Master distributes the items $\{a\}$ to W_1, $\{b\}$ to W_2 and $\{c\}$ to W_3. Each Worker computes NOP_2 by pruning the candidate patterns that does not satisfy the *maxOR* constraint. Consider W_1, where item a can be joined with every other item. Thus, the set of possible combinations are $\{\{a, b\}, \{a, c\}, \{a, d\}\}$. Among these patterns, the pattern $\{a{,}d\}$ is pruned because $OR(a, d) = ITD(a) \cap ITD(d)|/|ITD(d)| = 1.0$, which is greater than *maxOR* (0.8). Therefore, *NOPs* in form of *PPC* set at Worker W_1 will be $\{\{a{:}\{b, c\}\}$. The Worker calculates CS of each generated new pattern and attaches it as shown in Fig. 2. The coverage support of $\{a, b\}$, $CS(a, b) = ITD(a) \cup ITD(b)|/|TDB| = 0.83$. The Master collects all the second level partial *NOPs* from Workers and forms $NOP_2 = (\{a : \{b, c\}, b : \{c, d\}, c : \{d\}\})$. Note that, NOP_2 here is already in PPC form and is equivalent to $\{\{a, b\}, \{a, c\}, \{b, c\}, \{b, d\}, \{c, d\}\}$. Further, we obtain

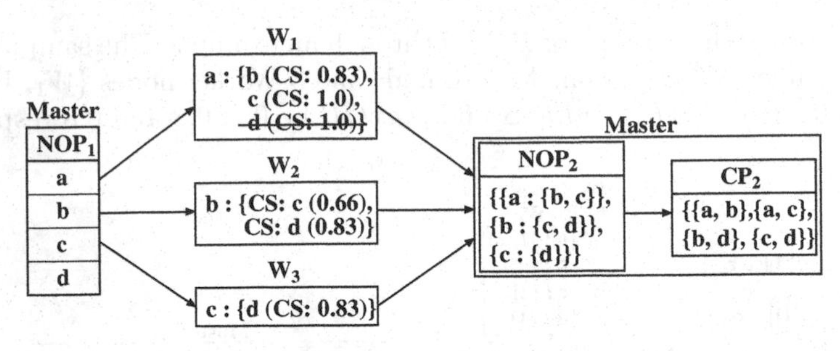

Fig. 2. Second level iteration

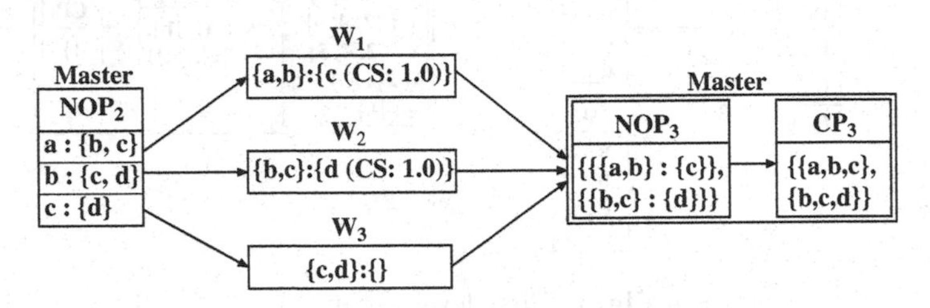

Fig. 3. Third level iteration

CP_2 = $\{\{a,b\},\{a,c\},\{b,d\},\{c,d\}\}$ by comparing the CS of each pattern in NOP_2 with $minCS$ constraint.

Level 3. Figure 3 depicts the third level iteration of *DCPM*. The Master distributes the $PPCs$ $\{a{:}\{b,\ c\}\}$, $\{b{:}\{c,\ d\}\}$ and $\{c{:}\{d\}\}$ to W_1, W_2 and W_3 respectively. Consider W_1 with the PPC $\{a{:}\{b,\ c\}\}$. Here, b is joined with c to form the candidate pattern $\{a,b,c\}$. The W_1 computes OR and CS for the generated pattern $\{a,b,c\}$ using $OR(a,b,c) = (ITD(a) \cup ITD(b)) \cap ITD(c)|/|ITD(c)|$ $= 0.66$ and $CS(a,b,c) = (ITD(a) \cup ITD(b)) \cup ITD(c)|/|TDB| = 1.0$. Since, OR of $\{a,b,c\}$ is 0.66, which is less than 0.8, the pattern $\{a,b,c\}$ forms NOP and is sent to the Master. The Master collects all the 3rd level partial $NOPs$ from Workers and forms $NOP_3 = \{\{a,b\} : \{c\},\{b,c\} : \{d\}\}$. Similarly, CP_3 $= \{\{a,b\} : \{c\},\{b,c\} : \{d\}\}$ is extracted by comparing the CS of each pattern in NOP_3 with $minCS$ constraint. Further, no new 4th level candidate patterns can be generated because each cluster has only one entry in it's LLI, hence the *DCPM* terminates.

5 Performance Evaluation

We conducted our experiments in the ADA cluster [1] (at IIIT Hyderabad), which consists of 42 Boston SYS-7048GR-TR nodes equipped with dual Intel Xeon E5-2640 v4 processors, providing 40 virtual cores per node. The aggregate theoretical peak performance of ADA is 47.62 TFLOPS. We have conducted our experiments on a cluster of 24 virtual nodes. Each virtual node was allocated with 2 GB of memory.

We have used three datasets, namely BMS-POS dataset [10], Mushroom dataset [8] and T10I4D100K dataset [2]. The BMS-POS dataset is a click-stream dataset of an e-commerce company. This dataset consists of 515,596 transactions and 1,656 distinct items. The Mushroom dataset is a dense dataset having 8,124 transactions and 119 distinct items. Moreover, the T10I4D100K is a synthetic dataset generated by a dataset generator. This synthetic dataset has 100,000 transactions and 870 distinct items.

We conduct the experiments by varying the parameters relative frequency threshold ($minRF$), coverage support threshold ($minCS$), overlap ratio threshold ($maxOR$) and the number of machines (NM). Table 2 summarizes the parameters of our performance study. The performance metrics of our study include total execution time (ET) in seconds for extracting coverage patterns from the transactional dataset and Data Shuffled (DS), which represents the total amount of data shuffled between the machines during the execution. ET is the total time elapsed between program initiation at the presentation of the dataset as input and termination at the delivery of the final coverage patterns subject to the $maxOR$, $minRF$ and $minCS$ constraints. DS is then computed by calculating the total amount of data transmitted and received by the Master node at each level. Given N Worker nodes and K MB of total data to be broadcast by the Master, the total data shuffled for a broadcast to all Worker nodes is $(K \times N)$ MB.

Table 2. Parameters used in our experiments

Dataset	Parameter	Default value	Variations
Synthetic	$minRF$	0.04	[0.037–0.05], step-size: 0.01
	$minCS$	0.3	[0.5–1], step-size: 0.1
	$maxOR$	0.6	[0.05–0.45], step-size: 0.01
	No. of Machines (NM)	16	[4–24], step-size: 4
BMS-POS	$minRF$	0.04	[0.03–0.1], step-size: 0.01
	$minCS$	0.5	[0.5–1], step-size: 0.1
	$maxOR$	0.4	[0.05–0.45], step-size: 0.01
	No. of Machines (NM)	16	[4–24], step-size: 4
Mushroom	$minRF$	0.04	[0.04–0.2], step-size: 0.02
	$minCS$	0.4	[0.5–1], step-size: 0.1
	$maxOR$	0.35	[0.05–0.45], step-size: 0.01
	No. of Machines (NM)	16	[4–24], step-size: 4

As a reference, for the purpose of meaningful comparison, we have implemented the following approaches:

- **Coverage Pattern Mining algorithm (*CPM*):** This is the *CPM* algorithm, implemented on a single node with 16 GB main memory in the ADA cluster.

- **Distributed *CPM* with MapReduce (*DCPM-MR*)**: We implemented the MapReduce-based *CPM* proposed in [19], in the following manner. First, we distribute the TDB among the Worker nodes and compute relative frequency values for all items in one phase of MapReduce. Further, we calculate one-size non-overlap patterns NOP_1 and broadcast the frequency values of all items to the Worker nodes. Second, we compute two-size non-overlap patterns NOP_2 and coverage patterns CP_2 using one phase of MapReduce. Third, at any k^{th} level, non-overlap patterns NOP_k and coverage patterns, CP_k are computed using two phases of MapReduce. We implemented the approach using Apache Spark Framework.
- **Distributed *CPM* with ITD (*DCPM-ITD*)**: We implemented the MapReduce based *CPM* proposed in [19] by introducing the notion of ITD and optimizing the second MapReduce phase. In this method, we compute the ITD of the given dataset at the start of execution and broadcast it to all Worker nodes. In *DCPM-ITD*, the first MapReduce phase for candidate pattern generation remains the same as the *DCPM-MR*. However, in the next step for computation of CS and OR of the generated candidate patterns, the generate candidate patterns are distributed equally among all the Worker nodes. Each Worker computes the CS and OR for all the candidate patterns it receives and sends them back to the Master. The Master node then computes the $NOPs$ and CPs with respect to the constraints. Note that the data is shuffled without employing any clustered approach. We implemented the approach using Apache Spark Framework.
- ***DCPM***: We implemented the *DCPM* approach as discussed in Sect. 4.2. *DCPM* is implemented using Apache Spark architecture [29] and utilizes the default task scheduler of the framework.

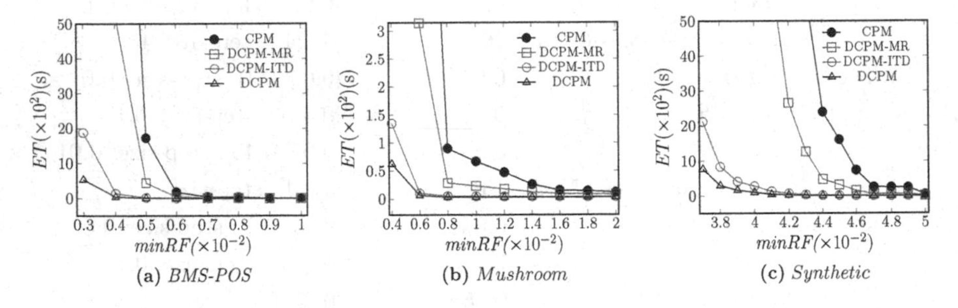

Fig. 4. Effect of varying *minRF*

5.1 Effect of Variations in MinRF

Figure 4 depicts the results of *minRF* versus *ET* for BMS-POS, Mushroom and Synthetic datasets. The results show that execution time decreases with the increase in *minRF* for all approaches. This is due to the decrease in the number of size-one frequent itemsets (items satisfying *minRF*).

Compared to *CPM*, *DCPM-MR* and *DCPM-ITD*, the proposed *DCPM* algorithm takes less time to extract coverage patterns from the datasets. In comparing *DCPM* with *DCPM-ITD*, the execution time of *DCPM* is reduced because *DCPM* does not have the MapReduce step for explicit generation of all the possible candidate patterns. In comparing *DCPM* to *DCPM-MR*, the reduction in execution time is due to inverse transactional data modelling and clustered distribution of non-overlap patterns. Further, in comparing the *DCPM-ITD* and *DCPM-MR*, the reduction in execution time is due to inverse transactional data modelling. *CPM* significantly underperforms w.r.t. three approaches because it is a serial approach and executed on a single machine. Observe that *DCPM* is 37 times faster than *DCPM-MR* and 3.3 times faster for the BMS-POS dataset than *DCPM-ITD* when *minRF* is 0.04.

The performance results in Fig. 4b and Fig. 4c exhibit similar trends for the Mushroom and Synthetic datasets respectively.

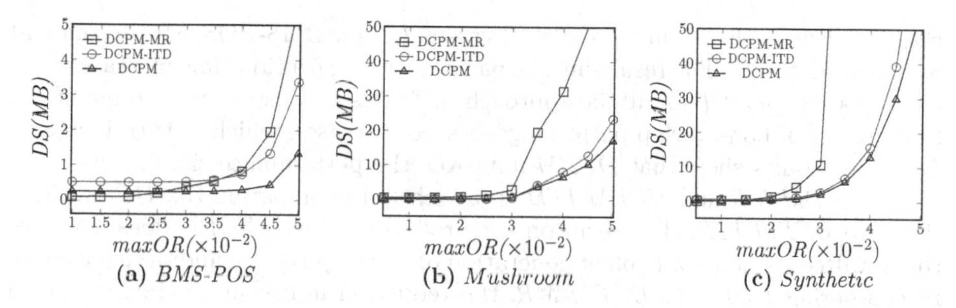

Fig. 5. Effect of varying *maxOR* on data shuffled

5.2 Effect of Varying *maxOR on Data Shuffled*

In this experiment, we show the results about the extent of data shuffled between the nodes. Figure 5 depicts the variation in data shuffled between nodes of varying *maxOR* threshold for BMS-POS, Mushroom and Synthetic datasets. The results in Fig. 5a show that data shuffled increases with an increase in *maxOR* for all the approaches. This is because when *maxOR* increases, the number of non-overlap patterns generated increases. Hence, the amount of data shuffled between the Master and Worker nodes increases.

Overall, the results show that data shuffled under *DCPM* is the least. More data is shuffled in *DCPM-ITD* over *DCPM* because of the clustered representation of candidate patterns in *DCPM*. The amount of data shuffled is significantly less in *DCPM* over *DCPM-MR* because all *NOPs* are distributed to all nodes under *DCPM-MR*, whereas in *DCPM*, only partial list of candidate patterns are distributed to each node in a compact form. As *maxOR* increases, the number of candidate patterns at each level increases, thus leading to a lot of data shuffling in *DCPM-MR*.

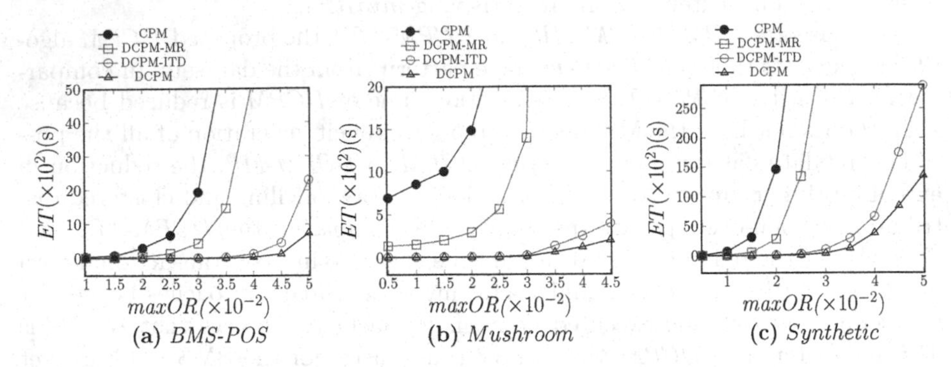

Fig. 6. Effect of varying *maxOR*

5.3 Effect of Variations in *maxOR*

Figure 6 depicts the results of *maxOR* versus *ET* for BMS-POS, Mushroom and Synthetic datasets. The results in Fig. 6a show that execution time increases with an increase in *maxOR* for all the approaches. This is because as *maxOR* increases, the number of non-overlap patterns generated increases, which in turn increases *ET*. The results show that *DCPM* improves the performance significantly over *CPM*, *DCPM-MR* and *DCPM-ITD* approaches. In comparing the *DCPM* algorithm to *DCPM-ITD*, the execution time reduces because *DCPM* does not have the MapReduce step for explicit generation of all the possible candidate patterns. In comparing *DCPM* to *DCPM-MR*, the reduction in execution time is due to inverse transactional data modelling and clustered distribution of non-overlap patterns. Further, the lesser execution time of the former is due to the efficient computation of *CS* and *OR* values using ITD.

We observed similar trends in the results for the Mushroom and Synthetic datasets as depicted in Fig. 6b and Fig. 6c respectively.

5.4 Effect of Variations in NM

Figure 7 depicts the result of *NM* (number of machines) versus *ET* for BMS-POS, Mushroom and Synthetic datasets. The results in Fig. 7a show that execution time decreases with an increase in *NM* and gradually reaches saturation value for all the approaches. For all the approaches, *ET* is the summation of both computation time and communication time. The *ET* of *DCPM* decreases with the increase in *NM* as the computation time decreases significantly due to parallel computation. It can be observed that *ET* is not significantly reduced for both

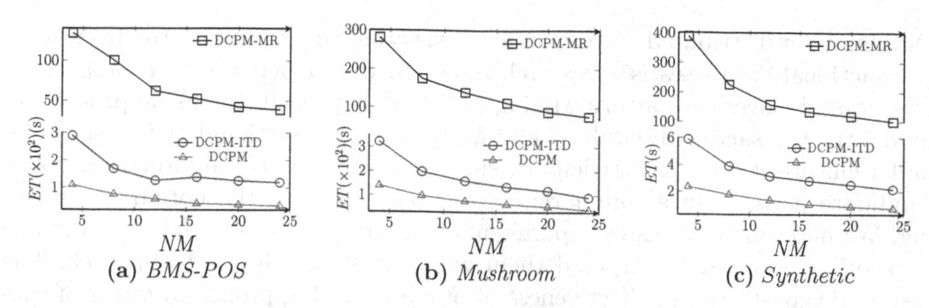

(a) *BMS-POS* (b) *Mushroom* (c) *Synthetic*

Fig. 7. Effect of varying *NM*

DCPM and *DCPM-ITD* as compared to *DCPM-MR* with the increase in NM. This is due to less computation in both *DCPM* and *DCPM-ITD* as compared to *DCPM-MR*. We can observe similar trends in the results for the Mushroom and Synthetic datasets as depicted in Fig. 7b and Fig. 7c. The values differ due to differences in dataset sizes and distribution of items in the transactions.

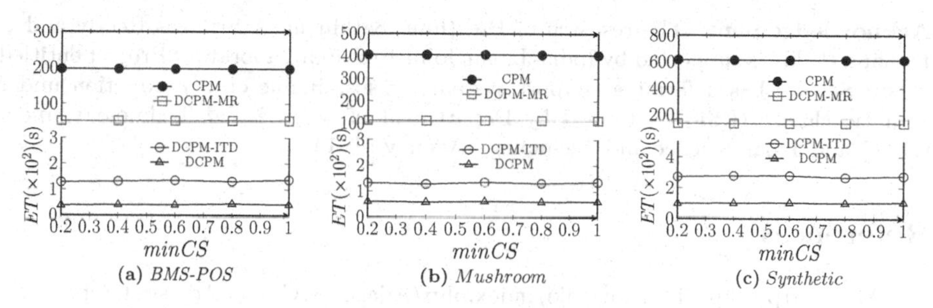

(a) *BMS-POS* (b) *Mushroom* (c) *Synthetic*

Fig. 8. Effect of varying *minCS*

5.5 Effect of Variations in *minCS*

Figure 8 depicts the results of *minCS* versus *ET* for BMS-POS, Mushroom and Synthetic datasets. The results in Fig. 8a show that execution time remains almost constant with an increase in *minCS* for all the approaches. This is because as *minCS* increases, the total number of frequent items and non-overlap patterns generated remain unaffected because of fixed *maxOR* and *minCS*. Thus, the execution time *ET* does not show much variation for any value of *minCS* in all the approaches. It can be observed that the *DCPM* algorithm outperforms other approaches for all datasets.

6 Conclusion

In pattern mining, developing fast and efficient parallel algorithms for handling large volumes of data becomes a challenging task. The mining of coverage patterns from transactional databases is one of the important data mining tasks.

Developing a distributed algorithm for extracting coverage patterns from large transactional databases is a research issue. In this paper, we have proposed a Distributed Coverage Pattern Mining (*DCPM*) approach. In this approach, we model the transactional database into an Inverse Transactional Database (ITD) and replicate it to each participant site, and transmit the information about candidate patterns in a summarized form by employing the notion of clustering. We performed extensive experiments using two real-world datasets and one synthetic dataset and compared them with the state-of-the-art approach. The results demonstrate the effectiveness of our proposed approach in terms of execution time and data shuffled across nodes.

As a part of future work, we plan to investigate the performance of *DCPM* by considering the issues of load balancing, skewed data-sets, memory utilization and so on. We also plan to investigate the issue of developing a distributed pattern-growth approach. Furthermore, we plan to extend *DCPM* and investigate scalable coverage pattern based algorithms, which have been proposed to improve the performance of banner advertising, search engine advertising and so on [4,9].

Acknowledgement. The research of Preetham Sathineni, A Srinivas Reddy and P Krishna Reddy is supported by India-Japan Joint Research Laboratory Project entitled "Data Science based farming support system for sustainable crop production under climatic change (DSFS)", funded by Department of Science and Technology, India (DST) and Japan Science and Technology Agency (JST).

References

1. ADA. http://hpc.iiit.ac.in/wiki/index.php/Ada_User_Guide. Accessed Sept 2021
2. Agrawal, R., Imieliński, T., Swami, A.: Mining association rules between sets of items in large databases. In: Proceedings ACM SIGMOD, vol. 22, pp. 207–216 (1993)
3. Budhiraja, A., Ralla, A., Reddy, P.K.: Coverage pattern based framework to improve search engine advertising. Int. J. Data Sci. Analytics **8**(2), 199–211 (2018). https://doi.org/10.1007/s41060-018-0165-3
4. Budhiraja, A., Reddy, P.K.: An approach to cover more advertisers in Adwords. In: Proceedings DSAA, pp. 1–10 (2015)
5. Budhiraja, A., Reddy, P.K.: An improved approach for long tail advertising in sponsored search. In: Candan, S., Chen, L., Pedersen, T.B., Chang, L., Hua, W. (eds.) DASFAA 2017. LNCS, vol. 10178, pp. 169–184. Springer, Cham (2017). https://doi.org/10.1007/978-3-319-55699-4_11
6. Chavan, K., Kulkarni, P., Ghodekar, P., Patil, S.: Frequent itemset mining for big data. In: International Conference on Green Computing and Internet of Things (ICGCIoT), pp. 1365–1368 (2015)
7. Chon, K.-W., Kim, M.-S.: BIGMiner: a fast and scalable distributed frequent pattern miner for big data. Clust. Comput. **21**(3), 1507–1520 (2018). https://doi.org/10.1007/s10586-018-1812-0
8. Dua, D., Graff, C.: UCI machine learning repository. http://archive.ics.uci.edu/ml. Accessed Sept. 2021

9. Gangumalla, L., Reddy, P.K., Mondal, A.: Multi-location visibility query processing using portion-based transactional modeling and pattern mining. Data Min. Knowl. Disc. **33**(5), 1393–1416 (2019). https://doi.org/10.1007/s10618-019-00641-3
10. Goethals, B., Zaki, M.: Frequent itemset mining implementations repository (2012). http://fimi.cs.helsinki.fi
11. Gui, F., et al.: A distributed frequent itemset mining algorithm based on spark. In: International Conference on Computer Supported Cooperative Work in Design, pp. 271–275 (2015)
12. Guo, J., Ren, Y.G.: Research on improved A Priori algorithm based on coding and MapReduce. In: Web Information System and Application Conference, pp. 294–299 (2013)
13. Kavya, V.N.S., Reddy, P.K.: Coverage patterns-based approach to allocate advertisement slots for display advertising. In: Bozzon, A., Cudre-Maroux, P., Pautasso, C. (eds.) ICWE 2016. LNCS, vol. 9671, pp. 152–169. Springer, Cham (2016). https://doi.org/10.1007/978-3-319-38791-8_9
14. Lin, J.C.W., Ahmed, U., Srivastava, G., Wu, J.M.T., Hong, T.P., Djenouri, Y.: Linguistic frequent pattern mining using a compressed structure. Appl. Intell. **51**(7), 4806–4823 (2021)
15. Liu, B., Hsu, W., Ma, Y.: Mining association rules with multiple minimum supports. In: Proceedings ACM SIGKDD, pp. 337–341 (1999)
16. Ogihara, Z.P., Zaki, M.J., Parthasarathy, S., Ogihara, M., Li, W.: New algorithms for fast discovery of association rules. In: Proceedings KDD, pp. 283–286 (1997)
17. Qiu, H., Gu, R., Yuan, C., Huang, Y.: YAFIM: a parallel frequent itemset mining algorithm with spark. In: IEEE International Parallel Distributed Processing Symposium Workshops, pp. 1664–1671 (2014)
18. Qiufeng, H., Qiang, L., Shiya, H., Yingcong, C.: Research on distributed parallel Eclat optimization algorithm. In: Proceedings International Conference on Artificial Intelligence and Big Data, pp. 149–154 (2020)
19. Ralla, A., Siddiqie, S., Reddy, P.K., Mondal, A.: Coverage pattern mining based on MapReduce. In: Proceedings ACM IKDD CoDS-COMAD, pp. 209–213 (2020)
20. Rathee, S., Kashyap, A.: Adaptive-miner: an efficient distributed association rule mining algorithm on spark. J. Big Data **5**(1), 1–17 (2018)
21. Rathee, S., Kaul, M., Kashyap, A.: R-Apriori: an efficient Apriori based algorithm on spark. In: Proceedings of the Workshop on Ph.D. Workshop in Information and Knowledge Management, pp. 27–34 (2015)
22. Sethi, K.K., Ramesh, D.: HFIM: a spark-based hybrid frequent itemset mining algorithm for big data processing. J. Supercomput. **73**(8), 3652–3668 (2017)
23. Singh, P., Singh, S., Mishra, P.K., Garg, R.: RDD-Eclat: approaches to parallelize Eclat algorithm on spark RDD framework. In: Proceedings International Conference on Computer Networks and Communication Technologies, pp. 755–768 (2020)
24. Srinivas, P.G., Reddy, P.K., Bhargav, S., Kiran, R.U., Kumar, D.S.: Discovering coverage patterns for banner advertisement placement. In: Tan, P.-N., Chawla, S., Ho, C.K., Bailey, J. (eds.) PAKDD 2012. LNCS (LNAI), vol. 7302, pp. 133–144. Springer, Heidelberg (2012). https://doi.org/10.1007/978-3-642-30220-6_12
25. Gowtham Srinivas, P., Krishna Reddy, P., Trinath, A.V., Bhargav, S., Uday Kiran, R.: Mining coverage patterns from transactional databases. J. Intell. Inf. Syst. **45**(3), 423–439 (2014). https://doi.org/10.1007/s10844-014-0318-3
26. Wu, J.M.T., Srivastava, G., Wei, M., Yun, U., Lin, J.C.W.: Fuzzy high-utility pattern mining in parallel and distributed Hadoop framework. Inf. Sci. **553**, 31–48 (2021)

27. Yahya, O., Hegazy, O., Ezat, E.: An efficient implementation of Apriori algorithm based on Hadoop-MapReduce model. Int. J. Rev. Comput. **12**, 59–67 (2012)
28. Yang, S., Xu, G., Wang, Z., Zhou, F.: The parallel improved Apriori algorithm research based on spark. In: International Conference on Frontier of Computer Science and Technology, pp. 354–359 (2015)
29. Zaharia, M., et al.: Apache Spark: a unified engine for big data processing. Commun. ACM **59**(11), 56–65 (2016)
30. Zhang, Z., Ji, G., Tang, M.: MREclat: an algorithm for parallel mining frequent itemsets. In: 2013 International Conference on Advanced Cloud and Big Data, pp. 177–180 (2013)

Outcomes of Speech to Speech Translation for Broadcast Speeches and Crowd Source Based Speech Data Collection Pilot Projects

Anil Kumar Vuppala(✉), Prakash Yalla, Ganesh S. Mirishkar, and Vishnu Vidyadhara Raju V

Speech Processing Laboratory, Language Technologies Research Centre, Kohli Center on Intelligent Systems, International Institute of Information Technology, Hyderabad, India
{anil.vuppala,prakash.yalla}@iiit.ac.in, {mirishkar.ganesh,vishnu.raju}@research.iiit.ac.in

Abstract. Speech-to-Speech Machine Translation (SSMT) applications and services use a three-step process. Speech recognition is the first step to obtain transcriptions. This is followed by text-to-text language translation and, finally, synthesis into text-speech. As data availability and computing power improved, these individual steps evolved. However, despite significant progress, there is always the error of the first stage in terms of speech recognition, accent, etc. Having traversed the speech recognition stage, the error becomes more prevalent and decreases very often. This chapter presents a complete pipeline for transferring speaker intent in SSMT involving humans in the loop. Initially, the SSMT pipeline has been discussed and analyzed for broadcast speeches and talks on a few sessions of Mann Ki Baat, where the source language is in Hindi, and the target language is in English and Telugu. To perform this task, industry-grade APIs from Google, Microsoft, CDAC, and IITM has been used for benchmarking. Later challenges faced while building the pipeline are discussed, and potential solutions have been introduced. Later this chapter introduces a framework developed to collect a crowd-sourced speech database for the speech recognition task.

Keywords: Speech recognition · Machine translation · Speech synthesis · Crowd-source database

Supported by Technology Development for Indian Languages (TDIL), Ministry of Electronics and Information Technology (MeitY), Government of India, for allowing us to work on the "Speech to Speech Translation & performance measurement platform for Broadcast Speeches and Talks" pilot project and "Crowd Sourced Large Speech Data Sets To Enable Indian Language Speech - Speech Solutions".

S. N. Srirama et al. (Eds.): BDA 2021, LNCS 13147, pp. 341–354, 2021.
https://doi.org/10.1007/978-3-030-93620-4_24

1 Introduction

Nowadays, information processing systems have become an integral part of human life. These systems generally take information in one form and process (transform) it into another form. Human-computer interaction (HCI) is predominantly used by considering speech as a source of input as speech signal is a unique and natural way of communication among human beings. Due to the advancement in technology, few researchers were curious about the mechanisms involved in the mechanical realization of human speech abilities, to the desire to automate the simple tasks that are inherently required for HCI. So the motive of the HCI is to listen to human speech and carry out their commands accordingly.

The magnitude of innovations being observed in the fast pace of the deep-learning era made real-time communication possible with a ray of light. By leveraging it, these days, portable devices are being enabled with voice or text-based services as it helps in connecting the global market without any hassle, such as removing the language barriers (wherein speech from one language is taken as an input and transformed into other languages). The dream to have an automatic speech-to-speech machine translation (SSMT) system comes into action if trust exists in deep learning, cognitive computing, and big data models. As of now, most of the SSMT systems [1–6] are compromising with transcription and translation quality. On this grounds, most of the research groups and industries have drawn their interest to mitigate the issue for the smooth functioning of the SSMT pipeline. The SSMT system could be used more broadly in live streaming, customer service management, a wherein person-to-person conversation is needed. These applications will bridge the gap between one language to another, and it would also help in a multilingual society like India.

In building the SSMT system, there involve multiple challenges in all the three blocks which are namely,

- **Automatic Speech Recognition**
  - It is a process of transforming a speech signal to its corresponding text.
- **Machine Translation**
  - It is process in which a text is translated from one natural language to another.
- **Speech Synthesis (text-to-speech)**
  - It converts a given text into speech signal.

From the above mention three blocks, speech synthesis has a decent amount of maturity in the commercial space across the languages. Machine translation (MT) and Speech recognition is still an unsolved problem due to various factors. Coming to ASR, it is mainly due to limited vocabulary size, speaking rate, environmental variations (external noises), different dialects etc.

In this work we have demonstrated the capability of technologies for performing speech to speech translation in Indian context in an industry grade technology stack and capture performance benchmarks to enable industry application grade technology development. Speech to speech translation is performed

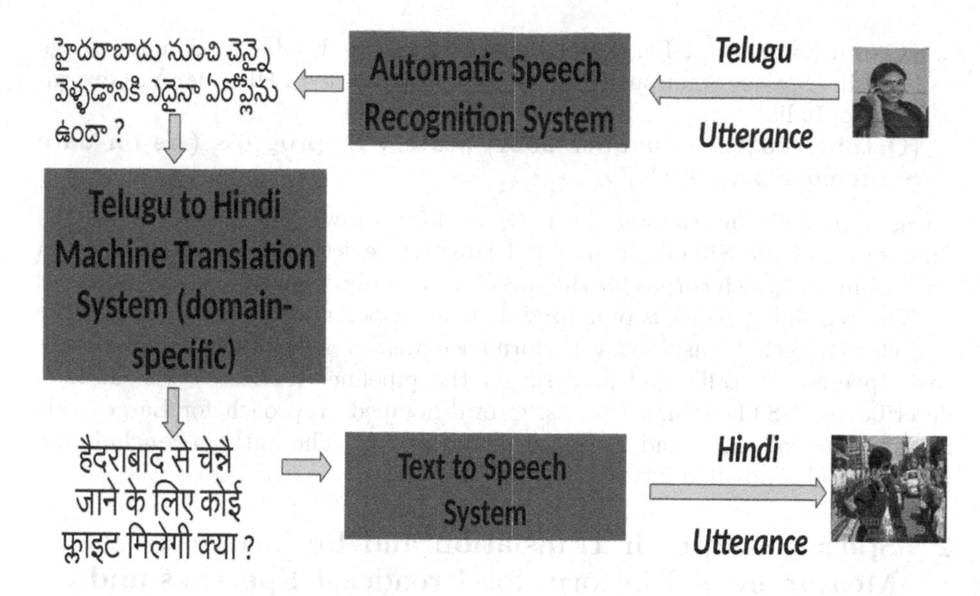

Fig. 1. High-level block schematic of speech to speech machine translation system

using speech recognition, machine translation and text to speech. The typical block diagram for SSMT system is shown in Fig. 1. It is a waterfall architecture, where ASR, MT, and SS systems are joined together to form an SSMT system [7]. The output obtained from an ASR system is given as input to an MT system, and the output obtained from the MT system is given as an input to a SS system. Given accurate ASR and accurate MT systems, this architecture may be sufficient to achieve the goal of SSMT systems. However, due to the limitations of the current state-of-art technology in ASR and MT areas, there are errors or ambiguities in the output of these components, which are propagated to its successive components.

It is pretty evident from Fig. 1 that the primitive block is an ASR system. An ASR system's task is to convert speech in the source language to its corresponding text. The performance of the SSMT system majorly depends on the ASR system; if the output of the ASR system is erroneous, then it is likely to propagate throughout the pipeline, which results in an incorrect speech-to-speech translation. Therefore, the speech recognition [8] component must be concerned less with transcription fidelity than semantic fidelity, while the MT component must try to capture the meaning or intent of the input sentence [9–15] without being guaranteed a syntactically legal sequence of words. In this chapter, we have broadly discussed about two of the projects which we have worked on, which are namely,

1. "Speech to Speech Translation & performance measurement platform for Broadcast Speeches and Talks funded by TDIL MeitY, Government of India. (**June 2019 to March 2020**) **Status: Completed**"

2. "Crowd Sourcing of Large Speech Data Sets To Enable Indian Language Speech - Speech Solutions (Pilot Project) funded by TDIL MeitY, Government of India.
(October 2020 to October 2021) Status: In progress (As on date September 29, 2021) "

So as a part of this chapter, initially, we have drawn our experience in the functioning of the SSMT pipeline and later strategies involved in collecting a crowdsourced speech corpus for the speech recognition task have been discussed.

The remaining paper is organized as follows: Section 2 briefly describes the Speech to Speech Translation & Performance Measurement Platform for Broadcast Speeches & Talks and analysis on the pipeline. In Sect. 3, the authors describe the CSTD-Telugu Corpus: Crowd-Sourced Approach for Large-Scale Speech data collection and its experimental results. The authors conclude the paper and give possible future directions in Sect. 4.

2 Speech to Speech Translation and Performance Measurement Platform for Broadcast Speeches and Talks

In this work, we demonstrate the capability of technologies for performing speech to speech translation in Indian context in an industry grade technology stack and capture performance benchmarks to enable industry application grade technology development. Speech to speech translation is performed using speech recognition, machine translation and text to speech. In multilingual society like India, speech to speech translation has plenty of potential applications like educational videos translation, overcoming the language barrier for communication and learning etc. In Indian context it is challenging because of lack of proper speech recognition, machine translation and text to speech systems for Indian languages. In this project we will convert Prime Minister Maan Ki Baat Hindi speech to English and Telugu automatically. The same will be extended to other speech to speech application needs expressed by broadcast industry e.g. sports commentary, multi-lingual interviews etc.

Observation and Analysis. Speech-to-speech translation systems have been developed for converting one language speech into another language speech, with the goal of helping people who speak different languages to communicate with each other. As mentioned earlier, such systems have usually been broken into three separate components they are as follows:

1. Automatic Speech Recognition (ASR)
   - to transcribe the source speech to text The performance evaluation metric for an ASR system is calculated in terms of Word Error Rate (WER).

$$WER(\%) = \frac{S + I + D}{N} \times 100$$

where Insertion - Ins (I), Deletion - Del (D), Substitution - Sub (S)

2. Machine Translation (MT)
   - to translate the transcribed text into the target language. In this project we have considered human evaluation by judges on a scale of 4 point scale.[1]
     1. **Unacceptable:** Certainly incomprehensible and/or little or no information is accurately transferred.
     2. **Possibly Acceptable:** Perhaps comprehensible (given enough context and/or time to work); Some information has certainly been transferred.
     3. **Acceptable:** Not perfect (stylistically or grammatically odd), but certainly perceptible and with perfect transfer of all important information.
     4. **Ideal:** Not an accurate translation, but grammatically correct and accurately transmitted with all information.
3. Speech synthesis (Text-to-Speech)
   - to generate speech in the target language from the translated text. In this project we have considered human evaluation by judges on a scale of 5 point scale. (See footnote 1)
     1. **Bad:** Totally unacceptable, unintelligible, annoying.
     2. **Poor:** Sounds very unnatural, weird, many issues like noise, shaky, muffling but intelligible.
     3. **Fair:** Robotic, some minor issues like noise, shaky, muffling, but overall is acceptable.
     4. **Good:** Natural and close to human beings.
     5. **Excellent:** Very natural and sounds like a human being.

Dividing the task into such a cascade of systems has been very successful, powering many commercial speech-to-speech translation products. Following are the set of observations which we have observed while integrating:

- **Steps followed in Human Evaluation**

  S1. The Audio is passed through the respective ASR system. As discussed prior the main job of an ASR is to convert the given audio into corresponding text. So once text is obtained we calculate the Word Error Rate (WER) from the ground truth (reference) transcripts.

  * Following are statistics of sessions which we have considered for ASR, MT and synthesis analysis for respective APIs (Google, MSR, CDAC, and IITM):

  Session June 2019

Audio duration (HH:MM:SS)	00:30:59
Number of sentences	213
Number of words	4516

[1] This metric has been provided by Microsoft - IDC Hyderabad.

Session July 2019

Audio duration (HH:MM:SS)	00:25:34
Number of sentences	390
Number of words	3588
Session August 2019	
Audio duration (HH:MM:SS)	00:31:34
Number of sentences	425
Number of words	4348

Consider the following examples:
Example1

Actual utterance / Reference

‘मन क बात’ हमेशा क तरह, मेर तरफ से भी और
आपक तरफ से भी एक प्रतीक्षा रहती है ।

ASR Output

मन क बात हमेशा क तरह, मेर तरफ से भी और
आपक तरफ से भी एक DEL रहती है ।

DEL - Deletion

So, in the above utterance the number of deletion is 1

So now we have to make a note of the number of deletions, number of substitutions and number of insertions from the ASR outputs.

S2. Later the text from the ASR output is fed to the translation system and the translation output is given to human's for evaluation. And the scores have been given by them. So the number of participants in this exercise are 15 and the average of all these 15 participants has been reported. The performance evaluation is done to the scale of 4.

S3. The translated text is taken and fed to the respective speech synthesis engines for the generation of text to speech. So once we had the synthesized output we performed subjective evaluation with 15 participants and asked to grade the outputs to the scale of 5 and mean of which is reported here.

The platform which is developed has been integrated using multiple API's and it is observed the stability of it mainly depends on their server (Parent node) (Tables 1 and 2).

Actual utterance / Reference

हमेशा की तरह, मेरी तरफ से भी और आपकी तरफ से भी एक प्रतीक्षा रहती है। इस बार भी मैंने देखा कि बहुत सारे पत्र, comments, phone मिले हैं –ढेर सारी कहानियां हैं, सुझाव हैं, प्रेरणा हैं

MSR ASR Output

हमेशा की तरह, मेरी तरफ से भी , और आपकी तरफ से भी, एक प्रतीक्षा रहती है | इस बार भी, मैंने देखा कि भाव सारे पत्र, कॉमेंट्स, फोन मिले **DEL** , ढेर सारी कहानियां हैं | सुझाव है | प्रेरणा है |

Google ASR Output

हमेशा की तरह मेरी तरफ से भी और आपकी तरफ से भी एक प्रतीक्षा रहती है इस बार भी मैंने देखा कि **DEL** सारे पत्र कॉमेंट्स फोन **DEL** मिले **DEL** ढेर सारी कहानियां हैं सुझाव है प्रेरणा है

CDAC ASR Output

हमेशा की तरह मेरी तरफ से भी और आपकी तरफ से भी एक प्रतीक्षा रहती है इस बार भी मन देखा क **DEL** बहुत सारे प **DEL** कम स पखं **SUB DEL** मले **DEL** ढेर सार **DEL** कहा नयाँह सझु|व है ◌**DEL**रणा है

Table 1. Summary of all the API's used and metrics used in calculating the performance evaluation is as follows:

S. No	Sessions	API's	Metrics			
			WER (%)	MT (/4)		TTS (/5)
				Hi-Te	Hi-En	
1.	June 2019	Google	9.4	3.3	3.1	3.1
		MSR	6.0	3.4	3.41	
		CDAC	9.2	–	–	–
		IITM	–	–	–	2.7
2.	July 2019	Google	9.54	3.2	3.1	3.1
		MSR	6.2	3.4	3.4	3.41
		CDAC	9.52	–	–	–
		IITM	–	–	–	2.7
3.	August 2019	Google	9.2	3.3	3.2	3.1
		MSR	6.1	3.41	3.4	3.41
		CDAC	9.52	–	–	–
		IITM	–	–	–	2.7

2.1 Challenges Faced While Building the SSMT Pipeline

The two verticals, namely, speech processing and text processing, need to be carefully handled while performing any task related to transcription or translation. Among both, text processing has wider exposure and is better understood.

Table 2. Summary of all the APIs used and latency involved in each component and the number which is reported below have been tested on 100 Mbps internet speed.

S. No	Sessions	API's	Latency (mm:ss)			Total
			ASR	MT	TTS	
1.	June 2019	Google	00:06	00:08	00:14	00:28
		MSR	00:08	00:08	00:20	00:36
		CDAC	00:06	–	–	–
		IITM	–	–	00:26	–
2.	July 2019	Google	00:06	00:08	00:14	00:28
		MSR	00:08	00:08	00:20	00:36
		CDAC	00:06	–	–	–
		IITM	–	–	00:26	–
3.	August 2019	Google	00:06	00:08	00:14	00:28
		MSR	00:08	00:08	00:20	00:36
		CDAC	00:06	–	–	–
		IITM	–	–	00:26	–

Pointing this out, text translation itself is an arduous task wherein it has to handle the nuances of the target language. In literature, few of the groups have tried to automatically translate spoken language into its target language text and reported that apart from loss in the context information, it also involved difficulty while handling semantics, domain context, disfluency (repairs, prolongations, false starts, etc.), dialog effect and few more uncertainty.

Issues: As discussed above, the difficulties involved in both speech recognition and text translation, people have made a few attempts to solve and bring naturalness into the pipeline. Few of them are,

Speaking Style (Read Speech vs Conversational). In practice, the articulation of speech, place a significant role in building speech-based products. In read speech mode, it is found that there will not be hesitations or disfluencies as most of it is prompted carefully. Whereas, in the case of conversational speech, such kind of behavior will not be observed. Given such information to the system, while building, it produces high accuracy. For example, News channels are considered as read speech and meeting scenarios comes under conversational speech.

Pacing (Consecutive vs. Simultaneous). Pauses are very common in the natural mode of communication, which help segregate the utterances spoken while producing the translations by the systems. This type of mechanism comes under consecutive mode. Therefore it eases the process of translation. In the SSMT pipeline, recognition and translation engines are expected to perform in synchronous mode while the speaker speaks. This type of mode is said to be a simultaneous mode.

Speed and Latency (Real-Time vs. Delayed Systems). Problems may persist in the SSMT pipeline depending on the speed requirements and the waiting time, which in general is called latency. In the synchronous mode, an optimal threshold value should be considered so that the speaker is in line with the output of the system. Consider the scenario where the lecture is being delivered or live streaming program, where the SSMT output is indeed expected to have low latency. But in standalone cases(post-hoc, viewing, or browsing), there is no need to worry about the speed and latency of the system.

Microphone Handling. In general, speakers tend to use microphones close to the mouth, which yields a clear speech signal like mobile phones. It is noticed that performance degradation is observed when speakers are far away from the microphone as it captures external or background noise.

Human Factors and Interfaces: Speech translation facilitates human communication in all ways, so a human interface is required. In a perfect world, we need to hear and comprehend that the discussion accomplices communicate in our language and don't have an interpretation program: the errand of the interface is to make the language hindrance as straightforward as could be expected. We need greatest speed and least obstruction from one viewpoint while keeping up with most extreme precision and effortlessness on the other. These are competitive goals. Better interface solutions help balance them; But no definitive solutions are expected in the near future, as even human commentators generally spend considerable time on clarity dialogues. As long as the exact accuracy is unclear, effective error recovery mechanisms are desirable. The first step is to enable users to identify errors in speech recognition and translation. Once the errors are found, mechanisms are needed for correction and then for adaptation and improvement. Literate users can detect errors generated by the speech recognition engine on the device screen. Text-to-speech playback of ASR results is used (but rarely used so far) for illiterate users or to initiate blind use. Some systems may allow users to type the wrong word or write it by hand to correct ASR errors. Facilities may be provided instead for voice-based correction (although these are also rarely used). The whole input may be repeated, but the same errors may be repeated, or new ones may be triggered. At last, multimodal goals can be upheld, for instance, in the realistic interface with manual determination of the mistake and voice correction. In any case, if a fragment or elocution can be revised, it very well may be shipped off machine interpretation (essentially in frameworks where ASR and MT parts are exact). Then, at that point, the distinguishing proof and adjustment of the interpretation results can be worked with. In ASR or MT, blunders are irritating, however repeating mistakes are extremely irritating. In a perfect world, frameworks ought to gain from their missteps with the goal that mistakes over the long haul and being used are limited. On the off chance that AI or some neural model is made accessible, clients should exploit any remedies provided. Then again, intelligent update components might be given.

Neural Speech to Speech: SSMT is to decipher discourse from one language into another. The neural model is useful for separating correspondence hindrances between individuals who don't share a typical language. In particular, it is feasible to prepare models to achieve the undertaking straightforwardly without depending on transitional message portrayal. It is rather than customary SSMT frameworks, which can be comprehensively grouped into three sections: Automatic Speech Recognition (ASR), Text-to-Text Machine Translation (MT), and Text-Speech (TTS) combination. Course frameworks have the likely issue of blunders between parts, e.g., distinguishing proof mistakes prompting more huge interpretation blunders. SSMT models keep away from this issue through preparing to address task-to-end. They additionally enjoy upper hands over course frameworks as far as diminished computational necessities and lower inductance delay, as just one disentangling step is needed rather than three. In addition, direct models can naturally retain translingual and non-linguistic information during translation. Finally, direct conditioning of the input speech makes it easier to learn to create clear pronunciation of words that do not require translation, such as names. End-to-end training can be given to the direct speech-to-speech translation format. Multi-task training, especially with speech-to-text tasks, facilitate training without pre-defined settings to influence high-level representations of source or target information in the form of transcripts. However, intermediate text representation is not used during inference. End-to-end architecture can be developed on a sequence network based on a vocoder that converts attention-based sequences (the ability to translate speech into speech) and target spectrograms into time-domain waveforms.

In this work, we have integrated API's from different vendors and developed a platform for evaluating the performance measure on broadcast speeches. As a part of it we have found that though Microsoft API's have a bit large latency when compared to the CDAC and Google it is producing the accurate outcomes. So among the lot Microsoft stands first next CDAC-Pune (for ASR) and later Google (Not adopted for MKB data). Later few challenges related to pipeline has been discussed. The demo of this project can be found here[2].

3 CSTD-Telugu Corpus: Crowd-Sourced Approach for Large-Scale Speech Data Collection

The availability of speech databases for Indian languages is minimal. Most of the Indian languages are spoken widely throughout the global still. There exist no annotated speech databases for building reliable speech technology products. To bridge this gap, people in the literature have adopted crowdsourcing for collecting such data collaboratively. In this project, we describe an experience of Telugu speech database collection for building automatic speech recognition tasks. It was done in collaboration with Ozontel Technologies private limited, Pacteraedge, and CIE IIIT-Hyderabad. In other sections, we explain the platform we have developed and strategies adopted for collecting the corpus.

[2] https://drive.google.com/file/d/1Xu0ELaHtgXRXulwf-FqV_6qZHv6BCghR/view.

3.1 Overview of the Pipeline

The framework has been developed so that any person can log into it by providing the basic information and start contributing his/her speech to the platform. In this process, people can either use their laptops or mobile phones as a medium for recording. The main advantage of crowdsourcing is that anybody can contribute from anywhere, sitting across the globe. So in this project, our collaborators have reached and pooled an audience for the data collection. The setup used for this crowd-sourced data collection is shown in Fig. 3. An upper bracket of 90 min is allotted for a speaker to provide his/her speech. Before starting the recording, each of them is provided with the guidelines to be followed like, and they should speak clearly, distinctly, naturally with few filler words. Once the

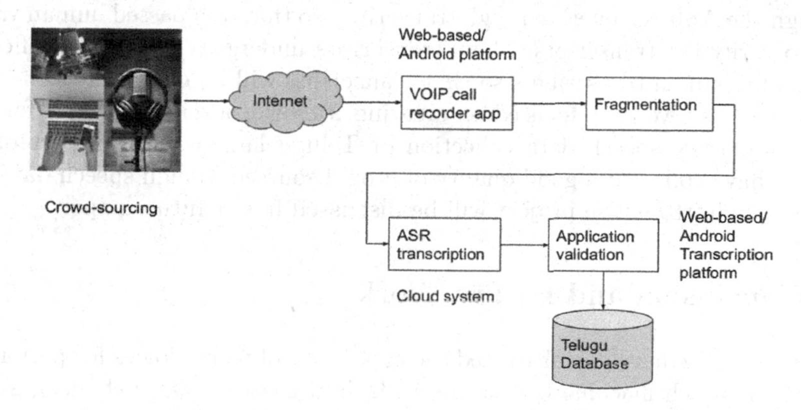

Fig. 2. Crowd-sourced set-up for Telugu speech data collection.

Filters Active - 1 | Collapse All | Show All | Clear All

Campaign Code		Status		Review	
SSAMP1	20	approved	1	False	1
SSAMP2	10	asr_done Jobs Pending (0/1)	27	None	28
		asr_done Jobs Pending (1/1)	1	True	1
		reviewed	1		

Show 10 entries | Search:

Name	Campaign Code	Status	Audio	Auto Transcript	User Transcript	Review	Actions
testing_001.wav	SSAMP1	reviewed	0:00 / 0:03	[illegible]	[illegible]	False	✔ ✘
testing_002.wav	SSAMP1	approved	0:00 / 0:01	[illegible]	[illegible]	True	
testing_003.wav	SSAMP1	asr_done Jobs Pending (0/1)	0:00 / 0:04	[illegible]		None	✘
testing_004.wav	SSAMP1	asr_done Jobs	0:00 / 0:02	[illegible]		None	✘

Reject Fragment

127.0.0.1:5500/rejectfragment?id=8ebb9bf2-f664-47ee-a12a-25caa7c5edaf

Fig. 3. Admin dashboard of the platform.

speech is captured through the VOIP interface, the platform make sure that the recorded speech is 16 Khz of sampling rate and 16 bit (Fig. 2).

The platform is built in such a way where users can select his/her topic of interest, which would be displayed on the user's screen. It also can handle multiple users simultaneously. Once the recording is done, the speech data is sent to the back-end server, and necessary formatting is done before sending it for further processing. Once the data is formatted, it is sent for fragmentation so that it could be used for building an ASR system building. The fragmentation algorithm is written in such way that it identifies non-speech region and chops the audio files with respective it. It also makes sure that the selected fragment is within the range of 3–15 seconds and fragments which doesn't comes under this category (specified duration) are discarded. The accepted fragments are passed through the ASR to generate rough transcripts so that it is passed human validators to verify the transcripts. These transcripts undergo a two level verification process by human transcribers so that transcripts will be error free.

In this task we have focused on creating a crowdsourcing platform for handling large-scale speech data collection of Telugu language. As a part of this task we have collected a good quality of crowd sourced Telugu speech database. Experience of the entire project will be discussed in out future paper.

4 Conclusion and Future Work

In this work, we have demonstrated the capability of technologies for performing speech to speech machine translation in Indian context. Later challenges faced while building the SSMT pipeline are broadly discussed. The platform which we have developed to collect a crowdsource speech corpus for speech recognition task is briefly explained.

Further progress awaits the maturity of vital components of any speech translation system - speech recognition, machine translation, speech synthesis, and practical infrastructure. In cases of demand, facilities are required to quickly switch between or interleave between automatic and human interpreters and tools to assist those interpreters. We also need feedback tools to assure customers that more human (and expensive) intervention is helpful. Serious R&D for speech-to-speech machine translation has to be continued worldwide, both with and without government sponsorship.

Acknowledgements. We would like to acknowledge the Technology Development for Indian Languages (TDIL), Ministry of Electronics and Information Technology (MeitY), Government of India, for allowing us to work on the pilot project for collecting large-scale Telugu corpus. We also thank Indian Institute of Technology-Madras Speech Lab, Microsoft IDC-Hyderabad, CDAC-Pune for providing there respective APIs for benchmarking the platform which we have developed. We would also like to thank our collaborators Ozonetel, Pacteraedge, and Swecha for Crowd Sourcing of Large Speech Data Sets To Enable Indian Language Speech - Speech Solutions (Pilot Project).

References

1. Anumanchipalli, G.K., Oliveira, L.C., Black, A.W.: Intent transfer in speech-to-speech machine translation. In: 2012 IEEE Spoken Language Technology Workshop (SLT), pp. 153–158. IEEE, December 2012
2. Zhang, R., Kikui, G., Yamamoto, H., Soong, F.K., Watanabe, T., Lo, W.K.: A unified approach in speech-to-speech translation: integrating features of speech recognition and machine translation. In COLING 2004: Proceedings of the 20th International Conference on Computational Linguistics, pp. 1168–1174 (2004)
3. Vilar, D., Xu, J., Luis Fernando, D.H., Ney, H.: Error analysis of statistical machine translation output. In: LREC, pp. 697–702, May 2006
4. Hashimoto, K., Yamagishi, J., Byrne, W., King, S., Tokuda, K.: Impacts of machine translation and speech synthesis on speech-to-speech translation. Speech Commun. **54**(7), 857–866 (2012)
5. Matusov, E., Kanthak, S., Ney, H.: On the integration of speech recognition and statistical machine translation. In: Ninth European Conference on Speech Communication and Technology (2005)
6. Frederking, R.E., Black, A.W., Brown, R.D., Moody, J., Steinbrecher, E.: Field testing the tongues speech-to-speech machine translation system. In: LREC, May 2002
7. Tomokiyo, L.M., Peterson, K., Black, A.W., Lenzo, K.A.: Intelligibility of machine translation output in speech synthesis. Presented at the Ninth International Conference on Spoken Language Processing (2006)
8. Salesky, E., Sperber, M., Black, A.W.: Exploring phoneme-level speech representations for end-to-end speech translation (2019). arXiv preprint arXiv:1906.01199
9. Carbonell, J.G., Lavie, A., Levin, L., Black, A.: Language technologies for humanitarian aid (2005)
10. Levin, L., et al.: The Janus-III translation system: speech-to-speech translation in multiple domains. Mach. Transl. **15**(1), 3–25 (2000)
11. Waibel, A., et al.: Speechalator: two-way speech-to-speech translation in your hand. In: Companion Volume of the Proceedings of HLT-NAACL 2003-Demonstrations, pp. 29–30 (2003)
12. Schultz, T., Alexander, D., Black, A.W., Peterson, K., Suebvisai, S., Waibel, A.: A Thai speech translation system for medical dialogs. In: Demonstration Papers at HLT-NAACL 2004, pp. 34–35 (2004)
13. Suebvisai, S., Charoenpornsawat, P., Black, A., Woszczyna, M., Schultz, T.: Thai automatic speech recognition. In: Proceedings (ICASSP'05) IEEE International Conference on Acoustics, Speech, and Signal Processing 2005, vol. 1, pp. I–857. IEEE, March 2005
14. Wilkinson, A., Zhao, T., Black, A.W.: Deriving phonetic transcriptions and discovering word segmentations for speech-to-speech translation in low-resource settings. In: INTERSPEECH, pp. 3086–3090 (2016)
15. Scharenborg, O., et al.: Building an ASR system for a low-resource language through the adaptation of a high-resource language ASR system: preliminary results. In: Proceedings of ICNLSSP, Casablanca, Morocco (2017)
16. Davis, S., Mermelstein, P.: Comparison of parametric representation for monosyllabic word recognition in continuously spoken sentences. IEEE Trans. Acoust. Speech Signal Process. **28**, 357–366 (1980)
17. Rabiner, L.: A tutorial on hidden Markov models and selected applications in speech recognition. Proc. IEEE **77**, 257–286 (1989)

18. Hastie, T., Tibshirani, R., Friedman, J.: The Elements of Statistical Learning - Data Mining, Inference, and Prediction. SSS, Springer, New York (2009). https://doi.org/10.1007/978-0-387-84858-7
19. Freitas, J., Calado, A., Braga, D., Silva, P., Dias, M.: Crowdsourcing platform for large-scale speech data collection. In: Proceedings FALA (2010)
20. Jyothi, P., Hasegawa-Johnson, M.: Acquiring speech transcriptions using mismatched crowdsourcing. In: Proceedings of the AAAI Conference On Artificial Intelligence, vol. 29 (2015)
21. Butryna, A., et al.: Google crowdsourced speech corpora and related open-source resources for low-resource languages and dialects: an overview. ArXiv Preprint ArXiv:2010.06778 (2020)
22. Arora, K., Arora, S., Roy, M., Agrawal, S.: Multilingual crowdsourcing methodology for developing resources for under-resourced Indian languages
23. Chopra, M., Medhi Thies, I., Pal, J., Scott, C., Thies, W., Seshadri, V.: Exploring crowdsourced work in low-resource settings. In: Proceedings of the 2019 CHI Conference on Human Factors in Computing Systems, pp. 1–13 (2019)
24. Prasad, K., Virk, S., Nishioka, M., Kaushik, C.: Crowd-sourced technical texts can help revitalise Indian languages. In: Proceedings Of LREC 2018, Workshop WILDRE4, pp. 11–16 (2018)
25. Jonell, P., Oertel, C., Kontogiorgos, D., Beskow, J., Gustafson, J.: Crowdsourced multimodal corpora collection tool. In: Proceedings of the Eleventh International Conference on Language Resources and Evaluation (LREC 2018) (2018)
26. Arora, S., Arora, K., Roy, M., Agrawal, S., Murthy, B.: Collaborative speech data acquisition for under resourced languages through crowdsourcing. Procedia Comput. Sci. **81**, 37–44 (2016)
27. Abraham, B., et al.: Crowdsourcing speech data for low-resource languages from low-income workers. In: Proceedings of the 12th Language Resources and Evaluation Conference, pp. 2819–2826 (2020)
28. Srivastava, B., et al.: Interspeech 2018 low resource automatic speech recognition challenge for Indian languages. In: SLTU, pp. 11–14 (2018)
29. Bell, L., Boye, J., Gustafson, J.: Real-time handling of fragmented utterances. In: Proceedings NAACL Workshop on Adaptation in Dialogue Systems, pp. 2–8 (2001)
30. Yang, Z., Liu, W., Jiang, W., Hu, P., Chen, M.: Speech fragment decoding techniques using silent pause detection. In: Chinese Conference on Pattern Recognition, pp. 579–588 (2012)
31. Povey, D., et al.: The Kaldi speech recognition toolkit. In: IEEE 2011 Workshop on Automatic Speech Recognition and Understanding (2011)
32. Gales, M.: Maximum likelihood linear transformations for HMM-based speech recognition. Comput. Speech Lang. **12**, 75–98 (1998)

Author Index

Printed in the United States
by Baker & Taylor Publisher Services

Printed in the United States
by Baker & Taylor Publisher Services